Chemical
Problem-Solving
by Dimensional Analysis

Use this card to mask the
answer column as you
complete each page.

Use this card to mask the
answer column as you
complete each page.

Use this card to mask the
answer column as you
complete each page.

SECOND EDITION

Chemical Problem-Solving by Dimensional Analysis

A Self-Instructional Program

Arnold B. Loebel

Merritt College, Oakland, California

Houghton Mifflin Company · Boston

Dallas   Geneva, Illinois
Hopewell, New Jersey   Palo Alto   London

To Elizabeth L. Birns

Printed in the U.S.A.
Library of Congress Catalog Card Number: 77-78565
ISBN: 0-395-25516-3

Contents

Chapter 12
Solution Concentration I:
Molarity, Formality, Molality
215

Chapter 13
What is Equivalency? A
Bookkeeping Trick   236

Chapter 14
Solution Concentration II:
Normality   252

Chapter 15
Stoichiometry Involving
Solutions   268

Preface to the Second Edition

The very widespread acceptance of the first edition of this totally programmed problem-solving book clearly showed the need for a text in this unique format which will serve the needs of a highly varied audience. The book has been used in situations which were never considered when it was first published. It has been used as the sole text in remedial courses for students with some previous chemistry background. It has been used as the sole text in courses preparatory to General Chemistry or given concurrently with General Chemistry, pre-General Chemistry or terminal chemistry courses for allied health students. It also has been used as a supplementary text for all the above courses. It has even found its way into college bookstores, where it is available on recommendation from instructors to students with problem-solving difficulties.

The totally programmed format has made it possible for the text to be used in both an auto-tutorial situation—in which the student is rewarded with pass-fail credit for completing a certain number of modules—and in a normal lecture course. The ability of the book to work under all these varied situations has been gratifying indeed.

Even more gratifying have been the letters I have received from students using the text, saying what a help it has been to them in their studies. These letters often included a reticent mention of some error which appeared in the text. I hope that, in this second edition, all errata have been removed.

The section in Chapter 1 on the slide rule has been removed from this second edition. The emergence of very inexpensive pocket calculators has made the slide rule as antiquated as the horse and buggy. However, the alacrity with which these calculators produce eight-digit answers has necessitated a reinforcement of the section in Chapter 1 on significant figures. Much of this section has been completely rewritten, with strong warnings to the students on the dangers of the calculators.

The first edition was written to cover the types of calculations generally found in the first semester of a General Chemistry course. Usually pH is not covered until the second semester. However, the book has been used in such a wide variety of situations that it became apparent that pH should be included. Since pH does not fall into a dimensional-analysis form, it has been added as an appendix. To help the student, the appendix is totally programmed, as is the body of the text. If the instructor is concerned only with whole-number pH values, he or she may stop the student midway through the appendix.

Putting in a section on pH made it necessary to include logarithms. These have been placed in Chapter 1, Mathematics Review. The coverage is quite complete, but the use of logarithms for multiplication and division has been disregarded. The students' pocket calculators serve this function. However, the use of logarithms for determining powers and roots is included, as well as natural logarithms, for students' future reference. If the instructor wishes to eliminate any of these parts, no problems will result.

Many teachers have commented that they would like this text to go further into chemical calculations. Consequently, I added a lengthy chapter on thermodynamic calculations, and included in it those topics for which dimensional analysis is a viable tool: specific heats, latent heats, thermochemical reactions, and calorimetry. Hess' Law is brought out in a rather unique way. This chapter—Chapter 17—is quite independent of all material after stoichiometry, and can be dealt with at any point after Chapter 8.

In the first edition, calculations involving solutions (Chapters 10 through 14) preceded calculations involving gas laws (Chapters 15 and 16). Most beginning chemistry textbooks have the reverse order. Therefore, in this second edition, I have rearranged the chapters to conform to this customary sequence. Gas laws now appear in Chapters 10 and 11, and solutions are in Chapters 12 through 16.

I want to thank Professor Larry Krannich, of the University of Alabama, whose careful reading of the first edition was an immense help in uncovering the errata about which "grateful students" shyly write. Professor Krannich's interest in the book since its first publication in 1974 has been truly appreciated. I also want to thank Professor Curtis Sears, of Georgia State University, for our conversations, which have been very helpful in crystallizing my ideas as to the direction that the revisions should take. I am also grateful to Professor Bernard Coyle, City College of San Francisco; Professor John H. Nelson, University of Nevada; and Professor David W. Brooks, University of Nebraska. Their assistance in reviewing the manuscript for the Second Edition has proved helpful.

Preface to the First Edition

This programmed book covers the types of chemical problems common to the first semester of freshman chemistry or a full-year college chemistry survey course. It should add an extra dimension to a beginning course, but it does not stray into the provinces of a good chemistry textbook. Students who have completed high school chemistry but who have been unable to pass a qualifying examination into general college chemistry have found this particularly useful in preparatory course work in chemical mathematics.

The book offers pretests and achievement tests for each module. Thus it is ideal for self-paced individual study since it allows the student to check his progress as he or she goes. As an extra aid, Appendices I and II cover chemical nomenclature. As it was being written, the manuscript was thoroughly class tested, and these appendices developed from a series of nomenclature quizzes.

Chapter 1 brings students up to the mathematical level required for the rest of the book. Readers are subsequently introduced to dimensional analysis, the only method used here for solving problems. (Gas law problems are an exception; they do not lend themselves to this procedure.) At first glance, this approach may seem nondidactic. It literally cajoles students away from the logical method of problem-solving in steps, traditionally conceived as the way of getting students to "think." However, sound motivation for "thinking" is also present in the dimensional-analysis method which has the added advantage of giving readers on-the-spot check points for testing how well they have learned a particular point. This is practical and remarkably successful.

Divisions are not included in the setup of problems. Instead, the invertibility of conversion factors, i.e., the use of their reciprocals, is brought out from the beginning. Students believe that they are always multiplying. This has eliminated many student errors. The chapters follow the order of most general chemistry courses, although the chapters related to solutions (10 through 14) precede the chapters on gases (15 and 16). The sequence can be reversed at the instructor's discretion since there is virtually no carry-over from one group to the other. There are also two chapters on stoichiometry (8 and 9). The first contains relatively simple material, while the second contains more complex problems with various involved conversions. If an instructor feels that the complicated stoichiometry chapter is unnecessary or overly burdensome, it can be eliminated.

Chapters 11 and 12 involve the concept of equivalency. The current trend is to downgrade the use of equivalency and normality, particularly in freshman chemistry. Consequently, these two chapters can be omitted, as the carry-over to Chapter 13 is minimal. Where the concept does occur, it is noted in the text so students will know that the problems are optional.

Numerical answers to pretests and a scoring procedure follow on the next page. Students can grade themselves, and each student's score will reveal whether he or she is ready to go directly to the problem set at the end of the chapter, or whether he or she should work through certain, or all, sections of the chapter. Problems in the pretest are used as examples in the body of each chapter.

Following each chapter are problem sets, arranged in order of increasing complexity and sometimes divided into groups corresponding to various sections of the chapter. Answers and dimensional-analysis setups for all problems follow the problem set. Thus students can immediately see where they are correct and where they may have made mistakes. Relatively complex problems are marked with an asterisk to indicate they are optional. Better prepared and more motivated students thus have a challenge, and the text takes on broader flexibility for use by students with a range of background skills.

The appendix on nomenclature is programmed in the same manner as the text. The level is relatively low, but it covers most compounds encountered in the first semester of freshman chemistry. IUPAC notation has been used, although lip service is paid to the old names such as ferric chloride and cobaltous nitrate.

I would like to thank the many students who have used this text and whose feedback has helped to develop it into its present form. Also I would like to thank Professor Ethelreda Laughlin of Cuyahoga Community College whose careful scrutiny has helped eliminate many errors, and Professor Mildred Johnson of the City College of San Francisco whose hawk-like eye has uncovered innumerable impossibly concentrated solutions and unlikely chemical reactions.

A.B.L.

To the Student

This book has a very special design. It is a self-instructional program, and you must work with it in a very specific way. If you do, you will find that in a short time you will be able to solve chemical problems of dazzling complexity with so little effort that it will seem as though someone else were doing the work. It will be no more difficult to figure out the number of seconds in three hours than to figure out the number of milliliters of a 6.0 N solution of potassium permanganate required to completely react with 1.0 liter of a 3.7 weight percent solution of sodium oxalate whose density is 1.08 g/ml. (If this problem looks impossible, wait until you get to Chapter 13; it will be a snap!)

The book is divided into eighteen chapters, each of which covers only a single topic. Early in the chapter is a group of problems which cover the material in that chapter. This is called a pretest. If you think that you can handle the material in the chapter, take the pretest. On the next page you will find the answers to these problems and a guide to scoring yourself. Your score will tell you either that you can skip immediately to the problem set at the end of the chapter, or that you had better work through certain sections of the chapter first.

Each chapter is presented with the text on the inside and a separate column on the outside with occasional words or numbers. This is the special design of a self-instructional program. Even if you have worked with this type of a program before, you will still benefit by reading the next few words of instruction here.

In the text, on the inside of the page, you will see blanks, _____, in the middle of sentences or after questions. In the column on the outside, on the same line as the _____, you will find the word or number which fills the _____. The one thing you must do—and without this it won't work at all—is to write the response for each and every _____ without looking at the response in the outside column. You can do this best by covering the column on the outside with the card located at the front of the book. Read the text material, and when you get to a _____, write down the appropriate word or number either in the _____ or on a piece of scratch paper. (You are going to need scratch paper anyway, since as you progress in the book the responses to the _____'s will involve building up the solution to the problems in the order of dimensional analysis, and you must keep a running solution in order to proceed to the next _____.) Then uncover the response for the _____ and check the answer you wrote down to see that it is the same. It will almost always be the same. If, by some strange chance it isn't, just go back a few _____'s and try it over again until you are sure you understand.

The main thing to remember is that you cannot simply read through the program as you would a textbook. You must work

through and write down the responses to every _____. You are not taking a test or simply filling in blanks when you are working with a program; you are learning by doing. The program works so that each response gives you sufficient understanding to make the next response, and on and on and on.

When you get to the end of the chapter you will be ready to do a real problem set. There are special instructions on some of these which will tell you that some problems need not be done. You can check your answers as soon as you have completed a problem set. All the answers are given on the following page, and the problems are worked out so that you can examine your solutions. You often will be told how many you should have done correctly and to which section of the chapter you should refer to find the method of solving a certain type of problem that you couldn't manage.

Remember that you are not just taking tests, you are learning by doing. You can work along at your own rate, and you can grade yourself as you go. In this way, you will be able to check your learning each step along the way. You will be surprised at how easy this is to do. So tear out the card, cover up the responses, and get to work.

Arnold

Chemical
Problem-Solving
by Dimensional Analysis

Chapter 1
Mathematics Review

PRETEST

1. Express 56,000 in scientific notation (that is, as a number between 1 and 10 times 10 raised to a power).

2. Express 0.0031 in scientific notation.

3. Perform the operation $3.5 \times 10^{-6} \times 2.0 \times 10^{2}$.

4. Perform the operation $\dfrac{8.4 \times 10^{-4}}{2.0 \times 10^{2}}$.

5. Perform the operation $6.7 \times 10^{18} + 3.00 \times 10^{19}$.

6. Perform the operation $(3.0 \times 10^{-8})^{3}$.

Using the four-place table of logarithms in Appendix III, determine the following.

7. $\log 10^{3}$

8. $\log 55.5$

9. $\log 0.000558$

10. antilog -9.2291

11. $e^{0.25}$

12. $\sqrt[5]{35}$

13. Express 2.4321640 to three significant figures.

14. Perform the operation

$$6.0 \times 4.3 \times \frac{4.3271}{3.}$$

15. Perform the operation $4.321 + 6.5$ and express your answer to the correct number of significant figures.

PRETEST ANSWERS

1. 5.6×10^4
2. 3.1×10^{-3}
3. 7.0×10^{-4}
4. 4.2×10^{-6}
5. 3.67×10^{19} or 36.7×10^{18}
6. 2.7×10^{-23} or 27×10^{-24}
7. 3
8. 1.7443
9. -3.2534
10. 5.90×10^{-10}
11. 1.284
12. 2.036
13. 2.43
14. $4. \times 10^1$
15. 10.8

If you had all of them right, you're a whiz. You can skip Chapter 1 and do the problem set at the end immediately.

If you had all but one right in numbers 1–6, you probably can skip Section A in Chapter 1. If you missed two or more out of questions 1–6, you should work through Section A. If you got all of numbers 7–12 correct, you can skip Section B in Chapter 1. If you just missed number 11, take a look at the part of this section on natural logarithms. If you missed only number 12, check the last part of Section B, Determination of Powers and Roots. If you did numbers 13, 14, and 15 all correctly, you can skip Section C.

The whole purpose of this chapter is to give you the tools you will need to get a number answer once you have set up chemical problems by "dimensional analysis." There is nothing very complicated in this chapter. It is divided into three sections which, though quite different, all have to do with handling numbers. Remember that the way to use this text is to cover the outside of the page with a piece of paper so that you cannot see the answers that go in the blanks. Then you read the inside, and when you get to a _____ you write down what goes in it. Then slide your piece of paper down the outside and check that you have written down the correct word or number. We are going to start with what is called "exponential notation."

SECTION A   Exponential Notation

There is a very convenient way of showing in mathematical
shorthand that you have multiplied the same number repetitively.
You write a small, raised number after the digit which is being
multiplied by itself. This raised number is called an exponent or a
power. For example,

$$2 \times 2 \times 2 = 2^3.$$

You have multiplied three 2's together and the exponent is 3. You
can also say that you have 2 to the third power.
  Using the same system,

$$2 \times 2 \times 2 \times 2 = \underline{}$$ 2^4

You have multiplied _____ 2's together, and the exponent is _.   four, 4

You can also say that you have 2 to the _____ power or write ____.   fourth, 2^4
  Using the same system,

$$5 \times 5 \times 5 = \underline{},$$ 5^3

which is 5 to the _____ power. Similarly,   third

$$7.5 \times 7.5 \times 7.5 \times 7.5 \times 7.5 \times 7.5 = \underline{}.$$ 7.5^6

You have multiplied ___ 7.5's together, so you can also say you   six

have 7.5 to the _____ power.   sixth

  If you multiply two 2's together, that is, 2 to the _____ power,   second

you get 4. If you have only one 2, it equals _. Of course, if   2

you only have one of a number you say it is to the ___ power,   first

which you can write exponentially as _. Any number raised to the 2^1
first power is the number itself. Therefore,

$$6.6^1 = \underline{}$$ 6.6

In this case there is only ___ 6.6 and you can say it is raised to the   one

____ power. If you had 10^1, this would be equal to __ since any   first, 10

number to the ____ power is equal to itself.   first
  If you had 10 to the second power, which can be written

exponentially as ____, this would mean that you have multiplied 10^2

_ 10's together,   2

$$10 \times 10$$

which equals _____. If you have 10^3, which is 10 to the _____ power,   100, third

you would have multiplied _____ 10's together,   three

$$10 \times 10 \times 10,$$

1000

which equals _____. Let's take a look at the values of these powers of 10 and see if any pattern shows up.

$$10 = 10^1 = \quad 10$$

$$10 \times 10 = 10^2 = \quad 100$$

$$10 \times 10 \times 10 = 10^3 = \quad 1000$$

$$10 \times 10 \times 10 \times 10 = 10^4 = 10,000$$

one

first, two

second, 1000

third

4

The number 10 contains _____ zero and is equal to 10 raised to the _____ power. The number 100 contains _____ zeros and is equal to 10 raised to the _____ power. The number _____ contains three zeros and is equal to 10 raised to the _____ power. If you write the number 10,000 exponentially, you will have an exponent of ___ since there are four zeros. What about the number

10^{12}

1,000,000,000,000.? Written exponentially, it would be _____.

Now you can see why we bother to write numbers exponentially. It is a very neat way of writing very large (or, as you shall see, very small) numbers in a small space.

Now take a wild guess at how you could write the number 1

no

10^0, zero

expressed as 10 to a power. The number 1 has ___ zeros, so you should write this exponentially as ____, which is 10 to the ____ power. This looks a little strange. However, it does not mean that 10 is multiplied by zero, which would be equal to zero. It simply means that no 10's are multiplied together at all. *Any number to the zero power equals* 1.

The next thing we will do is see how you can express exponentially numbers which have digits other than 1 and 0 in them. Say you have the number 400, which equals 4×100. You

10^2

know from the above that you can write 100 exponentially as ____,

10^2

so instead of writing 4×100, you can write $4 \times$ ____.

10^3

Using this method, 6000. can be written as $6 \times$ ____, and 50,000.

5×10^4

can be written as _____.

Here is a sneaky but very simple way of doing this whole thing. Actually, what we have been doing in all this is simply moving the decimal point. The first example shown above was 400. which equals 4.00×10^2. The decimal point is moved two places to the left,

4̖0̖0̖.
2 1

and then you multiply by 10 raised to the second power.

In the second example, 6000., you move the decimal point three places to the left

6 0 0 0.

3 2 1

and get 6.000×10^3. The exponent of 10 is equal to the number of

_____ you move the decimal point to the left. Then all you have to do is to write down the number and count the number of places you move the decimal point. Put in the decimal at the new location and multiply by 10 raised to a power which equals the number of

places moved. Try one yourself: $9,000,000. = 9.000000 \times \underline{\quad}$.

What about 470? This equals $47.0 \times \underline{\quad}$, but it also equals

$4.70 \times \underline{\quad}$. It even equals $.470 \times \underline{\quad}$. You can put the decimal point anywhere you want. All you have to do is count the number

of places you move the _____ to the left and multiply by 10 raised to that power.

How do you decide to which place it is best to move the decimal point? It has been found most convenient, when nothing else affects your decision, to move the decimal point far enough so that you have a number between 1 and 10 times 10 to the appropriate power. This is called "scientific notation." Therefore, 470 expressed in scientific notation would be 4.70×10^2 because 4.70 is between 1 and 10. The number 3476 expressed in scientific notation would

be _____ because 3.476 is between _____.

Now it is time for a little practice to make sure you have all of this under your belt. Write the following numbers in scientific notation.

$$43. = \underline{\qquad}$$

$$7569. = \underline{\qquad}$$

$$432,506.43 = \underline{\qquad}$$

$$9.0 = \underline{\quad}$$

The last one is a little tricky. Actually it is 9.0×10^0. Since 10^0 equals 1 and multiplying by 1 doesn't change anything, you need not bother to write 10^0. However, some workers feel that it is not true scientific notation unless the 10^0 is shown.

Now let's see if you can do it backwards. When written as a number not in exponential notation, the number 6.54×10^5 will be

_____.

If you got the correct answer, and feel fairly confident of yourself, you can skip the next short section and go directly to the section on negative exponents. If you had a little trouble, read through what is immediately below.

---

places

10^6

10^1

10^2, 10^3

decimal point

3.476×10^3, 1 and 10

4.3×10^1

7.569×10^3

4.3250643×10^5

9.0

654,000

Consider $5. \times 10^2$. This is a short way of writing $5. \times 10 \times 10 = 5. \times 100 = 500$. The easy way of getting to 500 is to write $5. \times 10^2$ with a lot of zeros after the 5.

$$5.0000000 \times 10^2$$

The exponent on the 10 tells you how many places the decimal was moved to the left. All you have to do is to move the decimal back to the right that many places.

$$5 \,.\, 0 \underset{1}{0} \underset{2}{0}\, 0\ 0,$$

and the number becomes 500.

Now look at the number 3.5678×10^3. The 10^3 tells you that the

three

decimal point has been moved _____ places to the left. Therefore,

three

you must move it back to the right _____ places. The digital

3567.8

number which is equal to 3.5678×10^3 must therefore be _____.

What about 6.5×10^4? The decimal point must be moved back to

fourth

the right four places because the 10 is raised to the _____ power. Since there are not that many numbers to move the decimal point,

zeros

you can add some _____, which don't change the value of the number. Thus you would have

$$6.5000000 \times 10^4,$$

four

and moving the decimal point back to the right ____ places gives

65,000

you the number _____.

$$6 \,.\, \underset{1}{5}\ \underset{2}{0}\ \underset{3}{0}\ \underset{4}{0}\, 0\ 0\ 0$$

You should now be able to do the first example. When written as

654,000

a nonexponential number, the number 6.54×10^5 is _____.

Negative Exponents

There is another convenient trick with exponential notation. If you divide by a number multiplied repetitively,

$$\frac{5}{2 \times 2 \times 2},$$

which you could write exponentially as

$$\frac{5}{2^3},$$

you can move the number written exponentially to the top (numerator) if you change its sign,

$$5 \times 2^{-3}.$$

The minus sign simply shows that you are dividing by the number raised to that power. Therefore, using the same procedure,

$$\frac{47}{5 \times 5 \times 5 \times 5} = \frac{47}{5^4} = 47 \times \underline{\quad},$$

or

$$\frac{2}{10^2} = 2 \times \underline{\quad}$$

and

$$\frac{7}{10^{-5}} = 7 \times \underline{\quad}.$$

If you shift a number written exponentially from the bottom to the top of the fraction, you change the sign. In the last example, you had a -5 on the bottom (denominator) and you changed the sign from $-$ to $+$ when you shifted it to the top. It is customary to omit $+$ signs; they are understood.

Do a few for practice.

$$\frac{11}{10^4} = \underline{\qquad}$$

$$\frac{5}{10^{-8}} = \underline{\qquad}$$

$$\frac{3}{2 \times 10^2} = \underline{\qquad}$$

(Notice that you can only shift the numbers written exponentially.)

$$\frac{6}{5 \times 10^{-2} \times 10^7} = \underline{\qquad}$$

What about this?

$$4 \times 10^{-3} = \frac{4}{\underline{\quad}}$$

That's right; if you shift a number written exponentially from the top to the bottom, you change its _____. It works both ways.

The reason that we bother with this at all is that it will allow you to express very small numbers in a convenient way. Consider $1. \times 10^{-2}$. This is really

$$\frac{1.}{10^2} = \frac{1.}{10 \times 10} = \frac{1.}{100}$$

Dividing this out, you get

.01

So

$$1. \times 10^{-2} = .01$$

5^{-4}

10^{-2}

10^5

11×10^{-4}

5×10^8

$\dfrac{3 \times 10^{-2}}{2}$

$\dfrac{6 \times 10^2 \times 10^{-7}}{5}$

10^3

sign

two | The decimal point has been moved ___ places to the left in going

−2 | from $1. \times 10^{-2}$ to .01. The power to which 10 is raised is __, so the number of places you move the decimal point is equal to the

10 | power to which the __ is raised.

We can put this the other way around: when you move a decimal

places | point to the right, the number of _____ that you move it gives you

power (exponent) | the negative _____ of the 10 by which you must multiply. Here is the sequence.

$$0.1 \quad = 1 \times 10^{-1}$$

$$0.01 \quad = 1 \times 10^{-2}$$

$$0.001 \ = 1 \times 10^{-3}$$

$$0.0001 = 1 \times 10^{-4}$$

10^{-9} | Following this pattern, 0.000000001 would equal $1. \times$ ____, since

nine | you had to move the decimal point ___ places to the right.

$$0 \cdot 0\ 0\ 0\ 0\ 0\ 0\ 0\ 0\ 1$$
$$1\ 2\ 3\ 4\ 5\ 6\ 7\ 8\ 9$$

There is a little memory trick to use in deciding whether the sign on the power of 10 is negative or plus when you move the decimal point left or right. When the number becomes **bigger**, you move the decimal point to the **right** and the power of 10 is **negative**. All the words have a **g**. Thus,

$$0.01 = 1. \times 10^{-2}$$

(The number 0.01 becomes 1, which is bi**gg**er, you move to the ri**g**ht, and the power is ne**g**ative.)

When the number becomes smaller, you move the decimal point to the left and the power of 10 is plus. All words have an **l**.

$$100. = 1. \times 10^2$$

(The number 100 becomes 1, which is smaller, you move to the left, and the power is plus.)

10^{-5} | The number 0.00001 can be written as $1. \times$ ____. Since 1. is

bigger | _____ than 0.00001, the sign on the power to which the 10 is

negative, right | raised is _____. The decimal point was moved to the _____.

10^6 | The number 1,000,000. can be written as $1. \times$ ___. Since 1. is

smaller | _____ than 1,000,000, the sign on the power to which 10 is raised

plus, left | is ____. The decimal point has been moved to the ___.

Here is a little practice. Simply note whether the power to which the 10 has been raised is plus or negative.

$$0.0001 = 1. \times 10\_^4 \quad \text{(bigger; negative; right)}$$

$$1000. \quad = 1. \times 10\_^3 \quad \text{(smaller; plus; left)}$$

$$0.1 \quad = 1. \times 10\_^1 \quad \text{(bigger; negative; right)}$$

$$10. \quad = 1. \times 10\_^1 \quad \text{(smaller; plus; left)}$$

−

+

−

+

So far we have only looked at numbers containing the digits 1 and 0; but in the same way that you were able to count the number of places moved to the left in large numbers to find the power of 10, you can count places in small numbers when you move the decimal point to the right to make them bigger. For example,

$$0.02 = 2. \times 10^{-2}.$$

You have moved the decimal point ＿＿ places to the ＿＿＿ and therefore have a −2 power.

two, right

$$0 \, . \, \underset{1 \quad 2}{0 \; 2}$$

Using the same procedure, if you want to write the number 0.00035 in scientific notation, you must move the decimal point

＿＿ places to the ＿＿＿. Therefore, the power of 10 will be ＿＿. The end result is that 0.00035 expressed in scientific notation is

four, right, −4

＿＿＿＿＿＿.

3.5×10^{-4}

You didn't forget what scientific notation is, did you? It is any

number between ＿＿ and ＿＿ times 10 to some power. If you moved your decimal point all the way to the end so that you have

1, 10

$35 \times$ ＿＿＿, then 35 is larger than 10. If you stop the decimal point

10^{-5}

in front of the 3, you have $0.35 \times$ ＿＿＿, and 0.35 is less than 1.

10^{-3}

To show yourself how well you can do this now, write the following numbers in scientific notation.

$$0.0000476 = \underline{\hspace{2cm}}$$

$$0.03 \quad = \underline{\hspace{1.5cm}}$$

$$35.67 \quad = \underline{\hspace{2cm}}$$

$$0.00004 \quad = \underline{\hspace{1.5cm}}$$

$$48671.5 \quad = \underline{\hspace{2.5cm}}$$

4.76×10^{-5}

3×10^{-2}

3.567×10^1

$4. \times 10^{-5}$

4.86715×10^4

(You probably have observed that when a number starts after the decimal point there is usually a zero written to the left of the decimal point. For example, we write 0.25, not .25. This is only a convention so that someone reading the number won't think that the decimal point is just a spot of dirt on the paper. To prevent any such doubts, a zero is put before the decimal.)

Let's see now if you can do these negative exponents backwards.

0.0000037

The digital number equal to 3.7×10^{-6} is _____.

If you got this last answer, you can skip the next section and go immediately to the section entitled Multiplying Numbers Expressed in Exponential Notation. If you didn't get the answer or don't feel too confident, work through the next short section.

If you have a number like 6.25×10^{-3}, the negative sign on the exponent means that the decimal place must have been moved to

right

the _____. The number of the exponent, 3, means that it has been

three

moved _____ places. Consequently, to get a decimal number you

three, left

must move the decimal point _____ places back to the ___. Since there are no numbers three places to the left, you just put in some extra zeros

$$0\ 0\underbrace{{,}0\ 0\ 6}\ .\ 2\ 5$$
$$3\ 2\ 1$$

and count off the number of places you need. The final result, then,

0.00625

is that the decimal number equal to 6.25×10^{-3} is _____.

Here is another example. The number 4.3615×10^{-8} has a negative 8 exponent on the 10. This means that the decimal point

eight, right

must have been moved _____ places to the _____ to get the number. Therefore, to convert it back to decimal notation you must

eight, left

move the decimal point back _____ places to the ___. So you put a

zeros

lot of _____ in front of the number,

0000000004.3615

_____,

and count off the number of places you need. The final result of this is that the decimal number equal to 4.3615×10^{-8} is

0.000000043615

_____.

Now you can try the very first one again. The decimal number

0.0000037

which is equal to 3.7×10^{-6} is _____. See how easy it is?

Multiplying Numbers Expressed in Exponential Notation

You are now going to see something remarkable. Say you want to multiply $10^2 \times 10^3$.

$$10^2 = 10 \times 10$$

10 × 10 × 10

$$10^3 = \underline{\hspace{3cm}}.$$

Therefore,

$$10^2 \times 10^3 = (10 \times 10) \times (10 \times 10 \times 10)$$

$$= 10 \times 10 \times 10 \times 10 \times 10.$$

But a much shorter way of writing five 10's multiplied together is

____. Therefore,

10^5

$$10^2 \times 10^3 = 10^5.$$

Can you see any relationship between $10^2 \times 10^3$ and 10^5? It

seems that all that you have done is to ____ the exponents.

add

We can check this to see if it is really so. Say you want to multiply $10^4 \times 10^5$.

$$10^4 = 10 \times 10 \times 10 \times 10$$

$$10^5 = \text{\underline{\hspace{5cm}}}.$$

$10 \times 10 \times 10 \times 10 \times 10$

Therefore,

$$10^4 \times 10^5 = 10 \times 10 \times 10 \times 10 \times 10 \times 10 \times 10 \times 10 \times 10.$$

A shorter way of writing nine 10's multiplied together gives you the final result,

$$10^4 \times 10^5 = 10^9.$$

Once again you can see that when you multiply a number raised to a power by the same number also raised to a power you just

____ the powers (exponents). You can do a few more yourself.

add

$$10^{11} \times 10^2 = \text{\underline{\hspace{1cm}}}$$

10^{13}

$$10^3 \times 10 = \text{\underline{\hspace{1cm}}} \quad \text{(The first power is understood.)}$$

10^4

$$10^4 \times 10^8 = \text{\underline{\hspace{1cm}}}$$

10^{12}

If you have numbers other than 10's raised to a power, you must multiply these separately. It is much easier if you group the digit numbers together and all of the 10's at the end. For example,

$$(3 \times 10^2) \times (4 \times 10^5)$$

can be rewritten as

$$3 \times 4 \times 10^2 \times 10^5.$$

Then you multiply the numbers as you would normally, multiplying the 10's to a power by adding the exponents.

$$3 \times 4 \times 10^2 \times 10^5 = 12 \times 10^7$$

Therefore, if you want to multiply $(5 \times 10^5) \times (6 \times 10^6)$, you can

rewrite this as ____. Then you multiply the numbers normally and the 10's to a power by adding exponents to give the

$5 \times 6 \times 10^5 \times 10^6$

30×10^{11}

3.0×10^1, 3.0×10^1

3.0×10^{12}

8×10^{24}

2.8×10^7

3.6×10^{13}

final answer _____. This answer is not in scientific notation, but you know that 30 can be written in scientific notation as

_____. Therefore, 30×10^{11} is equal to _____ $\times 10^{11}$.

Multiplying the powers of 10 by adding exponents gives _____.
   Try a few more yourself.

$$4 \times 10^{11} \times 2 \times 10^{13} = \underline{\hspace{2cm}}$$

$$7 \times 10^2 \times 4 \times 10^4 = \underline{\hspace{2cm}} \quad \text{(in scientific notation)}$$

$$2 \times 10^4 \times 3 \times 10^6 \times 6 \times 10^2 = \underline{\hspace{2cm}}$$

(It doesn't matter how many numbers you have multiplied together.)
   What will happen if the exponent is negative? Consider $10^3 \times 10^{-2}$. Recall that a negative exponent simply means that the exponential number was raised from the bottom of the fraction. You could rewrite this as

$$\frac{10^3}{10^2}$$

and breaking this down to 10's multiplied repetitively,

$$\frac{10 \times 10 \times 10}{10 \times 10}.$$

Now you can see that two of the 10's on the bottom will cancel two 10's on top,

$$\frac{\cancel{10} \times \cancel{10} \times 10}{\cancel{10} \times \cancel{10}},$$

leaving 10^1. This is just what you should expect. You have added the exponents 3 and -2 algebraically to give 1.
   Try a few yourself.

10^5

10^{-4}

10^{-5}

10^5

10^{-21}

$$10^{11} \times 10^{-6} = \underline{\hspace{1cm}}$$

$$10^{-9} \times 10^5 = \underline{\hspace{1cm}}$$

$$10^{-3} \times 10^{-2} = \underline{\hspace{1cm}}$$

$$10^{15} \times 10^{-4} \times 10^{-6} = \underline{\hspace{1cm}}$$

$$10^{-27} \times 10^3 \times 10^{-1} \times 10^4 = \underline{\hspace{1cm}}$$

If you were able to do all of the above, you can skip the next section and go on to Dividing Numbers Expressed in Exponential Notation. If you had any problem, the rules for algebraic addition are given below.

Algebraic Addition

If you add numbers of the same sign (either $+$ or $-$), the answer is the sum of the numbers and the sign is the same as that of the numbers. If you add numbers of different signs, you subtract the smaller from the larger, and the sign is the same as the sign of the larger. If you have more than two numbers, it is most convenient to add all those of the same sign separately. Then subtract the smaller sum from the larger sum and give your answer the sign of the larger.

Here are some examples.

$$2 + 6 = \underline{} \qquad\qquad 8$$

$$-2 + (-6) = \underline{} \qquad\qquad -8$$

$$5 + (-3) = \underline{} \qquad\qquad 2$$

$$3 + (-5) = \underline{} \qquad\qquad -2$$

$$6 + (-4) + (-3) = \underline{} \qquad\qquad -1$$

$$(-7) + (-4) + 3 + (-9) = \underline{} \qquad\qquad -17$$

Remember that a plus sign ($+$) is usually understood, so you don't bother to write it. A minus sign ($-$) is always written.

Dividing Numbers Expressed in Exponential Notation

To divide two numbers you set up a fraction with the number you are dividing on the top (numerator) and the number you are dividing it by on the bottom (denominator). Therefore, if you want to divide 18 by 3, you write this as the fraction ___.

$$\frac{18}{3}$$

If you want to divide 10 raised to a power by another 10 raised to a power, you set up the same kind of fraction. For example, 10^5 divided by 10^2 can be written as the fraction ___. However, you know that you can shift a number raised to a power from the

$$\frac{10^5}{10^2}$$

bottom of a fraction to the top simply by changing the ___ of the

sign

exponent. Doing this to $10^5/10^2$ gives _____. You now have a

$10^5 \times 10^{-2}$

multiplication that you perform by algebraically _____ the exponents, the final answer for the division of 10^5 by 10^2 is

adding

therefore ___.

10^3

It is really very simple. Try a few.

$$\frac{10^7}{10^6} = \underline{} \qquad\qquad 10^1$$

$$\frac{10^4}{10^9} = \underline{} \qquad\qquad 10^{-5}$$

$$\frac{10^{-1}}{10^3} = \underline{} \qquad\qquad 10^{-4}$$

$$\frac{10^7}{10^{-5}} = \underline{} \qquad\qquad 10^{12}$$

$$\frac{10^4}{10^4} = \underline{} \qquad\qquad 1 \,(= 10^0)$$

(Here the number on the bottom equals the number on the top, and they simply cancel each other out. If you remember this when you have a lot of exponential numbers on the top and bottom, it can speed up getting the solution.)

10^{-3}

$$\frac{\cancel{10^{-6}} \times 10^2 \times \cancel{10^{-3}}}{\cancel{10^{-3}} \times 10^5 \times \cancel{10^{-6}}} = \frac{10^2}{10^5} = \underline{}$$

You will also find that if you have more than one exponential number on the top or bottom, it may be faster if you algebraically add the exponents on the bottom and top separately and then shift the denominator to the numerator.

10^{10}

$$\frac{10^5 \times 10^{-3} \times 10^7}{10^4 \times 10^3 \times 10^{-8}} = \frac{10^9}{10^{-1}} = \underline{}$$

Actually, when you are solving real problems you will have numbers as well as 10's raised to various powers. You will find it much simpler if you rewrite the fraction grouping all the numbers together at the beginning and putting all the 10's at the end. Then you can handle the numbers in the normal fashion, making the multiplications and divisions and adding the exponents for the powers of 10. For example,

$$\frac{3 \times 10^3 \times 4 \times 10^{-5}}{2 \times 10^7 \times 6 \times 10^1} = \frac{3 \times 4}{2 \times 6} \times \frac{10^3 \times 10^{-5}}{10^7 \times 10^1}$$

$$= 1 \times 10^{-10}.$$

Thus, if you have the operation

$$\frac{1 \times 10^2 \times 2 \times 10^{-6}}{2 \times 10^8 \times 3 \times 10^{-4}},$$

first you rewrite to group all the numbers together to give

$\dfrac{1 \times 2}{2 \times 3} \times \dfrac{10^2 \times 10^{-6}}{10^8 \times 10^{-4}}$

_____. Then you solve the multiplication and division

0.33

of the numbers to give ____, and you solve the powers of 10 to

10^{-8}

give ____. When you carry out the calculation you get the answer

0.33×10^{-8}

_____. However, this isn't in scientific notation. You can

10^{-1}

change the 0.33 to $3.3 \times$ ____. This means that the answer now is

10^{-1}

$3.3 \times$ ____ $\times 10^{-8}$, or, finishing the multiplication, it is

3.3×10^{-9}

_____.

Try a few for practice.

$$\frac{3 \times 10^6 \times 8 \times 10^2}{6 \times 10^{-4}} = \underline{}$$

4×10^{12}

(If you got 4×10^4 for your answer, you forgot to change the sign on the 10^{-4} when you shifted it to the numerator.)

1.5

$$\frac{5 \times 10^{-3} \times 4 \times 10^{-5} \times 3 \times 10^1}{2 \times 10^{11} \times 2 \times 10^{-17}} = \underline{}$$

(If you got 15×10^{-1} you were right but didn't express your answer in scientific notation. If you got 15×10^{-13}, you forgot to change the sign of the exponent.)

Adding and Subtracting Numbers Expressed in Exponential Notation

If you want to add or subtract numbers that are multiplied by 10 to a power, you can only do this if the power of 10 is the same for all numbers. The reason for this will become very clear if you write the exponential numbers as ordinary numbers. Consider

$$
\begin{array}{r}
4. \times 10^{-1} \\
+ \ 3. \times 10^{2} \\
\hline
\end{array}
$$

If you write 4×10^{-1} as a digital number, it equals ___. If you write

3×10^{2} as a digital number, it equals ___. Then you add

$$
\begin{array}{r}
0.4 \\
+ \ 300. \\
\hline
300.4
\end{array}
$$

which equals 3.004×10^{2}. You see, then, that you simply cannot say $3+4$ is 7.

However, if the numbers are times the same power of 10, then they can be added and the result multiplied by that power of 10. Consider

$$
\begin{array}{r}
5. \times 10^{-1} \\
+ \ 2. \times 10^{-1} \\
\hline
\end{array}
$$

If you write 5×10^{-1} as an ordinary number, it equals ___. If you

write 2×10^{-1} as an ordinary number, it equals ___. Then you add

$$
\begin{array}{r}
0.5 \\
+ \ 0.2 \\
\hline
0.7
\end{array}
$$

which equals $7. \times 10^{-1}$.

Therefore, what you must do when you want to either add or subtract numbers which are times a power of 10 is to convert all the numbers to the same power of 10. This is quite easy now since you know that if you multiply by a 10 to a negative exponent, the number becomes bigger because the decimal point is moved to the right. When you multiply by 10 to a plus exponent, the number becomes smaller since you have moved the decimal point to the left.

If you want to add

$$
\begin{array}{r}
4. \times 10^{1} \\
+ \ 5. \times 10^{2} \\
\hline
\end{array}
$$

you can either convert $4. \times 10^{1}$ to something $\times 10^{2}$, or $5. \times 10^{2}$ to something $\times 10^{1}$.

0.4
300
0.5
0.2

Say you multiply $4. \times 10^1$ by 10^1. Since this is a plus exponent, the number becomes smaller and the decimal point moves to the left. You then have

$$0.4 \times 10^1 \times 10^1 = 0.4 \times 10^2$$
which you can add to $\quad \underline{5. \times 10^2}$
to give $\quad 5.4 \times 10^2$

Or you could have multiplied $5. \times 10^2$ by 10^{-1}. Since this is a negative exponent, the number becomes bigger and the decimal point moves to the right. You then have

$$50. \times 10^2 \times 10^{-1} = 50. \times 10^1$$
which you can add to $\quad \underline{4. \times 10^1}$
to give $\quad 54. \times 10^1 = 5.4 \times 10^2$

Let's look at another. You want to add

$$\begin{array}{r} 8.56 \times 10^{-3} \\ + \ 6.10 \times 10^{-1} \\ \hline \end{array}$$

Say you decide to convert everything to $\times 10^{-1}$. This means that

you will have to multiply 8.56×10^{-3} by _____. This exponent is

_____ so that you will have to move the decimal point _____ places

to the _____, and the number will become _____ having a value of

_____. The new 10 raised to a power will be $10^{-3} \times$ _____.

Multiplying these you have _____. You can then make the addition

$$\begin{array}{r} \underline{\phantom{6.1856 \times 10^{-1}}} \\ + \ 6.10 \times 10^{-1} \\ \hline \underline{\phantom{6.1856 \times 10^{-1}}} \end{array}$$

Try a few.

$$6. \times 10^5 + 9. \times 10^3 = \underline{}$$

$$4.1 \times 10^{-3} + 7. \times 10^{-2} = \underline{}$$

$$9. \times 10^1 - 3. \times 10^2 = \underline{}$$

(Remember the rules for algebraic addition!)

$$8.7 \times 10^2 + 7.1 \times 10^4 + 1.0 \times 10^1 = \underline{}$$

Raising Numbers Expressed in Exponential Notation to a Power

If you want to multiply a number expressed in exponential notation by itself repetitively, you can use the same shorthand method used for ordinary numbers. You write a small, raised

[margin answers]

10^2

$+2$, two

left, smaller

0.0856, 10^2

10^{-1}

0.0856×10^{-1}

6.1856×10^{-1}

6.09×10^5

7.41×10^{-2}

-2.1×10^2

718.8×10^2 or 7.188×10^4 or 7188×10^1

number following the number which is multiplied by itself telling how many times it is done. Thus

$$2.5 \times 10^2 \times 2.5 \times 10^2 \times 2.5 \times 10^2$$

can be written as

$$(2.5 \times 10^2)^3.$$

The raised 3 indicates that you have multiplied three 2.5×10^2's together. You always write parentheses before and after the number so that it is clear just what is being multiplied by itself.

If we group the numbers at the beginning and the 10's to a power at the end, we get

$$2.5 \times 2.5 \times 2.5 \times 10^2 \times 10^2 \times 10^2.$$

Of course, $10^2 \times 10^2 \times 10^2 = $ ___ by adding exponents when you multiply. But 10^2 multiplied together three times also can be

written by the shorthand notation as _____.

Can you see any relationship between $(10^2)^3$ and 10^6? Since

2 ____ 3 equals 6, when you raise a number already to a power

to another power, you simply _____ powers. Thus we can write

$$(10^2)^3 = 10^{(2 \times 3)}.$$

There isn't much we can do with the three 2.5's multiplied together except write $(2.5)^3$ and multiply them out when we need a real number. Therefore,

$$(2.5 \times 10^2)^3 = (2.5)^3 \times 10^6 = 15.6 \times 10^6.$$

Consider $(10^2)^4$. This can be written as

$$10^2 \_\_\_,$$

which is equal to

$$10\_$$

You try a few.

$$(10^6)^2 = \_\_\_$$

$$(4 \times 10^3)^2 = 4^2 \times (10^3)^2 = _____$$

(You take the 4^2 separately as 4×4.)

$$(10^{-3})^2 = \_\_\_$$

(The sign convention for multiplication is that if you multiply two numbers of the same sign, the sign of the answer is $+$. If you

10^6
$(10^2)^3$
times
multiply
$\times 4$
8
10^{12}
16×10^6 or 1.6×10^7
10^{-6}

multiply two numbers of different signs, the sign of the answer is −.)

8 × 10⁻²⁴

$$(2 \times 10^{-8})^3 = \underline{\hspace{2cm}} (2^3 = 2 \times 2 \times 2 = 8)$$

0.11 × 10⁻²² or 1.1 × 10⁻²³

$$(3 \times 10^{11})^{-2} = \underline{\hspace{1cm}} \left(3^{-2} = \frac{1}{3^2} = \frac{1}{3 \times 3} = 0.11\right)$$

0.0156 × 10³ or 1.56 × 10¹

$$(4 \times 10^{-1})^{-3} = \underline{\hspace{1cm}} \left(4^{-3} = \frac{1}{4^3} = \frac{1}{4 \times 4 \times 4} = 0.0156\right)$$

Determining the Roots of Numbers Expressed in Exponential Notation

It is possible to reverse the entire process of multiplying the same number together a certain number of times. Instead of asking what that will give, you ask, "What is the number which if I multiply it by itself a certain number of times will result in a given number?" This is called getting the root. An example that you will find familiar is,

What is the square root of 9?

Square is the common way of saying "to the second power," so the question really is,

What number multiplied repetitively two times gives 9? The answer, of course, is 3 since

$$3^2 = 9.$$

Another familiar example is,

What is the square root of 16?

two

16, 4

This is asking what number multiplied repetitively ____ times will

equal ___, and the answer is ___.

The common way of saying to the third power is "cube." Thus, if you are asked for the cube root, you are asked what number

three

multiplied by itself ____ times will give a certain value.

2

What is the cube root of 8? If you multiply three ___'s together

2

you get 8, so the cube root of 8 is ___.

Sometimes you will see roots written

$$\sqrt{} \text{ (square root)} \quad \sqrt[3]{} \text{ (cube root)}.$$

However, the most convenient way to write a root of a number is to write 1 over whatever root it is. Thus the square root of 4 could

8¹ᐟ³

be written $4^{1/2}$ and the cube root of 8 could be written as ____. In

12¹ᐟ¹⁵

the same way, the fifteenth root of 12 could be written as _____. The reason that this is such a convenient way of writing roots will become very obvious. If you have

$$(2^4)^2$$

you multiply the two exponents to get ___. If you want to know what the square root of 2^4 is, you write

$$(2^4)^{1/2},$$

and when you multiply $4 \times 1/2$ you get the new exponent,

$$2^2.$$

Let's check and see if this answer is correct. You want to see if 2^2

is the number which multiplied together ___ times equals 2^4.

Does it? ___.

Since this seems to work, let us try to calculate the cube root of 10^9 by the same method. This can be written as

$$(10^9)^{1/3}.$$

If you multiply 9 by 1/3 you get ___. So the cube root of 10^9 must

be 10^3. Check whether ___ 10^3's multiplied together equal 10^9.

Do they? ___.

What is the fifth root of 10^{10}? This can be written as (10^{10})___,

and multiplying these exponents gives ___. Using the same

methods, the eleventh root of 10^{-33} would be ___.

You probably now see what you are really doing. To get a root of a number expressed exponentially, you divide the exponent by the root you want to take. This is identical with multiplying by 1

over the root. Thus the twelfth root of 10^{48} would be 10___. You

have simply divided ___ by ___.

If you have a digital number times a 10 to a power you take the root of each separately. This is the same as when you raised numbers to a power. For example, the square root of 4×10^4 is written as

$$(4 \times 10^4)^{1/2}.$$

Then you take the 1/2 powers separately,

$$4^{1/2} \times (10^4)^{1/2},$$

which equals

$$2 \times 10^2.$$

Here is another example. The cube root of 8×10^{12} should be

written as ___. When you take the 1/3 powers separately,

you get ___. The cube root (1/3 power) of 8 is 2, and

10^{12} to the 1/3 power is ___. Thus the answer for the cube root

of 8×10^{12} is ___.

The cube root of 2.7×10^{-2} should be written as ___.

Answers (right margin):

2^8

two

Yes

3

three

Yes

1/5

10^2

10^{-3}

4

48, 12

$(8 \times 10^{12})^{1/3}$

$8^{1/3} \times (10^{12})^{1/3}$

10^4

2×10^4

$(2.7 \times 10^{-2})^{1/3}$

$2.7^{1/3} \times (10^{-2})^{1/3}$

When you take the 1/3 powers separately you get _____. However, you have a little problem here. When you try to take the cube root of 10^{-2}, you find that you cannot divide 3 into -2 and get a whole number. You have $10^{-2/3}$, which isn't very useful. To end up with a whole number exponent, the original power to which the 10 is raised must be exactly divisible by the root that you want to take. This isn't an impossible problem, since you can change the exponent by shifting the decimal place of the number. Thus,

$$2.7 \times 10^{-2} = 27 \times 10^{-3}.$$

(You multiply by 10^{-1}, which is negative, and the decimal point

right

moves to the ____.) Now you have an exponent which is exactly divisible by 3. You can rewrite the separated roots as

$$27^{1/3} \times (10^{-3})^{1/3}.$$

3×10^{-1}

Since the cube root of 27 is 3, the final answer will be _____.
Try this one:

$$(1.6 \times 10^7)^{1/2}.$$

You can see immediately that the exponent, 7, is not exactly

2

divisible by __. So you must change the exponent. It is most convenient to have your decimal number greater than 1, so you can

right

shift the decimal point to the ____ and get 16; but then you must

-1

multiply by 10 to the ____ power. (Remember, you get a bigger

right

number when you shift the decimal point to the ____, and the

negative

exponent will be _____.)
The rewritten number, then, will be

$$16^{1/2} \times (10^6)^{1/2},$$

4×10^3

which equals _____.
How would you rewrite

$$(3.2 \times 10^{-9})^{1/5}$$

$32^{1/5} \times (10^{-10})^{1/5}$

so that the exponent is exactly divisible? _____. If

2×10^{-2}

$2^5 = 32$, what is the solution? _____.
For the moment, we will not worry about how to get roots of decimal numbers. As a matter of fact, you will rarely have to worry about roots other than square and cube (1/2 and 1/3 powers), which you might be able to get from a calculator, if you have a very expensive one. Or, as you will see in the next section, on logarithms, you can easily determine any root using logarithms.

SECTION B Logarithms

There is an even shorter way to handle numbers written in exponential notation. Instead of writing a number raised to a power, you just write the power. When you do this, you say you have written the logarithm. Since we almost always write the number 10 raised to a power, when you write the power to which 10 is raised, you say you have written the *common logarithm* or the *logarithm to the base* 10 or simply the *logarithm*. There is one number other than 10 whose power is often written as a logarithm. This is a special number like π, and it is abbreviated as *e*. When you write the power to which *e* is raised, you say you have written the *natural logarithm* or the *logarithm to the base e* or the *Naperian logarithm*. We will have a little to say about these natural logarithms at the end of this section. But for the moment, we will consider common logarithms.

Logarithms have all the properties of exponents. Thus if you want to multiply two numbers, you add their logarithms. When you have a number raised to a power, you can simply multiply its logarithm by the power. If you want to express a very large or a very small number, writing its logarithm will save writing a lot of zeros. It is for these reasons that logarithms are very useful.

Logarithms of Numbers Containing Only One Digit Other Than Zeros

Since the common logarithm (or just logarithm) is the exponent to which 10 is raised, the logarithm of 10^2 is equal to 2. The abbreviation for logarithm to the base 10 (common logarithm) is log (or if you must be super-exact, $\log_{10}$). Thus you can write

$$\log 10^2 = 2 \quad \text{and} \quad \log 10^{-1} = -1$$

$$\text{or} \quad \log 10^5 = \underline{\hphantom{xxx}} \quad \text{or} \quad \log 10^{-3} = \underline{\hphantom{xxx}}$$

5, −3

As you see, the values of the logarithms may be either positive or negative. However, as with powers, you don't bother to write + signs. They are understood.

If you want to get the logarithm of a number that is not written in exponential notation, you have to convert it to exponential notation. Then you simply write the power to which 10 is raised. For example, if you want to get log 1000, you first write the 1000

as 10 to a power. This would be $1000 = \underline{\hphantom{xxx}}$. Now to get the

10^3

logarithm, all you have to do is write the $\underline{\hphantom{xxx}}$ to which 10 is raised.

power

$$\log 1000 = \log 10^3 = \underline{\hphantom{xxx}}$$

3

If you want to find log 0.00001, first you must write the number

in $\underline{\hphantom{xxxxx}}$ notation.

exponential

$$\log 0.00001 = \log \underline{\hphantom{xxx}}$$

10^{-5}

power	Now you can write the logarithm simply by writing the _____ to which 10 is raised.
−5	$\log 0.00001 = \log 10^{-5} =$ _____

You have probably noticed that the sign of the logarithm, + or −, depends on whether the number whose logarithm you are getting is larger or smaller than 1. To show you this, the following table lists a series of numbers from very large to very small. Some are shown in exponential notation. Others have their logarithms given. Complete the table.

	Number	Exponential Notation	Log
10^3	1000	_____	3
2	100	10^2	_____
10^1	10	_____	1
	1	10^0	0
10^{-1}	0.1	_____	−1
−2	0.01	10^{-2}	_____
10^{-3}	0.001	_____	−3

You can see from this list that numbers greater than 1 have a positive logarithm (the + is understood), and numbers less than 1

− (negative), zero	have a _____ logarithm. The logarithm of 1 itself is_____, since any number to the zero power equals 1. If you have forgotten this, check back to page 4.
0	Since log 1 = _____ and log 10 = 1, the logarithm of a number between 1 and 10 must be equal to something between 0 and 1. It is some decimal number. The logarithms of the whole numbers between 1 and 10 are shown in the following table.

Number	Logarithm	Number	Logarithm
1	0.000	6	0.778
2	0.301	7	0.845
3	0.477	8	0.903
4	0.602	9	0.954
5	0.699	10	1.000

There isn't a direct relationship between the logarithm and the number. Although 5 is halfway between 1 and 10, its logarithm is not halfway between 0 and 1. So you will always need a table of logarithms (or a calculator, if yours happens to have logarithms) to find the logarithms of numbers between 1 and 10. In these tables of logarithms, the column on the left side always lists the numbers

10	between 1 and _____ whose logarithms you are looking for. The values for the logarithms that appear in columns in the table always
1	have values between 0 and _____.

So far, we have considered only numbers that can be written as 10 to some power. But what happens if you want the logarithm of a number like 50? You can't just write this as 10 to some power. What you do is write the number in scientific notation. In this

example, you can write 50 as _____. In this way you have written it as some number between 1 and _____ multiplied by 10 to some power. And now you can take advantage of what happens when you multiply numbers written as 10 to a power. You simply add the powers. Since the logarithms are the powers (exponents),

5×10^1

10

when you multiply two numbers you can _____ their logarithms. In the example, 50 is 5×10^1 in scientific notation, and you can say

$$\log 50 = \log (5 \times 10^1) = \log 5 + \log 10^1$$

add

Then all you have to do is look up log 5 in a table (it is 0.699), and add it to $\log 10^1$.

$$\log 50 = \log (5 \times 10^1) = \log 5 + \log 10^1$$
$$= 0.699 + 1$$
$$= 1.699$$

Here is another example. Say that you want to find log 0.0003.

First you write the number in _____ notation. This gives you _____.

scientific

3×10^{-4}

$$\log 0.0003 = \log (3 \times 10^{-4})$$

Since 3 is multiplied by 10^{-4}, you can _____ their logarithms.

add

$$\log 0.0003 = \log (3 \times 10^{-4})$$
$$= \log 3 \_\_\_ \log 10^{-4}$$

$+$

Then you look up the value for log 3 in a table (it is 0.477), and you know that $\log 10^{-4} = $ _____. Writing all the steps, we have

-4

$$\log 0.0003 = \log (3 \times 10^{-4})$$
$$= \log 3 + \log 10^{-4}$$
$$= 0.477 + (\_\_\_)$$

-4

You must now add 0.477 to -4. This is an algebraic addition. You subtract the smaller number from the larger, and give the answer the sign of the larger. If you have forgotten this, check back to page 12.

$$\log 0.0003 = 0.477 + (-4)$$
$$= \_\_\_\_\_$$

-3.523

scientific notation	Here is another example of the same type. You want log 0.006. You start by writing 0.006 in _____.
6×10^{-3}	log 0.006 = log (_____)
	Since you have the logarithm of two numbers that are multiplied
add	together, you can _____ their logarithms.
	log 0.006 = log (6×10^{-3})
log 6 + log 10^{-3}	= _____
table	You can find log 6 by looking it up in a_____. You know that
−3	log 10^{-3} = _____. So
	log 0.006 = log (6×10^{-3})
	= log 6 + log 10^{-3}
0.778	= _____ + (−3)
	Making the algebraic addition, you have
−2.222	log 0.006 = _____
	Notice that, since you had the logarithm of a number that is less
− (negative)	than 1, the sign was _____.
	Here is one more for practice. You want log 400,000. Since the number is greater than 1, you know that the sign of the logarithm
+ (positive)	is _____. First you write the number in scientific notation.
4×10^5	log 400,000 = log (_____)
4, 0.602	Looking up log _____ in the table, you find that it is _____. You
5	know that log 10^5 is _____.
	So
	log 400,000 = log (4×10^5)
+ log 10^5	= log 4 _____
0.602, 5	= _____ + _____
5.602	= _____
	In this example, both the numbers you added had a + sign, so
+	you simply added them and gave your answer a _____ sign, which
understood	you don't write because it is _____. Notice that the number
+	400,000 is greater than 1, and therefore its logarithm has a _____ sign.

Logarithms of Numbers Containing
More Than One Digit Other Than Zeros

Say you want to find the logarithm of 55. When you write this in

scientific notation you have _____. But the table of logarithms
that we have been using contains only whole numbers between 1
and 10. How can you find log 5.5? You would have to have a
table that lists one digit after the decimal. Such tables do exist. For
example, such a table would show that numbers between 5.0 and
6.0 have the following values.

5.5×10^1

Number	Logarithm	Number	Logarithm
5.0	0.6990	5.6	0.7482
5.1	0.7076	5.7	0.7559
5.2	0.7160	5.8	0.7634
5.3	0.7243	5.9	0.7709
5.4	0.7324	6.0	0.7782
5.5	0.7404		

So, from this table, log 5.5 is _____. Now you can get the value
of log 55.

0.7404

$$\log 55 = \log (\underline{\hspace{2cm}})$$

$$= \log \underline{\hspace{1.5cm}} + \log 10^1$$

$$= \underline{\hspace{1.5cm}} + \underline{\hspace{1cm}}$$

$$= \underline{\hspace{1.5cm}}$$

5.5×10^1

5.5

0.7404, 1

1.7404

If you had a complete table for all numbers between 1.0 and 9.9,
it would be ten times longer than the one shown here. That's
pretty long, but with very small print it can fit on one page. But
what if you wanted to know log 5.55? This is a number with three
digits in it. You could make a table with all the numbers listed
in it from 1.00 to 9.99. This would be ten times longer still. To
save space, these tables list the first two digits down the left side
and the last digit across the top. Then instead of having one very
long list, you would have ten shorter lists side by side. For numbers
that have 5.5 for the first two digits, such a table would look like
this.

	0	1	2	3	4	5	6	7	8	9
5.5	.7404	.7412	.7419	.7427	.7435	.7443	.7451	.7459	.7466	.7474

.7443

5.55 × 10¹

5.55

.7443

1.7443

scientific

5.58 × 10⁻⁴

5.58, 10⁻⁴

.7466

−4

.7466, −4

−3.2534

Now you can find log 5.55. It is _____. And if you want log 55.5, you go through the following steps.

$$\log 55.5 = \log (\text{_____})$$

$$= \log \text{\_\_\_\_\_} + \log 10^1$$

$$= \text{\_\_\_\_\_} + 1$$

$$= \text{\_\_\_\_\_}$$

Here is another example. You want log 0.000558. You start by by writing it in _____ notation.

$$\log 0.000558 = \log (\text{_____})$$

Then you add the logarithms of the numbers that are multiplied together.

$$\log 0.000558 = \log \text{\_\_\_\_\_} + \log \text{\_\_\_\_\_}$$

You can find log 5.58 in the table above. It is _____. You know that log 10⁻⁴ equals _____.
So you can write

$$\log 0.000558 = \log 5.58 + \log 10^{-4}$$

$$= \text{\_\_\_\_\_} + (\text{\_\_\_\_})$$

$$= \text{\_\_\_\_}$$

If you didn't get the number shown, it means that you didn't subtract the smaller from the larger. If you didn't get a − sign it means that you didn't give your answer the same sign as the larger.

The type of table that enables you to find logarithms of numbers containing three digits always has the values of the logarithms expressed to four places to the right of the decimal point. Therefore, it is called a "four-place table of logarithms." You can find a complete four-place table in Appendix III at the back of this book. It covers two pages even though the print is very small. The print is somewhat larger in the *Handbook of Chemistry and Physics* (Cleveland, Ohio: The Chemical Rubber Company). It also covers two pages. Still more exact tables are available. There is a five-place table in the *Handbook of Chemistry* that is good for numbers like 5.555. It covers 20 pages! Since almost everyone has a pocket calculator today to do their multiplications and divisions, you will have very little use for such an exact table.

To use a four-place table of logarithms to find the logarithm of a number between 1 and 10, you go down the column on the left that contains the first two digits.

When you reach the line that gives the correct first two digits, you go across this line until you get to the column for your third digit. The number printed there is the logarithm for which you are looking.

For example, say you want the logarithm of 5.61. You go down the left-hand columns in the table in Appendix III until you reach the line with 5.6. Then you go across this line until you reach the column under 1. The number printed there, .7490, is the logarithm of 5.61. That's pretty easy.

Let's try one. We will use the four-place table in Appendix III to get log 2.46. First you go down the left-hand side until you

reach the line for the _____ values. Then you go across this line   2.4

until you reach the column for all third digits that are _____. The   6

logarithm for 2.46 that you find there is _____.   .3909

Here is another one. You want log 0.0080. Since the number is

less than 1, the sign on the log will be _____. You start by   − (negative)

writing the number in _____ notation.   scientific

$\qquad$ log 0.0080 = log (_____) 8.0×10^{-3}

Since you have two numbers multiplied together you can _____   add
their logarithms.

$\qquad$ log 0.0080 = log (8.0×10^{-3})

$\qquad\qquad\qquad$ = _____ + log 10^{-3} log 8.0

You know that log 10^{-3} equals _____. Now you must find log   −3
8.0 in the log table. You go down the left side until you reach the

line for the first two digits, _____. But what is the third digit?   8.0
Since no third digit appears in 8.0, you must assume that it is zero.

So the value for the logarithm will be in the 0 column. It is _____.   .9031
Now you can finish the problem.

$\qquad$ log 0.0080 = log 8.0 + log 10^{-3}

$\qquad\qquad\qquad$ = _____ + _____   .9031, −3

$\qquad\qquad\qquad$ = _____   −2.0969

To make sure that you understand this perfectly, check yourself on the following.

log 1.06	= _____	0.0253
log 84700	= _____	4.9279
log 0.03	= _____	−1.5229
log 0.00213	= _____	−2.6716
log 0.00000000895	= _____	−8.0482

If you had trouble with any of these, you had better go through all the examples in the text again.

Incidentally, you may sometimes find tables of logarithms that don't show the decimal point in the left-hand column. Instead of having 5.6, they have just 56. But you should understand that there is a decimal point after the first digit.

Antilogarithms

Say that you know the value of a logarithm and you want to know what number it represents. You simply reverse the process of finding the logarithm of a number, and you say you are finding the *antilogarithm*. The abbreviation for antilogarithm is antilog.

For example, say you know that the value of the logarithm of a number N is 3.

$$\log N = 3.$$

To find the value for the number N, you write

antilog 3. $= N$

3

You know that $\log 10^3 =$ _____, so it is easy to see that

antilog 3. $= 10^3$

All this means is that the number whose logarithm is 3 is 10^3.

10^{-1}

If you want antilog -1., you know that it is equal to _____. All you have done is write 10 to the power that the antilog asks for. This is all that you must do to get antilogarithms of numbers to the left of the decimal point. However, if you want an antilogarithm of a number that is less than 1, say .8545, you must use a table of logarithms. What you do is look through the values in the table until you find the number. Then you read off the value that would have given you this number. Looking through the four-place table in Appendix III, you see that log 7.15 is .8545. Therefore

7.15

antilog .8545 $=$ _____.

This isn't too hard a search, since the values for the logarithms become larger as you go farther down and farther to the right in the table.

Here is another example. You want antilog .6180. You look

4.1

through the table until you find this number. It is on the _____ line

5

and in the column that means the third digit is _____. Therefore

4.15

antilog .6180 $=$ _____.

If you want antilog .3054, you look through the logarithms in the

2.0

table and find that it is on the line that has _____ as the first two

2

digits. It is in the column that has _____ as the third digit. Therefore,

2.02

antilog .3054 $=$ _____.

You must have noticed that you have used two different ways of getting the antilogarithms of numbers. The first was antilog

10^3

3. $=$ _____. The number whose antilogarithm you wanted was

greater than 1 (to the left of the decimal), and you just wrote it as the power to which 10 is raised. The second was antilog 0.3053, and you had to look it up in a table of logarithms to find that is was 2.02. In this case, the number whose antilogarithm you wanted was less than 1 (to te right of the decimal). Thus, when you get antilogarithms of numbers, you write whatever is to the left of the

decimal as _____ to that power. Whatever is to the right of the

10

decimal you must _____ in a table of logarithms.

look up

Now what do you do if the number whose antilogarithm you want has numbers both to the left and to the right of the decimal? For example, say that you want antilog 6.776. You have to say that 6.776 = 6.+.776. It is the sum of two numbers, one to the left of the decimal and the other to the right. You can write

antilog 6.776 = antilog (6. + _____)

.776

Just as you add logarithms of numbers that are multiplied, you multiply numbers whose antilogarithms are added. You can write

antilog 6.776 = antilog (6. + .776)

= antilog 6. × antilog _____

.776

Now you can get antilog 6. just by writing 10 raised to the _____

6 (sixth)

power. And you can get antilog .776 by _____ it up in a log table.

looking

antilog 6.776 = antilog (6. + .776)

= antilog 6. × antilog .776

= _____ × 5.97

10^6

Usually when numbers are written in exponential notation, the decimal number comes first and the 10 to a power comes last. So you would write this as

antilog 6.776 = 5.97×10^6

Here is another example to make sure you have the whole process. You want antilog 3.356.

antilog 3.356 = antilog (_____)

3. + .356

= antilog 3. ____ antilog .356

×

You know antilog 3. = ____. To get antilog 0.356 you must ____

10^3, look

it up in a table. When you do, you find it is ____. So you can write

2.27

antilog 3.356 = antilog (3. + .356)

= antilog 3. × antilog .356

= 10^3 × ____

2.27

or written in the correct order, _____.

2.27×10^3

Do one yourself and see whether you come up with the right answer.

1.04 × 10⁸

    antilog 8.017 = _____

Now what do you do with antilogarithms of negative numbers?

−3

You can easily get the antilog (−3.). It is just 10 to the _____ power. But there are no negative numbers in the table of logarithms. You have no way of looking up antilog (−.676). What you must do is write the negative number whose antilogarithm you want as the algebraic sum of a negative number that is greater than one (to the left of the decimal point) and a positive number that is less than one (to the right of the decimal). For example, you can write −2.6 as (−3.+.4). You take one number greater than the number to the left of the decimal as the negative number (−3 for −2 in this example), and then you subtract what is to the right of the decimal to get the positive part (you subtract 0.6 in this example). Thus

−8.,   0.2

    −7.8   = _____ + _____

or

−7.,   0.55

    −6.45   = _____ + _____

or

−1.,   0.243

    −0.757 = _____ + _____

Now say that you want antilog −0.757. You break it up into the

greater

sum of a negative (−) number that is _____ than 1 and a

less

positive (+) number that is _____ than 1.

−1. + .243

    antilog −0.757 = antilog (_____)

multiply

Since you _____ the antilogarithms of sums of numbers, you can write

    antilog −0.757, = antilog (−1. + .243)

               = antilog −1. × antilog .243

10⁻¹

You know antilog −1 equals _____, but you must look up

1.76

antilog .243 in a log table. It is _____. Therefore

1.76 × 10⁻¹

    antilog −0.757 = _____.

Here is another example to see that you really have it. You want antilog −9.2291. You start by writing the number as the sum of a negative whole number and a positive number less than 1.

−10 + .7709

    antilog −9.2291 = antilog (_____)

multiply

You can _____ the antilogarithms of the numbers that are added.

antilog .7709

    antilog −9.2291 = antilog −10 × _____

You know that antilog −10 equals _____, but you must look up

antilog .7709 in a log table. It is _____. Therefore

     antilog −9.2291 = _____.

10^{-10}

5.90

5.90×10^{-10}

It is very important for you to be able to determine the
antilogarithms of negative numbers, since they always come up
when you are determining the concentration of acids from their
pH values. The whole idea of pH is covered in Appendix IV.

Here is one last problem that you will face when you are doing
antilogarithms. More often than not, the number you look up in the
log table will fall between two values in adjacent columns. For
example, say you are trying to get antilog .8473. You can find
.8470, and then the next number is .8476. The number you want is
between these two. Since the antilog .8470 is 7.03 and the
antilog .8476 is 7.04, the antilogarithm of your number, .8473,
must be somewhere between 7.03 and 7.04. But where? You will
have to guess, or as mathematicians say, "interpolate." Although
it isn't exactly right, we're going to assume that there is a direct
relationship (over this short range) between the number and its
logarithm. Thus .8473 is halfway between .8470 and .8476, so the
antilogarithm of your number must be halfway between 7.03 and
7.04. Well, halfway between 7.03 and 7.04 is 7.035. Therefore
you estimate that

    antilog .8476 = 7.035

Let's take a look at the mechanics of this interpolation process.
What you want to know is what part of the way between two
adjacent logarithms your number is. In our example it was halfway
or 0.5. Then you add this part to the lower of the two numbers,
which gave you the two adjacent logarithms. In this example, you
added the 5 to 7.03 and got 7.035. We will do a lot of examples,
and it will clear this up.

Let's look at the example we just did. The two adjacent
logarithms were .8476 and .8470. You subtract the smaller from
the larger.

    .8476 − .8470 = _____

.0006

Then you subtract the lower, .08470, from the number you want
to interpolate.

    .8473 − .8470 = _____

.0003

Thus your number is .0003 out of the .0006 total difference. You
can get a decimal number for this by dividing your difference by
the total difference.

$$\frac{.0003}{.0006} = \underline{}$$

0.5

So you put a 5 at the end of the value for the antilogarithm of the

7.03	smaller, .8470. Since the antilog .8470 is _____, your number
7.035	will have as its antilogarithm the value _____.
	Here is another example for some more practice. You want antilog .4716. Looking in the log table, you find that the nearest
.4713, .4728	you can get is between _____ and _____. The difference between these two adjacent logarithms is
.4713, .0015	.4728 − _____ = _____
	You subtract the smaller of these two from your number.
.4713, .0003	.4716 − _____ = _____
	So your number is three parts away out of a total of 15 parts. You divide to get the decimal
.2	$\dfrac{.0003}{.0015} =$ _____
2	Therefore you must put a _____ on the end of the antilogarithm of the smaller, .4713.
2.96	antilog .4713 = _____ (lower logarithm)
2.962	antilog .4716 = _____ (your logarithm)
	Try one more. You want antilog .5760. Looking in the log table,
.5752, .5763	you see that the nearest values you can get are _____ and _____. To get the total difference between these two adjacent values,
subtract	you _____ the smaller from the larger.
.0011	.5763 − .5752 = _____
	The difference between your value and the lower of the adjacent values is
.5760, .0008	_____ − .5752 = _____
	To get the parts away, you divide
.0008 .0011	$\dfrac{\rule{1cm}{0.4pt}}{\rule{1cm}{0.4pt}} = .727$
	Since this interpolation is only an approximation, you can drop everything except the first number, 7, and you put this digit on
lower	the end of the antilogarithm of the _____ of the two adjacent values.
	antilog .5752 = 3.76 (lower logarithm)
3.767	antilog .5760 = _____ (your logarithm)

Here is an example without any help. See whether you can work it out. You want antilog .2725. First you make the subtraction

_____ − _____ = _____   .2742, .2718, .0024

Then you make the subtraction

_____ − _____ = _____   .2725, .2718, .0007

Then you divide

$\dfrac{\rule{2cm}{0.4pt}}{\rule{2cm}{0.4pt}}$ = .2916   .0007

.0024

Since .2916 is very close to .3, you put a _____ on the end of the   3

antilogarithm of _____.   .2718

    antilog .2718 = _____   1.87

    antilog .2725 = _____   1.873

Here are a few to try by yourself. Work them out on a separate sheet of paper, and compare your answers with those in the margin.

    antilog .1007 = _____   1.261

    antilog .44   = _____ (remember .44 = .4400)   2.754

    antilog .609 = _____ (remember .609 = .6090)   4.065

(In this last example, you might have thought that the last digit was 4. When we get to significant figures, you will see why 5 is better.)

Now we are going to pull this all together. You will be finding the antilogarithm of a negative number that has digits both to the left and right of the decimal and that needs interpolation.

    antilog −6.43

The first thing you must do with a negative number is to make it the sum of a negative whole number and a positive number that

is _____ than 1.   less

    antilog −6.43 = antilog (_____)   −7 + .57

When you have the antilogarithm of a sum, you can _____ the   multiply

antilogarithms of the things that are added.

    antilog −6.43 = antilog (−7 + .57)

                 = antilog −7 _____ antilog 0.57   ×

You know that the antilog −7 is _____, but you have to look up 10^{-7}

.57 in a _____. Realize that .57 is the same as .5700. When   log table you make the interpolation as you did in the preceding part, you find that

3.715

    antilog .5700 = _____

Therefore you can write

3.715

    antilog -6.43 = 10^{-7} × _____

or in the usual order

3.715×10^{-7}

    antilog -6.43 = _____

This is about the worst kind of antilogarithm you will ever face.

Natural Logarithms

Recall that at the very beginning of this section on logarithms we said that there was another number besides 10 whose exponent you expressed as a logarithm. This number, which has the symbol e, has a value of 2.71828. Like π, it is not a whole number, and you could go on writing more digits at the end indefinitely. When you express logarithms "to the base e," you are expressing the exponent to which e is raised. These are called *natural logarithms*, and the abbreviation is ln (or if you are super-exact, $\log_e$).

    There are tables available that list numbers with their natural logarithms. However, you rarely need them because it is very simple to get from a natural logarithm to a common (to the base 10) logarithm. You simply multiply the common logarithm by 2.303 and you have the natural logarithm.

    ln N = 2.303 × log N

Say that you want to find ln 10^3. You can write

    ln 10^3 = 2.303 × log 10^3.

3.

But you know that log 10^3 = ____; therefore

    ln 10^3 = 2.303 × 3

6.909

        = _____

If you want ln 24.5, you write

2.303

    ln 24.5 = _____ × log 24.5

2.45×10^1

        = 2.303 × log (_____)

2.45, 10^1

        = 2.303 (log _____ + log _____)

.3892, 1

        = 2.303 (_____ + ____)

1.3892

        = 2.303 (_____)

3.1999

        = _____

It's just as easy as that.

More frequently what you will encounter is an equation with e raised to a power. You want to know what number this is equal to. What you actually want is the antilogarithm. For example, you want to know what $e^{0.25}$ is equal to. Since the natural logarithm is the power to which e is raised, you can write

$\ln N = 0.25$

(The natural logarithm equals the power.) You know from what we said before that

$\ln N = 2.303 \times \log N$

so you can say

$\ln N = 0.25 = 2.303 \times \log N$

or just

$0.25 = 2.303 \times \log N$

If you divide both sides of the equation by 2.303,

$$\frac{0.25}{2.303} = \frac{2.303}{2.303} \times \log N$$

You know that any number divided by itself equals 1, so you can write

$$\frac{0.25}{2.303} = \log N \quad \text{or} \quad .1086 = \log N$$

All you have to do now is find the antilogarithm of .1086, and you will know the value for N

antilog $.1086 = N = $ _____   1.284

Let's go through this once more, but you do the work. You want to know what number $e^{4.705}$ is equal to. You know that the

power to which e is raised is equal to its _____ logarithm.   natural
So you write

_____ $N = 4.705$ ln

To get from ln to log, you can write

$\ln N = $ _____ $\times \log N$ 2.303

So you can now write

_____ $= 2.303 \times \log N$ 4.705

and dividing both sides by _____, you get   2.303

$$\frac{4.705}{\rule{1cm}{0.4pt}} = \log N \quad \text{or} \quad \rule{1cm}{0.4pt} = \log N$$ 2.303,   2.0430

antilog

2 + .0430

×

1.104

All you must do now is find the _____ of 2.0430

N = antilog 2.0430 = antilog (_____)

= antilog 2. _____ antilog .0430

= ——— × 10^2

This is just about the whole story on natural logarithms.

Uses of Common Logarithms

Before the coming of the inexpensive pocket calculator, chemists used two tools to do their calculations: the slide rule and the table of logarithms. The slide rule is very fast for doing multiplications, divisions, squares, cubes, square roots and cube roots. However, the answer that you can read from a slide rule usually has only three digits. The chemist who needed greater precision could use a four- or even five-place table of logarithms. You see, the logarithm changes multiplication into simple addition, and division into subtraction. You can get powers and roots by multiplying or dividing the logarithms.

Now almost everyone has a pocket calculator. Even in its simplest form, the calculator does multiplications and divisions to an eight-digit answer, so hardly anyone bothers to use logarithms to multiply or divide. However, only the most expensive calculators can get powers or take roots of numbers. You can get the powers by repetitive multiplication, but frequently the answer "overflows" the capacity of the calculator. So, for those extremely rare chemical calculations that use roots and powers, you may still need logarithms.

Determining Powers and Roots

(Your instructor may suggest that you skip this section.)

In Section A of this chapter, pages 16 to 20, you learned that when you raised 10 to a power itself to a power (that is, multiplied it by itself repetitively), you could simply multiply the powers.

10^{-6}

Thus $(10^5)^3 = 10^{15}$. Or $(10^{-2})^3 =$ _____. When you wanted to get a root of 10 to a power, you simply divided the power by

10^3

whatever the root was. Thus $\sqrt[2]{10^6} = 10^3$ and $\sqrt[3]{10^9} =$ _____. In fact, you got these roots by multiplying the power to which the 10 was raised by 1/root, which is the same as dividing by the root.

power

Now, the common logarithm of a number is the _____ to which 10 is raised. Therefore, if you want a number to a power, you can just multiply the logarithm of the number by the power. If you

logarithm

want the root of a number, you simply divide the _____ of

the number by the root. Then you have the logarithm of your

answer. To get the answer itself you must take the _____.   antilogarithm
    Here is an example. You want to find the value of $(2.5)^{14}$. You
start by getting the logarithm of the number.

        log 2.5 = _____   .3979

Since you want the number to the 14 power, you must _____   multiply
the logarithm by 14.

        log $(2.5)^{14}$ = _____ × log 2.5   14

                 = 14 × _____   .3979

                 = _____   5.5706

You know that the logarithm of your number is 5.5706. To get the

number itself, you must take the _____ of 5.5706.   antilogarithm

        $(2.5)^{14}$ = _____ 5.5706   antilog

    If you get the antilogarithm, as you did in the preceding part of
this section, you find that

        antilog 5.5706 = _____ 3.721×10^5

So

        $(2.5)^{14}$ = 3.721×10^5

This isn't exactly correct because of the method you used in
interpolating. The correct answer is 3.72528×10^5. However, when
you get to Section C in this chapter, which is on significant figures,
you will see that 3.7×10^5 is as near to the answer as you are
permitted to write.

    Getting the roots of numbers using logarithms is the same as
taking numbers to a power. However, instead of multiplying the
logarithm by the power, you divide the logarithm by the root. Here
is an example. You want to know the value of $\sqrt[3]{2.5}$. You start by
getting the logarithm of the number.

        log 2.5 = _____   .3979

Then, to get the cube root, you divide the logarithm by _____.   3

$$\log \sqrt[3]{2.5} = \frac{0.3979}{\rule{2cm}{0.4pt}}$$ 3

             = _____   .1326

You now know what the logarithm of $\sqrt[3]{2.5}$ is. To get the value,

you must take the _____.   antilogarithm

        antilog .1326 = _____   1.357

    Once again there is a slight error due to the method of
interpolation. It is trivial, since the rules of significant figures, which

are coming up in Section C, say that the best you can write is just 1.4. The use of significant figures is going to be a revelation!

When you take roots of numbers expressed in scientific notation, you have to be careful with your signs. Here is an example. You

logarithm

want $\sqrt[4]{2.5 \times 10^{-3}}$. First you get the _____.

2.5, 10^{-3}

$$\log 2.5 \times 10^{-3} = \log \underline{\quad\quad} + \log \underline{\quad\quad}$$

.3979, (−3)

$$= \underline{\quad\quad} + \underline{\quad\quad}$$

−2.6021

$$= \underline{\quad\quad\quad}$$

divide

Since you want the fourth root, you must _____ the logarithm

4

by ____.

4

$$\log \sqrt[4]{2.5 \times 10^{-3}} = \frac{-2.6021}{\underline{\quad\quad}}$$

−.6505

$$= \underline{\quad\quad}$$

That was the place where you might have lost the − sign.

antilogarithm

Now to get the value itself, you must find the _____ of −.6505. Remember how we did negative numbers.

−1 + .3495

$$\text{antilog} -.6505 = \text{antilog} \,(\underline{\quad\quad\quad\quad})$$

×

$$= \text{antilog} -1 \underline{\quad\quad} \text{antilog} \,.3495$$

2.236 × 10⁻¹

$$= \underline{\quad\quad\quad\quad}$$

Once again the rules of significant figures tell you that you may only write this as 2.2×10^{-1}.

Determining pH

Actually, you are going to find the most use for logarithms when you determine the pH of acid and base solutions. This is because the p function means −log. There are other p functions, like pOH, pK_a, and pK_w. They mean −log H⁺, log OH⁻, −log K_a, and −log K_w. The pH is such an important topic that we cover it separately in Appendix IV. It is best if you do this appendix after you have done Chapter 12. However ,you can do it any time that your instructor suggests. We will not cover it here.

Uses of Natural Logarithms

Earlier in this section, we mentioned two uses that you have for natural logarithms. One is to find the value of e raised to a power. The other is the reverse, to find the natural logarithm of a number. These uses come about because there are several important relationships that have the general form

$$x = Ae^{-y/RT}$$

Depending on what you are interested in, x might be related to the speed of a chemical reaction, or to how easily a liquid evaporates, or even to the useful work that you can get out of a chemical reaction.
And y might be related to the heat that a reaction needs, or the heat it takes to evaporate a liquid, or the relative concentrations of reactants and products in a chemical reaction. T is always the temperature, and R is called the *gas constant* (see Chapter 10). The A is a constant that you are usually given. This is pretty complicated stuff, and it isn't the purpose of this book to teach you these relationships. All we want to do is to learn how to handle this kind of equation when it has numbers put into it. This means learning how to work with natural logarithms, since the equation

has _____ raised to a power. e

   You now know quite a bit about logarithms. You know that the

logarithm is the _____ to which a base is raised. You know that   power

if the base is 10, you are dealing with _____ logarithms. If the   common

base is e, you are working with _____ logarithms. You know   natural

that when two numbers are multiplied together, you can _____ their   add
logarithms. You know how to convert from natural to common logarithms. The relationship is

$$\ln N = \underline{\hspace{1cm}} \times \log N$$ 2.303

Let's see how putting all this information together will get the very complex equation we had at the start into a form that is easier to handle.

   Here is the equation again.

$$x = Ae^{-y/RT}$$

You start by taking the natural logarithms of both sides of the equation.

$$\ln x = \ln (Ae^{-y/RT})$$

Since $Ae^{-y/RT}$ is two things multiplied together, A and _____, $e^{-y/RT}$
you can add their logarithms.

$$\ln x = \ln A \underline{\hspace{1cm}} \ln e^{-y/RT}$$ $+$

But you know that the natural logarithm of e raised to a power is

power

equal to that _____. So

$$\ln e^{-y/RT} = \underline{\hspace{2cm}}$$

This simplifies the equation to

$$\ln x = \ln A + \left(-\frac{y}{RT}\right) \quad \text{or} \quad \ln x = \ln A \underline{\hspace{1.5cm}}$$

power

$-\dfrac{y}{RT}$

$-\dfrac{y}{RT}$

Then you want to convert the natural logarithms to common logarithms. Each time you have an "ln" you substitute

2.303

"_____ × log."

2.303 × log A

$$2.303 \times \log x = \underline{\hspace{3cm}} - \frac{y}{RT}$$

To get the equation into its simplest form, you divide each term by 2.303.

$$\frac{2.303}{2.303} \times \log x = \frac{2.303}{2.303} \times \log A - \frac{y}{2.303\,RT}$$

Now, since any number divided by itself equals 1, you can write

log A

$$\log x = \underline{\hspace{1.5cm}} - \frac{y}{2.303\,RT}$$

By going through this elaborate set of transformations, you are taking a relationship that had e raised to a power in it and fiddling around with it until you have something that is quite simple and contains only common logarithms. Then, if you are given values for any three of x, y, A, or T (you always know R), you can calculate the missing fourth. If you have a series of values for x and the corresponding T, you can plot this data graphically and determine the value of y from the slope.

SECTION C Significant Figures

As great as the new pocket calculators are, they can lead unsuspecting students to write some very silly answers to their calculations. The reason for this is that these calculators can produce answers with at least eight digits in them. Some of the more expensive ones even give you as many as twelve digits in the answer. More often than not, most of these digits don't mean anything. When you write them down, you are just writing a lot of numbers that you are not at all sure of.

Let's see how this comes about. Say that you want to measure the area of a piece of paper with a 12-inch ruler. Each inch is divided into sixteenths. The piece of paper is rectangular, so its area will be equal to the height times the width. You measure the height.

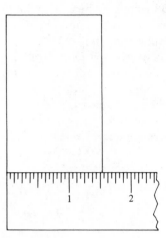

It looks as if it's about halfway between $1\frac{8}{16}$ and $1\frac{9}{16}$ inches. You guess that it is $1\frac{8.5}{16}$. Since a sixteenth of an inch is, in decimal notation, 0.0625 inch, 8.5 sixteenths is 8.5×0.0625 or 0.53125 inch. So the height is 1.53125 inches.

Next you measure the width of the piece of paper.

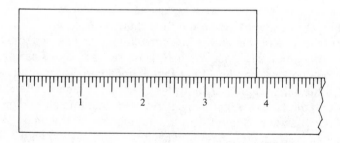

It looks as if it's a little more than $3\frac{13}{16}$ inches. You guess that it is two-tenths of the way between $\frac{13}{16}$ and $\frac{14}{16}$, and so you call it $3\frac{13.2}{16}$ inches. Since one-sixteenth of an inch is 0.0625 inch, 13.2 sixteenths is 0.825 inch. So your guess for the width is 3.825 inches.

Then you whip out your trusty calculator and get the area by multiplying 1.53125 times 3.825. With an eight-digit calculator you get 5.8570312 square inches. But writing down all these numbers is nonsense. As we will see, the last five are meaningless.

Say that you guessed a little wrong and said that the height, instead of $1\frac{8.5}{16}$, was $1\frac{8.6}{16}$. As a decimal number, this gives you a height of 1.5375 inches. Say that you also guessed a little wrong on the width. It was really $3\frac{13.3}{16}$ and not $3\frac{13.2}{16}$. In decimal notation this is 3.83125 inches. The area using these new values would be 5.8905468 square inches. Only the first two digits are the same as the area you got before, 5.8570312.

Now, let's go through the whole process again, but this time let's say that the true height and width are $1\frac{8.4}{16}$ and $3\frac{13.1}{16}$ inches, respectively. Expressing these in decimal notation, you have 1.525 inches and 3.81875 inches. The area these dimensions give you is 5.8235937 square inches. Once again, only the first two digits in the area are the same as those you got before.

Here are all these numbers in a table so that you can see what's going on.

Height (inches)	Width (inches)	Area (square inches)
$1\dfrac{8.4}{16}$	$3\dfrac{13.1}{16}$	5.8235937
$1\dfrac{8.5}{16}$	$3\dfrac{13.2}{16}$	5.8570312
$1\dfrac{8.6}{16}$	$3\dfrac{31.3}{16}$	5.8905468

You see, just by making a very small error in guessing how far between the sixteenth markings on the ruler the edge of the paper was, you have areas that vary from 5.82... up to 5.89... square inches. What is the sense in writing that the area is 5.8570312 square inches when you aren't really certain whether it is 5.82... or 5.89... square inches? Even the third digit is uncertain.

To take care of this problem, a convention has been adopted. This convention says that you may write as many digits as you are sure of, and then one final digit that is uncertain. These digits are called the *significant figures*. In this example, you can write the first two digits, 5.8, which are certain, and then one more digit, 6, that is not certain. So you would write the area of the piece of paper as 5.86 square inches and say that you have expressed it to *three significant figures*.

Now let's consider when you must use this significant-figure convention and how you use it. First you must realize that there are two different types of numbers. The first type of number comes either from a definition or by counting. Examples would be the 12 inches in a foot, the 60 minutes in an hour, or the 32 students (if you counted them) in a classroom. There is no uncertainty in this type of number, and if you use it in a calculation, you do not consider it when you determine the allowed significant figures in your answer. You might think of it as containing an infinite number of significant figures.

The second type of number is one that you obtain by measuring something with a tool or an instrument—like a ruler, a stop watch, a balance, and so forth. With this type of number, the certainty depends on how good an instrument you used, how careful you were, and how well you can read the instrument. There is always a limit on the certainty of the number you have read. Because of this, you must always obey the rules for significant-figures.

The number of significant figures that a measured quantity has depends on two things—the instrument and how large the measured object is.

Say that you weigh yourself on a bathroom scale, and the arrow points to a number between 146 and 147.

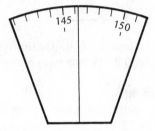

You are sure that you weigh 146 pounds plus some part of a pound, and you can guess that you are 0.3 pounds over 146 pounds. You therefore express your weight as 146.3 pounds. The

first three digits (1, 4, and 6) you are _____ of, and the last digit (3)   sure

you have _____. All of these digits are significant. You can say   guessed

that you have expressed your weight to ____ significant figures.   four
   If you weigh a grapefruit on the same scale and see

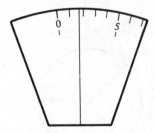

you are sure that the grapefruit weighs __ lb, but it weighs about   1

___ lb more than the 1 lb. (If we didn't guess exactly the same on   0.8 (my guess)
the last figure, this demonstrates the uncertainty.) To the correct

number of significant figures, then, you can say that the grapefruit

1.8, two

weighs ___ lb. This number contains ___ significant figures, 1 (of

sure, guess

which you are ____) and 8 (which is a ____).
Try weighing a plum on the same bathroom scale. You see

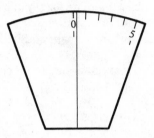

0.2

Its weight to the correct number of significant figures is ___. You must guess that it weighs 0.2 lb more than nothing. This number

one

contains ___ significant figure.
From the above three weighings, you can see that the heavier

greater (larger)

the sample you are weighing, the _____ the number of significant figures you can get from the same measuring device (for example, the bathroom scale).
Now let us weigh the grapefruit on a kitchen scale. For convenience the pounds have been divided into tenths rather than ounces.

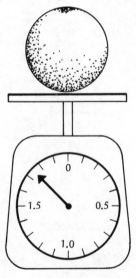

You are certain that the grapefruit weighs 1.7 lb and then a little more. You can guess it is 0.05 lb more than 1.7 lb. Therefore, to

1.75

the correct number of significant figures, its weight is ___ lb. This number contains _____ significant figures, while the previous weighing on the bathroom scale gave only two significant figures.

If you weigh the plum on the same kitchen scale,

it weighs, as best as you can read, exactly ___ lb. You are guessing that it does not weigh even a little more than .2 lb, so to show that you have guessed a last place of zero, you must write the weight

as ___ lb. Now you have _____ significant figures in the weight, a 2 of which you are certain and a 0 which you have guessed. The zero is a significant figure.

Remember that when you weighed the plum on a bathroom scale you only had one significant figure. Using a more sensitive measuring device, you can get more significant figures. You now know two ways of increasing the number of significant figures,

(1) _____ and (2) _____.

When you wrote the weight of the plum as .20 lb you could have put a zero in front of the decimal point, 0.20, to make sure that the decimal point isn't overlooked. This zero in front is not a significant figure; it is simply a place marker. As a matter of fact, all zeros written in front of a number are place markers and not significant figures. You could have written the weight in scientific notation as 2.0×10^{-1} lb. Here it is clear how many significant figures there are.

If you have a zero in the middle of a number, it is significant. For example the number 0.000034010 has five significant figures. The first five zeros are simply place markers and not significant; but the zero between the 4 and the 1 is significant, and the zero at the end is significant.

Thus, in the number 0.03570 there are ___ significant figures.

The _____ in front of the number don't count. The _____ at the end of the number is significant. Written in scientific notation the

number would be _____.

Answer column:

three

.2

.20, two

larger sample, more sensitive tool

four

zeros, zero

3.570×10^{-2}

four

The number 0.5670 has ____ significant figures since the zero

in front,   at the end

____ doesn't count. It is only a place marker. The zero _____ does count since it is part of the measurement.

five

The number 0.00035001 has ____ significant figures. It could be

3.5001×10^{-4}

written in exponential notation as _____.

five

The number 5680.0 has _____ significant figures. If you were to write this in scientific notation showing all the significant figures

5.6800×10^3,   significant

it would be _____. All the zeros at the end are _____.

Unfortunately, there can be some doubt as to whether final zeros are significant. If you haven't made the measurement yourself and you are told that a beaker contains 500 grams (g) of water, it might contain between 400 and 600 g (one significant figure), between 490 and 510 g (two significant figures) or between 499 and 501 g (three significant figures). However, if you are especially careful and only deal with numbers expressed in scientific notation, this problem will not come up. The three cases above would be $5. \times 10^2$ g (one significant figure), 5.0×10^2 g (two significant figures) and 5.00×10^2 g (three significant figures). You can say that all zeros after the decimal point are significant.

Multiplication and Division

When numbers expressed to the correct number of significant figures are multiplied or divided, the answer can have no more figures than the number with the fewest significant figures.

one

If you multiply $1. \times 1.345$, the 1. has ____ significant figure and

four

the 1.345 has ____ significant figures. The product, if you express it to the correct number of significant figures, can only have one significant figure.

three

If you divide 35.3/6.8, the 35.3 has ____ significant figures. The

two

6.8 has only ____ significant figures. The answer then can only be

two

expressed to ____ significant figures. This is the fewest number of significant figures.

The result of $(1.37 \times 0.3)/(0.0356 \times 1.5)$ can be expressed to only

one,   0.3

____ significant figure. The ____ has only one significant figure.

We can check this rule for determining the number of significant figures allowed in multiplication and division by using the example we had at the very beginning of this section. Recall that we calculated the area of a piece of paper from the measurements of its height and width. The height was measured to be $1\frac{8.5}{16}$ inches and the width was $3\frac{13.2}{16}$ inches. When we calculated the area, we showed that a missreading even as little as $\frac{9\div1}{16}$ gave answers that varied from 5.82 to 5.89 square inches. The first two digits, 5.8, are certain, but the third is not. At that point, we said that you could

write the answer only as 5.86, which has _____ significant figures.

    To check our rule for determining the number of significant figures, we have to express the total number of sixteenths in $1\frac{8\div5}{16}$ inches ($\frac{24\div5}{16}$) and in $3\frac{13\div2}{16}$ inches ($\frac{61\div2}{16}$). Both 24.5 and 61.2

have _____ significant figures. Therefore, the answer can have

only _____ significant figures. And it does. As we said at that point, any more figures are meaningless.

Addition and Subtraction

When you add numbers expressed to the correct number of significant figures, you must use a different set of rules for determining the correct number of significant figures allowed in the answer. In addition and subtraction, the location of the digits with respect to the decimal point controls the number of significant figures allowed in the answer. The answer may have more or fewer significant figures than the number that had the fewest significant figures. For example, if you add 0.5 + 9.74, the answer, expressed to the correct number of significant figures, is 10.2. The

10.2 has _____ significant figures, but the number 0.5 has

only _____.
    If you subtract 2.52 − 2.3, the answer, expressed to the correct

number of significant figures, is 0.2. The answer has only _____

significant figure, but the 2.3 has _____.
    Let's see how this comes about. In the first example, 0.5 + 9.74,

we'll assume that the last digit that was _____ could have been wrong by 2. Thus, the 0.5 could be as low as 0.3 or as high

as _____. The 9.74 could be as low as 9.72 or as high as _____. We add the two low values and separately add the two high values.

```
  0.3        0.7
  9.72       9.76
-------    -------
 10.02      10.46
```

Only the first two digits, 1 and 0, are the same. You are uncertain of the third digit, and therefore the third digit is the first uncertain digit. It is the last that you can show in your significant figure. The answer is therefore 10.2 and the digit 2 is uncertain.

Right column answers:

three

three

three

three

one

one

two

guessed

0.7,   9.76

In the second example, the subtraction $2.52 - 2.3$, let's assume that the last digit in each is wrong by only 1. Thus the 2.52 might be as low as 2.51 or as high as _____. The 2.3 might be as low as

2.53

2.4

2.2 or as high as _____. If we subtract the smaller possibility from the larger and the larger possibility from the smaller, we have

$$
\begin{array}{r} 2.53 \\ -2.2 \\ \hline 0.33 \end{array}
\qquad
\begin{array}{r} 2.51 \\ -2.4 \\ \hline 0.11 \end{array}
$$

You can see that even the first digit is not certain. Thus all that you can write for your answer is 0.2.

To handle additions and subtractions, you write the numbers to be added or subtracted so that all the decimal points are lined up one above another. For the two examples, this would be

$$
\begin{array}{r} 0.5 \\ 9.74 \\ \hline \end{array}
\qquad
\begin{array}{r} 2.52 \\ -0.3 \\ \hline \end{array}
$$

The last place that can appear in the answer is the column farthest to the right that has a digit in it for every number. You could draw a line down to show this.

$$
\begin{array}{r} 0.5| \\ 9.7|4 \\ \hline 10.2|4 \end{array}
\qquad
\begin{array}{r} 2.5|2 \\ -2.3| \\ \hline 0.2|2 \end{array}
$$

Anything to the right of the line cannot appear in your answer. The answers are 10.2 and 0.2.

Logarithms

Getting logarithms to the correct number of significant figures poses quite a problem. This is because the nature of logarithms varies with the value of the number whose logarithms you are considering. For example, suppose that you had the number 1.03, but you were uncertain of the last figure and it might be 1.02 or 1.04. The logarithms of the two uncertain possibilities are .0086 and .0170. These don't have any numbers in common. If, however, you had the number 9.03 and it might be as low as 9.02 or as high as 9.04, the two logarithms would be .9552 and .9562. In this case the first two numbers are the same.

All you are expected to do is watch the rules for significant figures by not putting any more figures after the decimal in a logarithm than you had in your original number. This is close enough.

Rounding Off

Now that you are able to decide to how many significant figures you can write an answer, you have to know how to cut your answer down to the right number of digits. This process is called "rounding off." It is very simple. There are only three rules, and they depend on the digits you want to remove.

(1) If the digit following the last digit you want to keep in your answer is less than 5 (that is, it is 0, 1, 2, 3, or 4), discard all the digits you do not want.

For example, round off 0.3752 to three significant figures. The

third significant figure is ___. (Recall that zeros preceding a number

are not significant.) The digit following the 5 is ___, which is less than 5, and so you simply discard everything which follows the 5,

giving the answer _____.

Try another example. If you round off 3.65748×10^3 to four

significant figures, the answer is _____. If you got the wrong answer it is possible that you didn't have the 7 as your fourth significant figure, which it is. Since the figure which follows the 7 is less than 5, you drop all the digits which follow the 7.

(2) If the digits following the last number you want to keep are greater than 5 (that is, 5 with any digit other than zeros following it, 6, 7, 8, and 9) you must add 1 to the number you want to keep and discard all the digits you do not want.

For example, round off 45.253 to three significant figures. The

third significant figure is the number ___. The digits which follow

it are 5 and ___ which are greater than just ___, so you add 1 to the 2 giving 45.3.

Here is another example. Round off 3.51866 to the fourth place

after the decimal point. The answer will be _____. Since the number following the last digit you wanted to keep is greater than

___ you have to add ___ to your last significant figure.

Try one more example. If you round off 4.3256 to three significant

figures, the third significant figure is the ___. The digits which follow

it are _____ than 5; in fact, they are 56. Consequently you must

___ 1 to the 2, giving the answer _____.

(3) If the digits following the last figure you want to keep are exactly 5 (that is, 5 followed by nothing but zeros), you add 1 to the last significant figure if it is an odd number and drop all the extra digits. If the last significant figure is an even number, you simply discard all the unwanted digits.

For example, round off 3.635 to three significant figures. The last

number you want to keep is ___. This is an odd number so you must

___ 1 to it, and the answer will be _____.

If you round off 0.50450 to three significant figures, the answer

5

2

0.375

3.657×10^3

2

3, 5

3.3187

5, 1

2

greater

add, 4.33

3

add, 3.64

0.504

even

0

4.560

will be _____. The reason you simply discard the numbers following the 4 is that 4 is an _____ number.

If you round off 4.56050 to four significant figures, the last digit you want to keep is __. It is considered an even number. Therefore the answer will be _____.

You have now finished reviewing all the mathematics which you will need for the types of problems covered in this text. On the next page you will find a set of questions which covers all of this material. Do this problem set and see if you can complete it in the suggested time.

PROBLEM SET

You will want to time yourself doing these problems. Do each section separately in the time noted. The answers will all be found on the following pages. Do not check your answers until you have finished a complete section. A method of scoring yourself appears with the answers. You need not worry about significant figures in Sections A and B.

Section A (10 minutes)

1. Express 5650000. in scientific notation.

2. Express 0.0000565 in scientific notation.

3. Express 0.5×10^{-3} in scientific notation.

4. The number 55×10^{-8} equals ___ $\times 10^{-7}$.

5. Express 5.3×10^2 as an ordinary number.

6. Express 6.73×10^{-2} as an ordinary number.

7. Perform the operation $4 \times 10^2 \times 2 \times 10^{-4}$.

8. Perform the operation $3.5 \times 10^{-1} \times 2 \times 10^{-7}$.

9. Perform the operation $\dfrac{2 \times 10^{-2} \times 3 \times 10^1}{6 \times 10^{-5}}$.

10. Perform the operation $(2 \times 10^{-4})^3$.

11. Perform the operation $\sqrt{2.5 \times 10^{-1}}$.

12. Perform the operation $0.5 \times 10^1 + 6 \times 10^{-1}$ and give the answer in scientific notation.

Section B (25 minutes)

Use the four-place table of logarithms in Appendix III to determine the following.

13. $\log 10^7 =$

14. $\log 0.0001 =$

15. $\log 5.35 =$

16. $\log 2500 =$

17. $\log 0.00087 =$

18. antilog $3 =$

19. antilog $7.9571 =$

20. antilog $2.608 =$

21. antilog $-3.8456 =$

22. $e^{3.7} =$

23. $\ln 7.65 =$

24. $(2.6 \times 10^3)^9 =$

25. $\sqrt[3]{0.000042} =$

Section C (12 minutes)

26. How many significant figures are in the number 450.0?

27. How many significant figures are in the number 0.0032?

28. How many significant figures are in the number 0.12040?

29. Express the number 0.000352 to two significant figures.

30. Express the number 0.357 to two significant figures.

31. Express the number 0.305 to two significant figures.

32. Express the number 0.145 to two significant figures.

33. Perform the operation

    $$\frac{3. \times 4.0 \times 3.254}{1.002 \times 2.05}$$

    and express your answer to the correct number of significant figures.

34. Perform the operation

    $$\frac{200.0 \times 20.0 \times 10.}{100.0}$$

    and express your answer to the correct number of significant figures.

35. Perform the operation $41.56 + 72. + 7.3$ and express your answer to the correct number of significant figures.

PROBLEM SET ANSWERS

Section A

You should get at least ten correct answers out of the twelve. If you do not, check your incorrect answers and see if the error was just carelessness. If you really don't understand why your answer was incorrect, refer back to that part of the text which has to do with the missed problems.

1. 5.650000×10^6

2. 5.65×10^{-5}

3. $5. \times 10^{-4}$

4. 5.5×10^{-7}

5. $530.$

6. 0.0673

7. $8. \times 10^{-2}$

8. $7. \times 10^{-8}$

9. $1. \times 10^4$

10. $8. \times 10^{-12}$

11. 5.0×10^{-1}

12. 5.6 (or just 6 to the correct number of significant figures)

Section B

You should get at least nine out of the first ten correct. If you didn't, you had better review the first three parts of Section B. If your errors were in the last four of these, 19 through 22, you only have to review the part on antilogarithms. You should be able to do both 23 and 24. If you missed either, redo the part of this section on natural logarithms. Questions 25 and 26 are covered in the section on the uses of common logarithms. If you get one of the two correct, that's good enough.

13. 7

14. -4

15. 0.7284

16. 3.3979

17. -3.0605

18. 10^3

19. 9.06×10^7

20. 4.055×10^2

21. 1.427×10^{-4}

22. 4.042×10^1 (4.0×10^1 to the correct number of significant figures)

23. 2.035

24. 5.432×10^{30} (5.4×10^{30} to the correct number of significant figures)

25. 3.476×10^{-2} (3.5×10^{-2} to the correct number of significant figures)

Section C

You should get at least nine correct answers out of the ten problems. If not, refer back to the text and redo the entire section.

26. four

27. two

28. five

29. 0.00035

30. 0.36

31. 0.30

32. 0.14

33. $2. \times 10^1$

34. 4.0×10^2

35. 1.21×10^2

(Notice how you can write these in scientific notation to the correct number of significant figures.)

Chapter 2
What is Dimensional Analysis ?

There will be no pretest for this chapter because you have had no specific preparation for the material in it. Nevertheless, it is the most important chapter in the whole book since it shows you the method you will be using to solve chemical problems.

Suppose I ask you "Do you have 5?" You probably would ask me, "Five what?" It might be 5 cents, or 5 fingers on a hand, or even 5 pounds of salt in 3 boxes. All numbers (with very, very few exceptions) are quantities of something. But only saying the number (quantity) isn't enough; you have to say the quantity of whatever you are talking about. The "whatever" is called the dimensions (or sometimes units) of the number.

Let us return to the question above, "Do you have 5?" If I were interested in money, I would say, "Do you have 5 cents?" "Cents" is the dimension of the number 5. If I were interested in weight, I

would ask, "Do you have 5 pounds?" "Pounds" is the _____   dimension
of the number 5. If I were interested in the length in yards, I would

say, "Do you have 5 ____?" Here yards is the dimension of the   yards

_____ 5.   number

Thus, for a number to mean anything, it must have a _____   dimension
written after it so you know what you are talking about.

SECTION A   Conversion Factors

The other possibilities mentioned above for what the 5 might be were 5 fingers per hand and 5 pounds of salt in 3 boxes. These bring us into what are called conversion factors. We will start with a very simple example of a conversion factor.

We all know that there are 60 seconds in one minute. This really means that 60 seconds equal 1 minute. Mathematically, this is

60 sec = 1 min.

If you divide both sides of an equality by the same thing, you still have an equality. Let's divide both sides by 1 min.

$$\frac{60 \text{ sec}}{1 \text{ min}} = \frac{1 \text{ min}}{1 \text{ min}}$$

But

$$\frac{1 \text{ min}}{1 \text{ min}} \text{ equals 1 (We say they "cancel.")}$$

so

$$\frac{60 \text{ sec}}{1 \text{ min}} = 1.$$

We call 60 sec/1 min a conversion factor and it equals 1. All conversion factors equal 1.

You are probably more accustomed to saying, "There are 60 seconds per minute." The "per" means divided by, and the number 1 is usually not written since it is understood.

What, then, is the conversion factor relating hours to days?

24 | Since there are ___ hours per day, you can write the conversion factor

24 hr |
$$\frac{\rule{2cm}{0.4pt}}{\text{day}}$$

1 | This conversion factor must equal ___, since all conversion factors

1 | equal ___.

$\dfrac{100\cancel{c}}{\$1}$ | The conversion factor relating cents (¢) to dollars ($) is _____.

1 | Since this is a conversion factor it must equal ___.

Let us go back to the first conversion factor which came from "60 seconds equals 1 minute,"

$$60 \text{ sec} = 1 \text{ min}.$$

Now what happens if you divide both sides of the equality by 60 sec?

$$\frac{60 \text{ sec}}{60 \text{ sec}} = \frac{1 \text{ min}}{60 \text{ sec}}.$$

The 60 sec/60 sec cancels and is equal to 1. Therefore,

$$1 = \frac{1 \text{ min}}{60 \text{ sec}}.$$

Since 1 min/60 sec also equals 1, it must also be a conversion factor. The two conversion factors, then, are

$$\frac{60 \text{ sec}}{\text{min}} \quad \text{and} \quad \frac{\text{min}}{60 \text{ sec}}.$$

They are the same except one is _____ compared to the other. The mathematical word for saying this is "inverted." Thus any conversion factor can be inverted and it will still be a conversion factor. Sometimes we say that one is the "reciprocal" of the other.

upside-down

Similarly, the inverted form of 24 hr/day is _____; it is still a

conversion factor and still equals __.

$$\frac{1 \text{ day}}{24 \text{ hr}}$$

1

The two conversion factors relating feet to inches would be

_____ and _____. Both of these equal __. This is because 12 in.

$$\frac{12 \text{ in.}}{1 \text{ ft}}, \quad \frac{1 \text{ ft}}{12 \text{ in.}}, \quad 1$$

_____ 1 ft.

equals

Many conversion factors have fixed definitions, such as, "There

are 16 ounces per pound," which can be written as _____. Others are presented in a statement such as, "A worker makes $7.50 per

$$\frac{16 \text{ oz}}{1 \text{ lb}}$$

hour," which you can write as _____, or as its reciprocal _____. You will learn very quickly how to spot these factors given in the problem.

$$\frac{\$7.50}{1 \text{ hr}} \quad \frac{1 \text{ hr}}{\$7.50}$$

You must be careful when you write the dimensions of a number. The dimension must be sufficiently explicit so that you cannot confuse it with the dimension of any other number. For example, if a problem says, "Onions cost 11¢ per pound," you can write the

conversion factor _____. However, if in the same problem you have, "Potatoes cost 6¢ per pound," then you must distinguish which vegetable you are talking about. Thus you write the complete conversion factor 11¢/1 lb onions for the onions, and then you

$$\frac{11¢}{1 \text{ lb}}$$

write _____ for the potatoes. The dimensions are then pounds of onions and pounds of potatoes.

$$\frac{6¢}{1 \text{ lb potatoes}}$$

It has become customary to leave out the number 1 in conversion factors. We usually write 60 sec/min rather than 60 sec/1 min. The reason that this can be done is that multiplying or dividing by 1 does not change the numerical value of what it is operating on. Therefore, from this point on all 1's in conversion factors will be omitted.

Conversion factors do not necessarily have to be for a single hour, or a single day, or a single foot or a single pound. For example, if a problem states that "An earth-mover can dig a trench 500 feet long in 3 hours," the conversion factors would be

$$\frac{500 \text{ ft}}{3 \text{ hr}} \quad \text{or} \quad \frac{3 \text{ hr}}{500 \text{ ft}}.$$

SECTION B   Solving Problems by Dimensional Analysis

Now we can get to the business at hand: how to solve a problem. It is really very simple. All problems ask a question whose answer will be a number *and its dimension*. All problems also have some information given in them which will also be a number *and its dimension*. You are going to consider the numbers and the dimensions separately. You will multiply the information given by conversion factors so that you cancel all dimensions except the dimension you want in your answer. You may multiply by as many conversion factors as you wish since all conversion factors equal

1, 1

__, and multiplying something by __ doesn't change its value.

Here is an example. How many seconds in 35 min? The question asked is "How many seconds?" Therefore, when you get an

seconds

answer it will have the dimension _____. The information given in the problem is 35 min.

You start by writing down the dimensions of the answer so that you can check yourself at the end.

? sec =

Then you supply the information given.

? sec = 35 min

You now want to multiply the 35 min by some conversion factor which will have the dimension minutes in the denominator (on the bottom) to cancel the unwanted minutes in the information given and have the dimensions of the answer (seconds) in the numerator (on top). How about the conversion factor 60 sec/min? Try it.

$$? \text{ sec} = 35 \text{ min} \times \frac{60 \text{ sec}}{\text{min}}$$

The unwanted dimension, minutes, cancels, and the only remaining dimension is seconds.

? sec = 35 × 60 sec

= 2100 sec

You know that this must be the correct answer because the only dimension left is the dimension of the question asked.

Try another problem. How many hours are in 6 days? The

How many hours?

question asked is "_____" The dimensions of the

hours

answer therefore will be _____, so you write down

hours

? _____ =

6 days

The information given is _____, so you write

6 days

? hr = _____

Now you want to multiply the information given by some

days

conversion factor which will have the dimension _____ in the

days

denominator (bottom) to cancel the unwanted dimension _____

in the information given. This should lead toward the dimensions of

the question asked. Such a conversion factor would be _____.
If you do this, you have

? hr = 6 days × _____,

and the unwanted dimension _____ cancels, leaving only the

dimension _____, which is the dimensions of the question asked.
This must be the answer. Its numerical value, once you have

multiplied the numbers together, is _____.
Multiplying by only one conversion factor usually will not get
you all the way to the dimensions of the answer. However, since
all conversion factors equal 1, you can multiply by as many as you
need. Consider the problem, "How many seconds in 3 years?" The

question asked is "_____" and the information

given is _____. So you write down

? sec = 3 yr

Now you need some conversion factor which has _____ in the

denominator to cancel the _____ in the information given and
which will move you in the direction of the dimensions of the
question asked.
What about 365 days/year?

$$? \text{ sec} = 3 \, \cancel{\text{yr}} \times \frac{365 \text{ days}}{\cancel{\text{yr}}}$$

The unwanted dimension years has been cancelled, but now you
have the unwanted dimension days left. You need a conversion
factor with days in the denominator to cancel that. Try

$$? \text{ sec} = 3 \, \cancel{\text{yr}} \times \frac{365 \, \cancel{\text{days}}}{\cancel{\text{yr}}} \times \frac{24 \text{ hr}}{\cancel{\text{day}}}$$

The unwanted dimension now is _____. To move toward seconds

you might next multiply by the conversion factor _____. This
gives

$$? \text{ sec} = 3 \, \cancel{\text{yr}} \times \frac{365 \, \cancel{\text{days}}}{\cancel{\text{yr}}} \times \frac{24 \, \cancel{\text{hr}}}{\cancel{\text{day}}} \times \frac{60 \text{ min}}{\cancel{\text{hr}}}$$

Finally, you must have a conversion factor with _____ in the
denominator to cancel out the unwanted dimension. This would be

_____.

$$? \text{ sec} = 3 \, \cancel{\text{yr}} \times \frac{365 \, \cancel{\text{days}}}{\cancel{\text{yr}}} \times \frac{24 \, \cancel{\text{hr}}}{\cancel{\text{day}}} \times \frac{60 \, \cancel{\text{min}}}{\cancel{\text{hr}}} \times \frac{60 \text{ sec}}{\cancel{\text{min}}}$$

Margin answers:

$\dfrac{24 \text{hr}}{\text{day}}$

$\dfrac{24 \text{ hr}}{\text{day}}$

days

hours

144 hr

How many seconds?

3 yr

years

years

hours

$\dfrac{60 \text{ min}}{\text{hr}}$

minutes

$\dfrac{60 \text{ sec}}{\text{min}}$

seconds

The only dimension left is _____, which is the dimension of the question asked. Thus, the correct answer must be

$$? \text{ sec} = 3 \times 365 \times 24 \times 60 \times 60 \text{ sec}$$

$$= 94{,}608{,}000 \text{ sec} = 9.4608 \times 10^7 \text{ sec}$$

Sometimes you will use a conversion factor in its inverted form. You might know the conversion factor from feet to miles. Since

$\dfrac{5280 \text{ ft}}{\text{mi}}$

there are 5280 feet per mile, the conversion factor is _____. In its

$\dfrac{\text{mi}}{5280 \text{ ft}}$

How many miles?

inverted form you would write _____. Now consider the problem of how many miles are in 7920 ft. The question asked is

"_____" so the dimensions of your answer must be

miles

_____. Since the information given is 7920 ft, you write

$$? \text{ mi} = 7920 \text{ ft.}$$

To cancel the unwanted dimension feet you need a conversion

feet

factor with _____ in the denominator. This would be the inverted form shown above.

$$? \text{ mi} = 7920 \text{ ft} \times \frac{\text{mi}}{5280 \text{ ft}}$$

The unwanted dimension feet cancels leaving only the dimension mile, which is the dimension of the question asked and therefore must be the answer.

$$? \text{ mi} = 7920 \times \frac{\text{mi}}{5280}$$

$$= 1.500 \text{ mi}$$

Very frequently the problem itself will have conversion factors in it. You may find it very helpful at first to write down all the conversion factors before you start a problem. Here is an example. "What is the cost of three shirts, if a box containing 12 shirts costs $27?"

The conversion factors given in the problem are

$$\frac{12 \text{ shirts}}{\text{box}} \quad \text{and} \quad \frac{\$27}{\text{box}}.$$

You must read the problem carefully to find these. It is stated that there are 12 shirts per box and the price is $27 per box.

The most important thing is to determine the dimensions of the answer to the question asked. In this problem the question asked is, "What is the cost?" The dimensions of cost are dollars. You must reword the question as, "How many dollars"? Now you can write

$$? \text{ dollars} =$$

The information given is _____, so you write

    ? dollars = 3 shirts.

The conversion factor which contains shirts as a dimension is

_____, and so you multiply by it.

$$? \text{ dollars} = 3 \cancel{\text{ shirts}} \times \frac{\text{box}}{12 \cancel{\text{ shirts}}}$$

Notice that it is used in the inverted form in order to _____ the dimension shirts in the information given. The unwanted dimension

now is ____, so now you multiply by the conversion factor ____ to give

$$? \text{ dollars} = 3 \cancel{\text{ shirts}} \times \frac{\cancel{\text{box}}}{12 \cancel{\text{ shirts}}} \times \frac{\$27}{\cancel{\text{box}}}.$$

The only dimension left is dollars, so your answer must be

$$? \text{ dollars} = 3 \times \frac{1}{12} \times \$27$$

$$= \$6.75$$

    All of the examples so far have been so simple that you could have solved them without bothering with dimensional analysis. Now here is a more difficult problem with some special points.

    What is the gas consumption in miles per gallon of an automobile if it uses 0.1 gal of gas in 100 sec when traveling 60 mi/hr?

    The first thing to do is reword the question so you know what the dimensions of the answer will be. "How many miles per gallon" can be reworded as, "How many miles equal 1 gallon?" Thus, the information given is 1 gal.

    The conversion factors given in the problem are

$$\frac{0.1 \text{ gal}}{100 \text{ sec}} \quad \text{and} \quad \frac{60 \text{ mi}}{\text{hr}}.$$

Starting with the dimensions of the answer and the information given, we have

    ? mi = 1 gal ×

The only conversion factor that you can write with gallons in the

denominator is _____, so you multiply by this.

$$? \text{ mi} = 1 \cancel{\text{ gal}} \times \frac{100 \text{ sec}}{0.1 \cancel{\text{ gal}}}$$

The unwanted dimension now is _____, but the next conversion factor you have related to miles has hours as its other dimension.

You must therefore convert from seconds to hours.

$$? \text{ mi} = 1 \,\cancel{\text{gal}} \times \frac{100 \,\cancel{\text{sec}}}{0.1 \,\cancel{\text{gal}}} \times \frac{\cancel{\text{min}}}{60 \,\cancel{\text{sec}}} \times \frac{\text{hr}}{60 \,\cancel{\text{min}}}$$

Now the unwanted dimension is hours, and you can multiply by the conversion factor given in the problem.

$$? \text{ mi} = 1 \,\cancel{\text{gal}} \times \frac{100 \,\cancel{\text{sec}}}{0.1 \,\cancel{\text{gal}}} \times \frac{\cancel{\text{min}}}{60 \,\cancel{\text{sec}}} \times \frac{\cancel{\text{hr}}}{60 \,\cancel{\text{min}}} \times \frac{60 \text{ mi}}{\cancel{\text{hr}}}$$

$$? \text{ mi} = 1 \times \frac{100}{0.1} \times \frac{1}{60} \times \frac{1}{60} \times 60 \text{ mi}$$

$$= 16.7 \text{ mi}$$

Incidentally, as we proceed in this text, a great many example problems will be presented conversion factor by conversion factor. The blanks you will be filling in will be these conversion factors. However, it will be an extremely good idea if, besides filling in the blanks, you also keep a running setup of the problem as we go along step by step, cancelling the unwanted dimensions as you go.

PROBLEM SET

For these problems set up the answer but don't bother to do all the multiplications and divisions. The answers follow directly. After you set up the answer immediately check to see if you have done it correctly. What you want to know is whether you have mastered the method of dimensional analysis.

1. How many seconds in 800 minutes?

2. How many minutes in 5 years?

3. How many years in 500 days?

4. How many dozens of eggs are there in 3500 eggs?

5. How many miles in 12,000 yards? There are 5280 ft/mi and 3 ft/yd.

6. How many decades are there in 9 centuries? There are 10 years in a decade.

7. What is the cost of 6 onions if 3 onions weigh 1.5 lb and the price of onions is 11¢ per lb?

8. How many hours will it take to drive to Los Angeles from San Francisco if an average speed of 52 mi/hr is maintained? The distance between the two cities is 405 miles.

9. What is the cost to drive from San Francisco to Los Angeles (still 405 miles) if the cost of gasoline is 64¢/gal and the automobile gets 18 mi/gal of gasoline?

10. How many oranges are in a crate if the price of a crate is $1.60 and the price of oranges is $0.20/lb? On the average there are three oranges per pound.

11. The price of a ream of paper is $4.00. There are 500 sheets of paper in a ream. If a sheet of paper weighs 0.500 oz, what is the price per pound of paper? There are 16 oz in a pound.

12. How many cars are there in a long freight train if it takes the entire train 2 min to pass a station as it travels 40 mi per hour? Each car is 50 ft long. There are 5280 ft in a mile.

PROBLEM SET ANSWERS

You should be able to do at least ten of these problems correctly. This chapter is so important that if you missed three or more problems, you had better go over the entire chapter again.

1. $? \text{ sec} = 800 \text{ min} \times \dfrac{60 \text{ sec}}{\text{min}}$

    $= 800 \times 60 \text{ sec}$

2. $? \text{ min} = 5 \text{ yr} \times \dfrac{365 \text{ days}}{\text{yr}} \times \dfrac{24 \text{ hr}}{\text{day}} \times \dfrac{60 \text{ min}}{\text{hr}}$

    $= 5 \times 365 \times 24 \times 60 \text{ min}$

3. $? \text{ yr} = 500 \text{ days} \times \dfrac{\text{yr}}{365 \text{ days}}$

$$500 \times \dfrac{1}{365} \text{ yr}$$

4. $? \text{ doz} = 3500 \text{ eggs} \times \dfrac{\text{doz}}{12 \text{ eggs}}$

$$= 3500 \times \dfrac{1}{12} \text{ doz}$$

5. $? \text{ mi} = 12{,}000 \text{ yd} \times \dfrac{3 \text{ ft}}{\text{yd}} \times \dfrac{\text{mi}}{5280 \text{ ft}}$

$$= 12{,}000 \times 3 \times \dfrac{1}{5280} \text{ mi}$$

6. $? \text{ decades} = 9 \text{ centuries} \times \dfrac{100 \text{ yr}}{\text{century}} \times \dfrac{\text{decade}}{10 \text{ yr}}$

$$= 9 \times 100 \times \dfrac{1}{10} \text{ decades}$$

7. $? \text{ cents} = 6 \text{ onions} \times \dfrac{1.5 \text{ lb}}{3 \text{ onions}} \times \dfrac{11\cancel{c}}{\text{lb}}$

$$= 6 \times \dfrac{1.5}{3} \times 11\cancel{c}$$

8. $? \text{ hr} = 405 \text{ mi} \times \dfrac{\text{hr}}{52 \text{ mi}}$

$$= 405 \times \dfrac{1}{52} \text{ hr}$$

9. $? \text{ cents} = 405 \text{ mi} \times \dfrac{\text{gal}}{18 \text{ mi}} \times \dfrac{64\cancel{c}}{\text{gal}}$

$$= 405 \times \dfrac{1}{18} \times 64\cancel{c}$$

10. The question asked is, "How many oranges are there in 1 crate?"

$$? \text{ oranges} = 1 \text{ crate} \times \dfrac{\$1.60}{\text{crate}} \times \dfrac{\text{lb}}{\$0.20} \times \dfrac{3 \text{ oranges}}{\text{lb}}$$

$$= 1 \times 1.60 \times \dfrac{1}{0.20} \times 3 \text{ oranges}$$

11. If you are having a little trouble, start by writing down all the conversion factors given in the problem. Then you can simply select the one you need from the collection.

$$? \text{ dollars} = 1 \, \cancel{lb} \times \frac{16 \, \cancel{oz}}{\cancel{lb}} \times \frac{1 \, \cancel{sheet}}{0.500 \, \cancel{oz}} \times \frac{\cancel{ream}}{500 \, \cancel{sheets}} \times \frac{\$4.00}{\cancel{ream}}$$

$$= 1 \times 16 \times \frac{1}{0.500} \times \frac{1}{500} \times \$4.00$$

12. $$? \text{ cars} = 1 \, \cancel{train} \times \frac{2 \, \cancel{min}}{\cancel{train}} \times \frac{\cancel{hr}}{60 \, \cancel{min}} \times \frac{40 \, \cancel{mi}}{\cancel{hr}} \times \frac{5280 \, \cancel{ft}}{\cancel{mi}} \times \frac{car}{50 \, \cancel{ft}}$$

$$= 1 \times 2 \times \frac{1}{60} \times 40 \times 5280 \times \frac{1}{50} \text{ cars}$$

This last problem was a little more difficult since the conversion factors were hidden in the language of the problem. If you were able to do it, you are well on your way to being able to solve the types of chemical problems you are going to face in this book. If you were not able to do it, don't worry. You are going to get a lot more practice!

Chapter 3
The Metric System

PRETEST

If you feel that you know the workings of the metric system of measurement, take this test. Try to work all the problems by dimensional analysis. A few conversion factors that you may not still have in your memory are given below. Many others that you should know are not. Try to complete this test in 45 minutes.

The answers follow immediately with a method of scoring yourself.

Conversion Factors

$\dfrac{2.54 \text{ cm}}{\text{in.}}$	$\dfrac{946 \text{ ml}}{\text{qt}}$	$\dfrac{454 \text{ g}}{\text{lb}}$	$\dfrac{4 \text{ qt}}{\text{gal}}$	$\dfrac{5280 \text{ ft}}{\text{mile}}$
$\dfrac{2 \text{ pt}}{\text{qt}}$	$\dfrac{16 \text{ fl oz}}{\text{pt}}$	$\dfrac{16 \text{ oz}}{\text{lb}}$	$\dfrac{2000 \text{ lb}}{\text{ton}}$	$\dfrac{1.61 \text{ km}}{\text{mile}}$

1. How many milliliters are in 16 liters?

2. How many yards are in a kilometer?

3. How many microliters are in 5 ml?

4. How many milligrams are in 3.00×10^{-4} ton?

5. How many angstroms are in 2×10^{-4} m?

6. The price of milk in France is 3.6 francs/liter. What is the price in dollars for 1 qt? There are 5 francs/dollar.

7. How long will it take a Porsche traveling at 120 km/hr to go 90 mi?

8. The speed of light is 3.0×10^{10} cm/sec. How many miles from the earth is the sun if it takes light 6.0 min to cover the distance?

9. If it takes 540 calories (cal) to vaporize 1.0 g of water at 100°C to steam, how many calories will it take to vaporize 6.0 pt of water? A liter of water weighs 0.96 kg.

10. How many dollars would it cost to drive a VW 5000 mi in Germany where gasoline costs 0.90 DM/liter? The car averages 12 km/liter and there are 2.3 DM/dollar.

11. What is the temperature in °C corresponding to 65°F?

12. What is the temperature in °F corresponding to -14°C?

(These last two are not generally solved using dimensional analysis, but do so here for practice.)

PRETEST ANSWERS

If you got the correct answer for eight out of the first ten problems, you are able to handle the metric system very well indeed. You can proceed immediately to the problem set at the end of Chapter 3. If, however, you were not able to do questions 11 and 12, you had better work the last section of Chapter 3, which shows you the conversions of temperatures.

If you missed more than two of the first ten problems, it will be worth your while to spend some time working through all of Chapter 3.

1. 1.6×10^4 ml

2. 1.09×10^3 yd

3. 5×10^3 μl

4. 2.72×10^5 mg

5. 2×10^6 Å

6. $0.68

7. 1.2 hr

8. 6.7×10^7 mi

9. 1.5×10^6 cal

10. $262

11. 18°C

12. 7°F

Don't be upset if you really did poorly on the pretest for this chapter; the problems were a little rough. Now you are going to get a chance to go through the chapter and get a little more practice using dimensional analysis.

SECTION A The Metric System

In this chapter we are going to examine the units of measurement used in all scientific work. These units are in what is called the metric system. Many countries use this system as their normal method of expressing measurements. As a matter of fact, only the United States and some countries that used to be British colonies still use what is called the British system of weights and measures. (Britain itself has gone metric, so many people now call it the U.S. system.) The metric system ,as you will see, is vastly simpler than the British system. A Congressional committee is investigating what will have to be done to shift weights and measures to the metric system in the United States. Someday you will be saying, "What a hot day; it is 32 degrees in the shade," or, "My car is making terrific mileage, 12 kilometers per liter of gas." At present, however, you will have to learn the metric system and also how to convert from the British system to the metric system.

The beauty of the metric system is that all the conversion factors are simply 1.00 times 10 raised to some power. What is more, the name of the dimension tells you to what power 10 is raised. This is how it works.

Prefix	Symbol	Meaning	Prefix	Symbol	Meaning
pico-	p	10^{-12}	deci-	d	10^{-1}
nano-	n	10^{-9}	kilo-	k	10^{3}
micro-	μ	10^{-6}	mega-	M	10^{6}
milli-	m	10^{-3}	giga-	G	10^{9}
centi-	c	10^{-2}	tetra-	T	10^{12}

Only about four of these prefixes are used commonly in introductory chemistry, so we are going to concentrate on them.

The basic unit of length in the metric system is the meter. A meter (m) is about a yard long, and it has 100 parts, each of which is called a centimeter (cm). Thus you have

$$1 \text{ cm} = 10^{-2} \text{ m.}$$

The conversion factor and its reciprocal form from this equality are

$$\frac{\text{cm}}{10^{-2} \text{ m}}, \frac{10^{-2} \text{ m}}{\text{cm}}.$$

Try to calculate the number of centimeters (cm) in 4.5 meters (m).

The dimensions of the answer will be _____, so you write

$$? \underline{\hspace{1cm}} =$$

The information given in the problem is _____, so you write

$$? \text{ cm} = \underline{\hspace{1cm}}.$$

The conversion factor which has _____ in the denominator to

centimeters

cm

4.5 m

4.5 m

meters

cancel the _____ in the information given is _____. You now write

$$? \text{ cm} = 4.5 \ \cancel{m} \times \frac{\text{cm}}{10^{-2} \ \cancel{m}},$$

and the only dimension left is _____, which is the dimension of the answer. The answer, then, must be

$$? \text{ cm} = 4.5 \times \frac{1}{10^{-2}} \text{ cm}$$

$$= 4.5 \times 10^2 \text{ cm}.$$

The meter can also be divided into 1000 parts, each of which is called a millimeter (mm). Thus

$$1 \text{ mm} = 10^{-3} \text{ m}.$$

The conversion factors from this equality are _____.
Try this somewhat more complicated problem: how many centimeters in 8×10^2 millimeters? The dimensions of the answer

will be _____, so you write

$$? \ \_\_\_ = .$$

The information given is _____, so you write

$$? \text{ cm} = _____.$$

The only conversion factor you have which has millimeters in the

denominator is _____, so you multiply by this factor:

$$? \text{ cm} = 8 \times 10^2 \ \cancel{mm} \times \frac{10^{-3} \text{ m}}{\cancel{mm}}.$$

This operation cancels the unwanted dimension millimeters, but

leaves a new unwanted dimension _____. A conversion factor

which will have the unwanted dimension _____ in the denominator and be related to the dimension of the answer to the

question asked is _____. Multiplying by this conversion factor gives

$$? \text{ cm} = 8 \times 10^2 \ \cancel{mm} \times \frac{10^{-3} \ \cancel{m}}{\cancel{mm}} \times \frac{\text{cm}}{10^{-2} \ \cancel{m}}$$

$$= 8 \times 10^2 \times 10^{-3} \times \frac{1}{10^{-2}} \text{ cm}$$

$$= 8 \times 10^1 \text{ cm}.$$

meters, $\dfrac{\text{cm}}{10^{-2} \text{ m}}$

centimeters

$\dfrac{\text{mm}}{10^{-3} \text{ m}}, \ \dfrac{10^{-3} \text{ m}}{\text{mm}}$

centimeters

cm

8×10^2 mm

8×10^2 mm

$\dfrac{10^{-3} \text{ m}}{\text{mm}}$

meters

meters

$\dfrac{\text{cm}}{10^{-2} \text{ m}}$

It should now start to become apparent why the metric system is so easy to handle. All the conversions simply amount to multiplying or dividing by 10 raised to some exponent. If you are not very confident handling this exponential notation, refer back to Section A in Chapter 1.

The meter is divided into still smaller parts. If you divide a meter into 1 million parts, you have what is usually called a micron. (It should be called a *micrometer*, but this looks like the name of a measuring device.) The symbol for 10^{-6} is the Greek letter μ (mu). Thus

$$1 \ \mu = 10^{-6} \text{ m. (Notice that the ''m'' is left out.)}$$

$$\frac{\mu m}{10^{-6} \text{ m}}, \ \frac{10^{-6} \text{ m}}{\mu m}$$

The conversion factors from this equality are _____.

The smallest length which will interest us, since it is in the size range of an atom, is the angstrom (abbreviation Å or A). Its name doesn't follow the rules for naming but the relationship is

$$1 \text{ Å} = 10^{-10} \text{ m.}$$

When you are dealing with very large lengths (which isn't too often in chemistry) the unit used is the kilometer (km). It is a thousand meters and the relationship is

$$1 \text{ km} = 10^3 \text{ m.}$$

Here is a problem: how many kilometers in 6.3×10^{35} Å? The

How many kilometers?

6.3×10^{35} Å

question asked is ''_____'' and the information

given is _____. You therefore write

$$? \text{ km} = 6.3 \times 10^{35} \text{ Å.}$$

A conversion factor which will cancel out the unwanted dimension

$$\frac{10^{-10} \text{ m}}{\text{Å}}$$

Å is _____, so you write

$$? \text{ km} = 6.3 \times 10^{35} \text{ Å} \times \frac{10^{-10} \text{ m}}{\text{Å}}.$$

meters

$$\frac{\text{km}}{10^3 \text{ m}}$$

You can now get from the unwanted dimension _____ to kilometers, the dimensions of the answer, by multiplying by the

conversion factor _____.

$$? \text{ km} = 6.3 \times 10^{35} \text{ Å} \times \frac{10^{-10} \text{ m}}{\text{Å}} \times \frac{\text{km}}{10^3 \text{ m}}$$

$$= 6.3 \times 10^{35} \times 10^{-10} \times \frac{1}{10^3} \text{ km}$$

$$= 6.3 \times 10^{22} \text{ km}$$

Now we can summarize these metric units of length in their conversion factor form.

Angstrom	$\dfrac{\text{Å}}{10^{-10}\text{ m}}$
Micron	$\dfrac{\mu}{10^{-6}\text{ m}}$
Millimeter	$\dfrac{\text{mm}}{10^{-3}\text{ m}}$
Centimeter	$\dfrac{\text{cm}}{10^{-2}\text{ m}}$
Kilometer	$\dfrac{\text{km}}{10^{3}\text{ m}}$

The basic unit of volume in the metric system is the liter (*l*), which is a little more than 1 qt. Since the letter *l* and the number 1 are identical or very similar in normal type, the abbreviation will not be used in this text. However, when you are writing, you can make the distinction whether you mean ℓ or 1.

There are only two commonly used metric units of volume other than the liter itself. One is a milliliter (ml), and from what you know about the prefix milli- it is obvious that

$$1 \text{ ml} = \underline{\hspace{1cm}} \text{ liter.}$$

The conversion factors from this equality are _____.

When the metric system was set up, it was done in such a way that 1 ml was the volume equivalent to a cube 1 cm on a side. That is,

$$1 \text{ ml} = 1 \text{ cm}^3.$$

There was a very slight error made at that time, but for all your calculations you can consider the equality above as correct.

The other common unit of volume is the microliter (μl). From the prefix micro- you know that

$$1 \text{ }\mu\text{l} = \underline{\hspace{1cm}} \text{ liter.}$$

Now we can summarize the metric units of volume in their conversion factor form

Milliliter	$\dfrac{\text{ml}}{10^{-3}\text{ liter}}$
Microliter	$\dfrac{\mu\text{l}}{10^{-6}\text{ liter}}$

Margin answers:

10^{-3}

$\dfrac{\text{ml}}{10^{-3}\text{ liter}}, \dfrac{10^{-3}\text{ liter}}{\text{ml}}$

10^{-6}

How many μl?

4.7 × 10⁻² ml

μl, 4.7 × 10⁻² ml

milliliters

$$\frac{10^{-3} \text{ liter}}{\text{ml}}$$

liter

$$\frac{\mu l}{10^{-6} \text{ liter}}$$

microliters

$$\frac{\text{ml}}{\text{cm}^3}, \quad \frac{\text{cm}^3}{\text{ml}}$$

How many milliliters?

Try a problem just for practice. How many μl in 4.7×10^{-2} ml? The question asked is "_____" and the information given is _____. You therefore write

$$? \underline{} = \underline{}.$$

The dimension you want to cancel is _____, and a conversion factor with this in the denominator is _____. Multiplying by this conversion factor leaves the unwanted dimension _____. The conversion factor which has this unwanted dimension in the denominator and which relates to the dimensions of the desired answer would be _____. Multiplying by this,

$$? \mu l = 4.7 \times 10^{-2}\,\cancel{ml} \times \frac{10^{-3}\,\cancel{\text{liter}}}{\cancel{ml}} \times \frac{\mu l}{10^{-6}\,\cancel{\text{liter}}}$$

leaves _____ as the only dimension, and so the answer must be

$$? \mu l = 4.7 \times 10^{-2} \times 10^{-3} \times \frac{1}{10^{-6}}\,\mu l$$

$$= 4.7 \times 10^{1}\,\mu l.$$

Recall that above it was stated that

$$1 \text{ ml} = 1 \text{ cm}^3.$$

The conversion factors that can be derived from this equality are

_____, _____. Actually, volumes can always be expressed as lengths cubed. Thus, in the British system of weights and measures we talk about quarts and gallons but also cubic yards and cubic feet.

There will be certain problems where you will have to use both the volume units based on liters and those based on some metric length cubed. The conversion factors relating milliliters to cubic centimeters will allow you to go from one to the other. Here is a problem to illustrate this. How many milliliters are contained in a cube which is 3.0 cm on each edge?

The first thing you must do is calculate the volume of the cube in cubic centimeters. The volume of a cube is its edge length raised to the third power.

$$\text{Volume} = (3.0 \text{ cm})^3.$$

Both the number and its dimension must be cubed.

$$\text{Volume} = (3.0)^3 \times (\text{cm})^3$$

$$= 27 \text{ cm}^3$$

Now you have the volume of the cube, which is the information given. The question asked is "_____." You therefore write

$$? \text{ ml} = 27 \text{ cm}^3.$$

In order to cancel the dimensions of the information given
and move toward the dimensions of the answer, you can multiply

by the conversion factor ＿＿.

$$? \text{ ml} = 27 \ \cancel{cm^3} \times \frac{ml}{\cancel{cm^3}}$$

$$\frac{ml}{cm^3}$$

The only dimension left is the dimension of the answer, so

$$? \text{ ml} = 27 \text{ ml}$$

must be the answer.
Now find the volume in liters of a cube which is 2 Å on each
edge. This problem may seem much harder, but it isn't; you simply
must use more conversion factors.
The first thing you must do is to determine the volume in
angstroms cubed (Å³). The volume of a cube is the edge length
cubed, so

$$\text{Volume} = (2 \text{ Å})^3 = 2^3 \text{ Å}^3 = 8 \text{ Å}^3.$$

(Notice that the number and the dimension of the number are
each cubed.)

The question asked is "How many liters?" and the information
given has been calculated as 8 Å³. Therefore,

$$? \text{ liters} = 8 \text{ Å}^3.$$

You have no conversion factor with Å³ as a dimension, but you do

have one with Å, which is ＿＿＿＿＿. If you cube this it still is a
conversion factor.

$$\frac{10^{-10} \text{ m}}{\text{Å}}$$

$$\left(\frac{10^{-10} \text{ m}}{\text{Å}}\right)^3 = \frac{10^{-30} \text{ m}^3}{\text{Å}^3}$$

Multiplying by this factor you have

$$? \text{ liters} = 8 \ \cancel{Å^3} \times \frac{10^{-30} \text{ m}^3}{\cancel{Å^3}}.$$

Now the unwanted dimension is ＿＿＿＿＿＿, and you need to

cubic meters (m³)

multiply this by some conversion factor containing ＿＿＿＿＿＿
in the denominator, which will lead you toward the dimensions of
the answer, liters. If you think about it you will realize that you
must work through the equality of milliliters and cubic centimeters.
You must therefore have a conversion factor containing cubic

cubic meters

meters and cubic centimeters. You can derive this factor by ＿＿＿＿
the conversion factor relating meters to centimeters.

cubing

$$\left(\frac{cm}{10^{-2} \text{ m}}\right)^3 = \frac{cm^3}{10^{-6} \text{ m}^3}$$

Multiplying by this factor you have

$$? \text{ liters} = 8 \,\cancel{\text{Å}^3} \times \frac{10^{-30} \,\cancel{m^3}}{\cancel{\text{Å}^3}} \times \frac{cm^3}{10^{-6} \,\cancel{m^3}}$$

Now you can get from cubic centimeters to the volume dimension, milliliters, by multiplying by the conversion factor ____.

$$\frac{ml}{cm^3}$$

$$? \text{ liters} = 8 \,\cancel{\text{Å}^3} \times \frac{10^{-30} \,\cancel{m^3}}{\cancel{\text{Å}^3}} \times \frac{\cancel{cm^3}}{10^{-6} \,\cancel{m^3}} \times \frac{ml}{\cancel{cm^3}}$$

It is very easy to convert from milliliters to liters.

$$? \text{ liters} = 8 \,\cancel{\text{Å}^3} \times \frac{10^{-30} \,\cancel{m^3}}{\cancel{\text{Å}^3}} \times \frac{\cancel{cm^3}}{10^{-6} \,\cancel{m^3}} \times \frac{\cancel{ml}}{\cancel{cm^3}} \times \frac{10^{-3} \text{ liter}}{\cancel{ml}}$$

The only dimension left is liters, so the answer must be

$$? \text{ liters} = 8 \times 10^{-30} \times \frac{1}{10^{-6}} \times 1 \times 10^{-3} \text{ liter}$$

$$= 8 \times 10^{-27} \text{ liter}.$$

This must seem very labored. What you should really remember is that when you want to raise a quantity to a power, you treat the number and the dimension separately. You raise the number to the power and you also raise the dimension to the power.

For practice, find the conversion factor relating square millimeters with square meters. The conversion factor relating millimeters to meters is _____.

$$\frac{10^{-3} \text{ m}}{mm}$$

You must square both the numerator and the denominator:

$$\left(\frac{10^{-3} \text{ m}}{mm}\right)^2 = \frac{(10^{-3})^2 \text{ m}^2}{mm^2} = \frac{10^{-6} \text{ m}^2}{mm^2}.$$

$$\frac{10^{-18} \text{ m}^3}{\mu^3}$$

Check yourself on this one. The conversion factor relating cubic microns to cubic meters is _____.

The unit of weight in the metric system is the gram (g). Approximately 30 g equal an ounce. There are only three commonly used units of weight other than the gram itself. One is the milligram (mg), and

$$10^{-3}$$

$$1 \text{ mg} = \text{____ g}.$$

For very small quantities there is the microgram (μg), and you now know that

$$10^{-6}$$

$$1 \,\mu\text{g} = \text{____ g}.$$

Finally, for larger weights there is the kilogram (kg), and

$$10^3$$

$$1 \text{ kg} = \text{____ g}.$$

You can now see how simple the metric system of naming things is. In the table below these units are summarized with their conversion factors.

Microgram	$\dfrac{\mu g}{10^{-6}\ g}$	
Milligram	$\dfrac{mg}{10^{-3}\ g}$	
Kilogram	$\dfrac{kg}{10^{3}\ g}$	

Do one problem for practice: how many kilograms in $10^{14}\ \mu g$? Putting down the dimensions of the answer and the dimensions of the information given, you have

$$?\ \underline{\quad} = \underline{\qquad}$$

kg, $10^{14}\ \mu g$

Multiplying this by a conversion factor to remove the unwanted dimension, you have

$$?\ kg = 10^{14}\ \mu g \times \underline{\quad}.$$

The unwanted dimension now is _____, and you must relate this

to _____, the dimensions of the answer. The conversion factor

which will do this is _____.

$\dfrac{10^{-6}\ g}{\mu g}$
grams

kilograms,
$\dfrac{kg}{10^{3}\ g}$

(Remember to keep writing down all the conversion factors on your own setup as we go along.) Multiplying by this factor, all the dimensions cancel except the dimension of the answer, and so you have

$$?\ kg = 10^{14}\ \cancel{\mu g} \times \frac{10^{-6}\ \cancel{g}}{\cancel{\mu g}} \times \frac{kg}{10^{3}\ \cancel{g}}$$

$$= 10^{14} \times 10^{-6} \times \frac{1}{10^{3}}\ kg = 10^{5}\ kg.$$

Go back to the pretest, and do problems 1, 3, and 5. See whether you come up with the correct answer. All the answers are expressed in scientific notation, which you should make a practice of doing.

Now you are going to do some very easy problems using the metric system.

If 2.0 m of a sample of steel wire weigh 0.65 kg, how many milligrams does 10 mm weigh? The conversion factor in the

problem is _____. The dimensions of the answer will be

_____. The information given is _____. Setting this up you

have

$$?\ mg = 10\ mm.$$

$\dfrac{0.65\ kg}{2.0\ m}$
milligrams,   10 mm

You now need a conversion factor with millimeters in the denominator which will allow you to proceed toward the

dimensions of the answer. Such a factor might be _____. Try it.

$$? \text{ mg} = 10 \text{ mm} \times \frac{10^{-3} \text{ m}}{\text{mm}}$$

$\dfrac{10^{-3} \text{ m}}{\text{mm}}$

This is fine. You can now use the conversion factor shown in the problem.

$$? \text{ mg} = 10 \text{ mm} \times \frac{10^{-3} \text{ m}}{\text{mm}} \times \frac{0.65 \text{ kg}}{2.0 \text{ m}}$$

To get from kilograms, the unwanted dimension, to milligrams, the dimensions of the answer, you simply multiply by the appropriate conversion factors.

$$? \text{ mg} = 10 \text{ mm} \times \frac{10^{-3} \text{ m}}{\text{mm}} \times \frac{0.65 \text{ kg}}{2.0 \text{ m}}$$

$$\times \frac{10^3 \text{ g}}{\text{kg}} \times \frac{\text{mg}}{10^{-3} \text{ g}}$$

Therefore,

$$? \text{ mg} = 10 \times 10^{-3} \times \frac{0.65}{2.0} \times 10^3 \times \frac{1}{10^{-3}} \text{ mg}$$

$$= 0.32 \times 10^4 \text{ mg} = 3.2 \times 10^3 \text{ mg}.$$

Now do one yourself. A European coal-burning locomotive uses 35 kg of coal to travel 3.0 km. How many grams of coal does it use

grams

6.0 cm

$\dfrac{35 \text{ kg}}{3.0 \text{ km}}$

7.0×10^{-1} g

to go 6.0 cm? The dimensions of the answer will be _____. The

information given is _____. The conversion factor given in the

problem is _____. Now perform your calculations. The final

answer is _____. Did you forget to change the answer into scientific notation? If so, you got 70×10^{-2} g.

SECTION B The British System of Weights and Measures

Unhappily, we are stuck with a very clumsy and nonscientific system of measurement in the United States. It is very confusing to have 16 oz in an avoirdupois pound, 12 oz in a Troy pound (used only for precious metals) and 32 fl oz in a quart. From years of use, however, you probably remember most of the conversion factors in our system. To refresh your memory, and to have them all in one place, the most common ones are tabulated on the following page.

Length	Volume	Weight
$\dfrac{12 \text{ in.}}{\text{ft}}$	$\dfrac{16 \text{ fl oz}}{\text{pt}}$	$\dfrac{16 \text{ oz}}{\text{lb}}$
$\dfrac{3 \text{ ft}}{\text{yd}}$	$\dfrac{2 \text{ pt}}{\text{qt}}$	$\dfrac{2000 \text{ lb}}{\text{ton}}$
$\dfrac{5280 \text{ ft}}{\text{mile}}$	$\dfrac{4 \text{ qt}}{\text{gal}}$	

Although it isn't any more difficult to set up a problem in British units than it is in metric units, the arithmetic, as you will see, is much more laborious. For example, let's try this problem: how many miles in 6.0×10^7 in.? The question asked is

"_____" so the dimensions of the answer will be

_____. The information given is _____, so you write

$$? \text{ miles} = 6.0 \times 10^7 \text{ in.}$$

The only conversion factor you have with inches is _____.

$$? \text{ miles} = 6.0 \times 10^7 \text{ in.} \times \frac{\text{ft}}{12 \text{ in.}}$$

The unwanted dimension ____ can be cancelled by multiplying by the conversion factor _____.

$$? \text{ miles} = 6.0 \times 10^7 \text{ in.} \times \frac{\text{ft}}{12 \text{ in.}} \times \frac{\text{mile}}{5280 \text{ ft}}$$

The only dimension left is ____, so the answer must be

$$? \text{ miles} = 6.0 \times 10^7 \times \frac{1}{12} \times \frac{1}{5280} \text{ mile}$$

$$= 9.5 \times 10^2 \text{ miles.}$$

Converting from the British System to the Metric System

Most problems that you will meet in introductory chemistry will involve changing measurements given in British units to the equivalent measurement in metric units, or vice versa. The conversion factors are shown in the following table.

Length	Volume	Weight
$\dfrac{2.54 \text{ cm}}{\text{in.}}$	$\dfrac{946 \text{ ml}}{\text{qt}}$	$\dfrac{454 \text{ g}}{\text{lb}}$
$\dfrac{1.61 \text{ km}}{\text{mile}}$		$\dfrac{\text{kg}}{2.2 \text{ lb}}$

How many miles?

miles, 6.0×10^7 in.

$\dfrac{\text{ft}}{12 \text{ in.}}$

$\dfrac{\text{foot}}{\text{mile}}$
$\overline{5280 \text{ ft}}$

miles

With these conversion factors it is very simple to convert any measurement in one system to the equivalent measurement in the other. Problem 2 in the pretest asks: how many yards are in a kilometer? Starting the problem as usual, you have

yd, 1 km

$$? \underline{\quad} = \underline{\quad}$$

$$\frac{\text{mile}}{1.61 \text{ km}}$$

You can multiply by the conversion factor _____ to cancel the unwanted dimension and, incidentally, to shift you to British system units.

miles

The unwanted dimension now is _____, and you can get this to the wanted dimension yards by multiplying by the two conversion

$$\frac{5280 \text{ ft}}{\text{mile}}, \frac{\text{yd}}{3 \text{ ft}}$$

factors _____ and ____. All the dimensions cancel except

yards

_____, so your answer must be

$$? \text{ yd} = 1 \times \frac{1}{1.61} \times 5280 \times \frac{1}{3} \text{ yd}$$

$$= 1.09 \times 10^3 \text{ yd}.$$

Let us look at Problem 4 in the pretest: how many milligrams in

How many milligrams?

3×10^{-4} ton? The question asked is "_____"

3×10^{-4} ton

and the information given is _____. You therefore write

mg, 3×10^{-4} ton

$$? \underline{\quad} = \underline{\quad}$$

The only conversion factor you have containing tons as a dimension

$$\frac{2000 \text{ lb}}{\text{ton}}$$

is _____. Multiplying by this factor, you have the unwanted

pounds

dimension _____ left. Now, however, you have a conversion factor which will have pounds as one dimension and a metric unit as the other. This will serve as your "bridge" between the two

$$\frac{454 \text{ g}}{\text{lb}}$$

systems. It is _____.

Multiplying by this "bridge" (conversion factor), you now have

grams

as the unwanted dimension _____, which is easily converted to

milligrams

_____, the dimensions of the answer, by using the conversion

$$\frac{\text{mg}}{10^{-3} \text{ g}}$$

factor _____. All the dimensions have now cancelled except the dimensions of the answer, so the answer must be

$$? \text{ mg} = 3 \times 10^{-4} \text{ ton} \times \frac{2000 \text{ lb}}{\text{ton}} \times \frac{454 \text{ g}}{\text{lb}} \times \frac{\text{mg}}{10^{-3} \text{ g}}$$

$$= 2.72 \times 10^5 \text{ mg}.$$

Try Problem 7 in the pretest, and you will see how very simple it is using dimensional analysis.

You want to find how long it will take a Porsche doing 120 km/hr to go 90 miles. Do the problem yourself and then check your setup with the one shown below.

$$? \text{ hr} = 90 \cancel{\text{ miles}} \times \frac{1.61 \cancel{\text{ km}}}{\cancel{\text{mile}}} \times \frac{\text{hr}}{120 \cancel{\text{ km}}}$$

$$? \text{ hr} = 90 \times 1.61 \times \frac{1}{120} \text{ hr}$$

$$= 1.2 \text{ hr}$$

There is one conversion factor given in Problem 7, and this is

_____ .

$$\frac{120 \text{ km}}{\text{hr}}$$

The only way in which problems can become more complicated is if more conversion factors are given in the question itself.

In Problem 6 in the pretest we are told that the price of milk in France is 3.6 francs per liter and asked what the price is in dollars for 1 qt. (There are 5 francs per dollar.) There are two conversion

factors in the question: _____ and _____ .

$$\frac{3.6 \text{ francs}}{\text{liter}}, \quad \frac{5 \text{ francs}}{\$}$$

Try to do the problem and see if you get the correct answer, which is $0.68. If you do not, check your setup with the one shown below.

$$? \text{ dollars} = 1 \text{ qt} \times \frac{946 \text{ ml}}{\text{qt}} \times \frac{10^{-3} \text{ liter}}{\text{ml}} \times \frac{3.6 \text{ francs}}{\text{liter}} \times \frac{\$1}{5 \text{ francs}}$$

Now do Problems 8, 9, and 10 in the pretest. The setups are shown below, but do not look at them until you have attempted the problems yourself.

Problem 8. $$? \text{ miles} = 6 \cancel{\text{ min}} \times \frac{60 \cancel{\text{ sec}}}{\cancel{\text{min}}} \times \frac{3 \times 10^{10} \cancel{\text{ cm}}}{\cancel{\text{sec}}} \times \frac{10^{-2} \cancel{\text{ m}}}{\cancel{\text{cm}}}$$

$$\times \frac{\cancel{\text{km}}}{10^3 \cancel{\text{ m}}} \times \frac{\text{mile}}{1.61 \cancel{\text{ km}}}$$

Problem 9. $$? \text{ cal} = 6 \cancel{\text{ pt}} \times \frac{\cancel{\text{qt}}}{2 \cancel{\text{ pt}}} \times \frac{946 \cancel{\text{ ml}}}{\cancel{\text{qt}}} \times \frac{10^{-3} \cancel{\text{ liter}}}{\cancel{\text{ml}}}$$

$$\times \frac{0.96 \cancel{\text{ kg}}}{\cancel{\text{liter}}} \times \frac{10^3 \cancel{\text{ g}}}{\cancel{\text{kg}}} \times \frac{540 \text{ cal}}{\cancel{\text{g}}}$$

Problem 10. $$? \$ = 5000 \cancel{\text{ mile}} \times \frac{1.61 \cancel{\text{ km}}}{\cancel{\text{mile}}} \times \frac{\cancel{\text{liter}}}{12 \cancel{\text{ km}}} \times \frac{0.90 \text{ DM}}{\text{liter}}$$

$$\times \frac{\$}{2.3 \text{ DM}}$$

SECTION C   Temperature Conversions

In the metric system temperatures are expressed in degrees Celsius, which can also be called degrees centigrade (°C). As you know, in the United States we express temperature in degrees Fahrenheit

(°F). It is quite easy to convert a temperature from one system to the other. The mathematics are so simple that there is no reason to use the formal dimensional analysis method to solve the problem.

In the British system the temperature of freezing water is taken to be 32°F, and the temperature of boiling water is assigned the value of 212°F. In the metric system the temperature at which water freezes is taken to be 0°C, and the temperature at which it boils is assigned a temperature of 100°C.

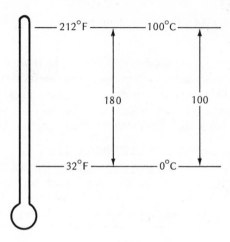

As you see, each °F is smaller than each °C. It takes 180 Fahrenheit degrees to equal only 100 centigrade (Celsius) degrees.

$$\frac{°F}{°C} = \frac{180}{100}$$

or simplifying

$$\frac{°F}{°C} = \frac{9}{5}$$

or

$$°F = \frac{9}{5}°C$$

When you convert from an actual temperature in one system to a temperature in the other, you will use this conversion factor to find by how many degrees the temperature in which you are interested differs from the freezing point of water. A temperature of 55°C is 55 Celsius degrees away from the freezing point of water. Remember that a temperature of 55°F is only 23 Fahrenheit degrees away from the freezing point of water since water freezes at 32°F.

Therefore, in making a conversion from one scale to the other, you will always use the number of degrees away from this reference point, the freezing point of water.

Number Fahrenheit degrees from reference point

$= $ temp °F $- 32$

Number Celsius degrees from reference point $=$ temp. °C

To convert temperatures you use the relation

$$\frac{\text{Fahrenheit degrees (°F)}}{\text{Celsius degrees (°C)}} = \frac{9}{5}.$$

Substituting the degrees away from the freezing point of water, you have

$$\frac{\text{temp °F} - 32}{\text{temp °C}} = \frac{9}{5}.$$

If you want to determine the temperature in °C, you will use the form

$$\text{temp °C} = \frac{5}{9}(\text{temp °F} - 32).$$

And if you want to determine the temperature in °F you will use the form

$$\text{temp °F} = \frac{9}{5}(\text{temp °C}) + 32.$$

Look at Problem 11 in the pretest; find the temperature in °C corresponding to 65°F. Using the above form, you have

$$°C = \frac{5}{9}(°F - 32)$$

$$= \frac{5}{9}(65 - 32)$$

$$= \frac{5}{9}(33)$$

$$= 18°C.$$

Problem 12 asks, what is the temperature in °F corresponding to
−14°C? Using the given form,

$$°F = \frac{9}{5}°C + 32$$

$$= \frac{9}{5}(-14) + 32$$

$$= -25 + 32$$

$$= 7°F.$$

(If you have difficulties with algebraic addition of negative
numbers, see Section A in Chapter 1. Also, if you had 6.8°F for the
last problem, you are not paying attention to the significant figures.
Check the addition of significant figures in Section C Chapter 1.)

There is one other temperature scale that you will use in gas law
calculations. This is an absolute temperature, and it is expressed in
degrees Kelvin (°K). The size of the degree is identical with the size
of a degree centigrade, but the freezing point of water is at 273°K.
To get a temperature in °K, you simply add 273 to the temperature
in °C. Thus, the boiling point of water (100°C) would equal

373

_____°K. We will return to the Kelvin scale in the chapter on gas
laws.

You are now ready to do the problem set on the metric system.

PROBLEM SET

As you do these problems, immediately check your answer. You will find both the setup and the numerical answers. The answers are expressed in scientific notation and to the correct number of significant figures. You will very rapidly see whether you are getting the hang of it. The entire set should not take more than 50 minutes to complete.

1. How many millimeters are there in 3.8 m?

2. How many grams are there in 4.5×10^9 μg?

3. An 8.2 kg sample contains how many grams?

4. How many milliliters are in a 9.5×10^{16} μl sample?

5. Express 4×10^{11} Å in kilometers.

6. A cube 8 cm on each side contains how many liters?

You may need the following conversion factors

$\dfrac{2.54 \text{ cm}}{\text{in.}}$	$\dfrac{12 \text{ in.}}{\text{ft}}$	$\dfrac{454 \text{ g}}{\text{lb}}$	$\dfrac{16 \text{ oz}}{\text{lb}}$	$\dfrac{946 \text{ ml}}{\text{qt}}$
$\dfrac{1.61 \text{ km}}{\text{mile}}$	$\dfrac{5280 \text{ ft}}{\text{mile}}$	$\dfrac{2.2 \text{ lb}}{\text{kg}}$	$\dfrac{2000 \text{ lb}}{\text{ton}}$	$\dfrac{4 \text{ qt}}{\text{gal}}$

7. How many centimeters in 3.0 miles?

8. How many gallons are there in 43 liters?

9. How many inches are in 6.0 m?

10. How many ounces are in 8.5 kg?

11. A square which is 2.5 m on a side has an area of how many square inches?

12. What is the cost of coal in dollars per ton if it costs $0.04/kg?

13. How many miles could you drive for $3.90 if the gas mileage of your car is 14 km/liter of gas and the price is $0.64/gal?

14. If 6.0 liters of mercury weigh 78 kg and cost $420, what is the price of mercury in dollars per pound?

15. A crane can pick up 3.0 tons of excavated earth in an hour. The crane operator's wages are $15 per hour. What, then, is the cost of picking up 85 kg of excavated earth?

16. What would it cost to send a rocket to Mars which is 7.9×10^7 km distant from Earth if the average speed of the rocket is 1600 miles/min and the solid propellant consumption averages 100 g every 1.5 sec and costs $500/lb?

17. What would it cost to inflate a weather balloon 14 ft in diameter with helium if the price of helium is $5.00/lb and 22.4 liters of helium weigh 4.0 g. The volume of a sphere is

$$\text{volume} = \frac{1}{6}\pi d^3, \quad \text{where } d \text{ is the diameter.}$$

π is one of the few dimensionless numbers and equals 3.1412.

18. If one afternoon you decide to dig a hole through the earth to China for a game of ping pong, how many centuries would it be before you got there if you could dig at the rate of 4 miles depth per day and the diameter of the earth is 12,700 km?

19. A temperature of $-9°F$ is equivalent to what temperature in $°C$?

20. A temperature of $4500°C$ is equivalent to what temperature in $°F$?

21. A temperature of $-15°C$ is equivalent to what temperature in $°K$?

22. A temperature of $80°F$ is equivalent to what temperature in $°K$?

23. When the United States shifts to the metric system, a high fever of $103.0°F$ will be given as how many $°C$?

PROBLEM SET ANSWERS

You can grade yourself on the problems in the following way: if you missed more than one problem in the group 1–6, you had better redo Section A. If you missed three or more problems in the next group, 7–18, you should redo Section B in the chapter. Problems 19–23 are covered in the short section on temperature, and you should get them all correct. If you did not, redo Section C.

1. $\text{? mm} = 3.8\ \cancel{m} \times \dfrac{mm}{10^{-3}\ \cancel{m}} = 3.8 \times 10^3\ \text{mm}$

2. $\text{? g} = 4.5 \times 10^9\ \cancel{\mu g} \times \dfrac{10^{-6}\ g}{\cancel{\mu g}} = 4.5 \times 10^3\ \text{g}$

3. $\text{? g} = 8.2\ \cancel{kg} \times \dfrac{10^3\ g}{\cancel{kg}} = 8.2 \times 10^3\ \text{g}$

4. $\text{? ml} = 9.5 \times 10^{16}\ \cancel{\mu l} \times \dfrac{10^{-6}\ \cancel{liter}}{\cancel{\mu l}} \times \dfrac{ml}{10^{-3}\ \cancel{liter}}$

   $= 9.5 \times 10^{13}\ \text{ml}$

5. $\text{? km} = 4 \times 10^{11}\ \cancel{Å} \times \dfrac{10^{-10}\ \cancel{m}}{\cancel{Å}} \times \dfrac{km}{10^3\ \cancel{m}}$

   $= 4 \times 10^{-2}\ \text{km}$

6. $\text{? liters} = 512\ \cancel{cm^3} \times \dfrac{\cancel{ml}}{\cancel{cm^3}} \times \dfrac{10^{-3}\ liter}{\cancel{ml}}$

   $= 5.12 \times 10^{-1}\ \text{liter}$

7. $? \text{ cm} = 3.0 \text{ miles} \times \dfrac{1.61 \text{ km}}{\text{mile}} \times \dfrac{10^3 \text{ m}}{\text{km}} \times \dfrac{\text{cm}}{10^{-2} \text{ m}}$

    $= 4.8 \times 10^5 \text{ cm}$

8. $? \text{ gal} = 43 \text{ liters} \times \dfrac{\text{ml}}{10^{-3} \text{ liter}} \times \dfrac{\text{qt}}{946 \text{ ml}} \times \dfrac{\text{gal}}{4 \text{ qt}}$

    $= 11 \text{ gal}$

9. $? \text{ in.} = 6.0 \text{ m} \times \dfrac{\text{cm}}{10^{-2} \text{ m}} \times \dfrac{\text{in.}}{2.54 \text{ cm}}$

    $= 2.4 \times 10^2 \text{ in.}$

10. $? \text{ oz} = 8.5 \text{ kg} \times \dfrac{2.2 \text{ lb}}{\text{kg}} \times \dfrac{16 \text{ oz}}{\text{lb}}$

    $= 3.0 \times 10^2 \text{ oz}$

11. $? \text{ in.}^2 = 6.25 \text{ m}^2 \times \dfrac{\text{cm}^2}{10^{-4} \text{ m}^2} \times \dfrac{\text{in.}^2}{(2.54)^2 \text{ cm}^2}$

    $= 9.69 \times 10^3 \text{ in.}^2$

12. $? \text{ dollars} = 1 \text{ ton} \times \dfrac{2000 \text{ lb}}{\text{ton}} \times \dfrac{454 \text{ g}}{\text{lb}} \times \dfrac{\text{kg}}{10^3 \text{ g}} \times \dfrac{\$0.04}{\text{kg}}$

    $= \$36.32$

If you are having difficulties, it would be a very good idea to start by writing down all the conversion factors given in the problems.

13. $? \text{ miles} = \$3.90 \times \dfrac{\text{gal}}{\$0.64} \times \dfrac{4 \text{ qt}}{\text{gal}} \times \dfrac{946 \text{ ml}}{\text{qt}} \times \dfrac{10^{-3} \text{ liter}}{\text{ml}}$

    $\times \dfrac{14 \text{ km}}{\text{liter}} \times \dfrac{1 \text{ mile}}{1.61 \text{ km}}$

    $= 2.0 \times 10^2 \text{ miles}$

14. $? \text{ dollars} = 1 \text{ lb} \times \dfrac{454 \text{ g}}{\text{lb}} \times \dfrac{\text{kg}}{10^3 \text{ g}} \times \dfrac{\$420}{78 \text{ kg}}$

    $= \$2.44 = \$2.4$ to the correct number of significant figures

15. $? \text{ dollars} = 85 \text{ kg} \times \dfrac{2.2 \text{ lb}}{\text{kg}} \times \dfrac{\text{ton}}{2000 \text{ lb}} \times \dfrac{\text{hr}}{3.0 \text{ ton}} \times \dfrac{\$15}{\text{hr}}$

    $= \$0.47$

16. $? \text{ dollars} = 7.9 \times 10^7 \text{ km} \times \dfrac{\text{mile}}{1.61 \text{ km}} \times \dfrac{\text{min}}{1600 \text{ miles}}$

    $\times \dfrac{60 \text{ sec}}{\text{min}} \times \dfrac{100 \text{ g}}{1.5 \text{ sec}} \times \dfrac{\text{lb}}{454 \text{ g}} \times \dfrac{\$500}{\text{lb}}$

    $= \$1.4 \times 10^8$ (140 million dollars!)

17. Volume $= \dfrac{1}{6}\pi(14\text{ ft})^3 = 1.5 \times 10^3\text{ ft}^3$

? dollars $= 1.5 \times 10^3\text{ ft}^3 \times \dfrac{(12)^3\text{ in.}^3}{\text{ft}^3} \times \dfrac{(2.54)^3\text{ cm}^3}{\text{in.}^3} \times \dfrac{\text{ml}}{\text{cm}^3}$

$\times \dfrac{10^{-3}\text{ liter}}{\text{ml}} \times \dfrac{4.0\text{ g}}{22.4\text{ liter}} \times \dfrac{\text{lb}}{454\text{ g}} \times \dfrac{\$5.00}{\text{lb}}$

$= 1.5 \times 10^3 \times 1728 \times 16.38 \times 1 \times 10^{-3} \times \dfrac{4.0}{22.4}$

$\times \dfrac{1}{454} \times \$5.00$

$= \$8.4 \times 10^1$

18. ? century $= 12{,}700\text{ km} \times \dfrac{\text{mile}}{1.61\text{ km}} \times \dfrac{\text{day}}{4\text{ miles}} \times \dfrac{\text{yr}}{365\text{ days}}$

$\times \dfrac{\text{century}}{100\text{ yr}} = 0.05\text{ century}$

(This problem is so easy that it should have been a rest after doing Problem 17. It is, however, at about the general level of difficulty of the problems you will face in chemistry.)

19. $°C = \dfrac{5}{9}(-9-32) = \dfrac{5}{9}(-41) = -23°C$

20. $°F = \dfrac{9}{5}(4500) + 32 = 8100 + 32 = 8132°F$

21. $°K = -15 + 273 = 258°K$

22. $°C = \dfrac{5}{9}(80-32) = \dfrac{5}{9}(48) = 27°C$

$°K = 27 + 273 = 300°K$

23. $°C = \dfrac{5}{9}(103.0-32) = 39.4°C$

Chapter 4
Density, Specific Gravity, and Percentage

In this chapter we are going to examine three conversion factors—density, specific gravity, and percentage. Density is a relationship of the weight of a substance to its volume and has the dimensions of grams/milliliter or grams/liter (gases). Specific gravity works out to be the ratio of the density of a substance to the density of water, $d(X)/d(H_2O)$. The percentage is the parts of a substance per 100 total parts of the system in which the substance is contained. Using these conversions, try the pretest below. It should take no longer than 20 minutes.

PRETEST

1. What is the density of benzene in grams/milliliter if 5.5 liters of benzene weigh 4.8 kg?

2. If the density of air is 1.3 g/liter, how many liters of air will weigh 2.0 lb?

3. Given that the density of mercury is 13.6 g/ml, then how many liters will 800 g of mercury occupy?

4. A cube of platinum which is 10 cm on each edge weighs 21 kg. What is the density of platinum?

5. What is the specific gravity of a solution of sulfuric acid if 100 ml solution weighs 150 g? The density of water at the temperature of interest is 0.996 g/ml.

6. A sample of calcite is 87.0% $CaCO_3$ by weight. How many grams of calcite are required to yield 125 g pure $CaCO_3$?

7. How many grams of iron are in 600 g iron ore if the iron ore is known to be 35.0% iron by weight?

8. A solution of HCl in water has a density of 1.1 g/ml. What is the weight percent of HCl in the solution if 150 ml of the solution contains 30 g HCl?

9. What is the percent purity of a sample of potassium permanganate if 21.0 g of the sample react with exactly 35.0 ml of a sodium oxalate solution? It is known that 10.0 ml of the sodium oxalate solution react with 5.70 g pure potassium permanganate.

10. How many grams of NaCl are in 500 ml of a solution which is 26.0 weight percent NaCl and has a density of 1.20 g/ml?

PRETEST ANSWERS

If you got nine out of the ten problems correct, you are probably ready to go directly to the problem set at the end of this chapter If you did Problems 1 through 4 correctly, you can skip Section A in the chapter, which is about density. If you got Problem 5 correct and feel confident about specific gravity, you can skip Section B, which is on this topic. Problems 6 and 7 are concerned with simple density calculations, and if you did these correctly, you know how to handle what is in Section C in the chapter and you may skip this. Problems 8 through 10 are a little more complicated and involve either percent purity or a combination of density and percentage. If you missed these, it would be a good idea to carefully read the end of Section C.

Many of these problems are worked out as examples in the body of the chapter.

1. 0.87 g/ml

2. 7.0×10^2 liter

3. 5.88×10^{-2} liter

4. 21 g/ml

5. 1.51

6. 144 g calcite

7. 210 g iron

8. 18%

9. 95%

10. 156 g NaCl

SECTION A Density

The density of a substance is defined as the mass per unit volume. Using the metric units of mass (grams) and volume (milliliters), we can say that the density is the number of grams which equal 1 ml. This equality is, of course, a conversion factor.

$$\text{Density} = \frac{\text{weight in grams}}{1 \text{ milliliter}}$$

The dimensions of this factor will be ___, or, as we usually say,

density is the number of grams ___ milliliter.

We use the units g/ml for densities of liquids and solids. However, the densities of gases are so small that it is more convenient to express them in g/liter. Thus for gases, the density is

the weight in grams per ___.

To determine the density of a substance, it is simply necessary to know the weight of a known volume. For example, if a 20 ml sample of alcohol weighs 16 g, what is the density of alcohol? To be consistent we can reword the question to ask how many grams

1 ml weighs. Now the question asked is _____. The

information given is ___, and in the problem the conversion factor is 20 ml per 16 g. Setting this up in the usual way, you have

$$? \_ = \_\_\_ \times \_\_\_$$

All dimensions except grams cancel, and you have as the final answer that the density is 0.8 g per ml.

It is possible in this particular calculation to take a short cut. Since density is always per 1.0 volume units (milliliters or liters), you could say that the question asked is "How many grams per milliliter?" and so the dimensions of the answer will be g/ml.

$$? \frac{g}{ml} =$$

Then the conversion factor given in the problem is in the dimensions of the answer, and you simply put it in.

$$? \frac{g}{ml} = \frac{16 \text{ g}}{20 \text{ ml}}$$

The dimensions are exactly those of the answer.

Try one this way. What is the density of oxygen gas if 11.2 liters weigh 16.0 g? Since you want density as your answer, the dimensions of the answer will be

$$? \_\_\_ =$$

(*Note:* it is per liter because oxygen is a gas.)

Then the conversion factor given in the problem is _____.

$\dfrac{g}{ml}$

per

liter

"How many grams?"

1 ml

g, 1 ml, $\dfrac{16 \text{ g}}{20 \text{ ml}}$

$\dfrac{g}{liter}$

$\dfrac{16.0 \text{ g}}{11.2 \text{ liters}}$

Putting this in you see that you have exactly the dimensions of the answer.

$$? \frac{g}{liter} = \frac{16.0 \ g}{11.2 \ liters}$$

$$= 1.43 \frac{g}{liter}$$

The only way that this type of problem can become more complicated is by having the weight or volume given in dimensions which must be converted to grams and milliliters (or liters). However, since you are now expert at converting metric and British dimensions, this should be very easy.

Try one. What is the density in grams per milliliter of antifreeze if

$\dfrac{g}{ml}$

$\dfrac{g}{ml}$

$\dfrac{2.3 \ lb}{qt}$

$\dfrac{2.3 \ lb}{qt}$

1.0 qt weighs 2.3 lb? The dimensions of the answer will be ___, so you write

$$? \underline{\quad} =$$

The conversion factor given in the problem is _____.

$$? \frac{g}{ml} = \underline{\qquad}$$

Now you must multiply by conversion factors which will cancel the unwanted dimensions and replace them with the dimensions of the answer. The conversion factor relating pounds to grams is

$\dfrac{454 \ g}{lb}$

$\dfrac{qt}{946 \ ml}$

_____. The conversion factor relating quarts to milliliters is

_____. If you have forgotten these conversion factors, check back into Section B, Chapter 3.

Multiplying by both of these factors, you have

$$? \frac{g}{ml} = \frac{2.3 \ \cancel{lb}}{\cancel{qt}} \times \frac{454 \ g}{\cancel{lb}} \times \frac{\cancel{qt}}{946 \ ml}$$

$$= 2.3 \times 454 \times \frac{1}{946} \frac{g}{ml}.$$

This leaves only the dimensions of the answer, which then is

$$? \frac{g}{ml} = 1.1 \frac{g}{ml}.$$

Incidentally, it is very common to write $\dfrac{g}{ml}$ as g/ml. This saves space for the printer since it gets everything on one line.

Now we shall see how you use density as a conversion factor, which it really is; it converts weight to volume. Here is a problem. If the density of carbon tetrachloride is 1.5 g/ml, what will 50 ml of carbon tetrachloride weigh? The question is, "What will it

weigh?" so the dimensions of the answer will be _____. The
information given is _____, and the conversion factor in the
problem is the density, _____. Writing these in the usual way, you
have

$$? \underline{\hspace{1cm}} = \underline{\hspace{1cm}} \times \underline{\hspace{1cm}}.$$

The only dimension which does not cancel is _____, so the
answer is _____.

It is very important that, besides just filling in the blanks, you
write down each conversion factor as you get it in your own setup.
Only by doing this will you start to get really good at setting up
these problems.

For a slightly more complicated example, let's do Problem 2 in
the pretest. If the density of air is 1.3 g/liter, how many liters of air
will weigh 2.0 lb?

The question asked is "_____" and so the
dimensions of the answer will be _____. The information given is
____, and the conversion factor in the problem is the density,
which is _____. Starting in the usual way, you write

$$? \underline{\hspace{1cm}} = \underline{\hspace{1cm}}.$$

The dimension of the information that you must cancel is _____.
However, the dimension for weight in the conversion factor in the
problem is _____. Consequently, before you can use the density,
you must convert _____ to _____. You know the conversion
factor which will do this; it is _____.

The unwanted dimension is _____, but this can be cancelled
using the density expressed so that the dimension _____ is in the
denominator. You write the conversion factor as _____. Multiplying
by this factor, the only dimension left is _____, which is the
dimension of the answer Therefore, the answer must be

$$? \text{ liters} = 2.0 \cancel{lb} \times \frac{454 \cancel{g}}{\cancel{lb}} \times \frac{\text{liter}}{1.3 \cancel{g}}$$

$$= 698 \text{ liters} = 7.0 \times 10^2 \text{ liters}.$$

Now try Problems 1, 3, and 4 in the pretest. The setups are shown
below, but don't look at them until you have done the problems.

Problem 1.　$? \dfrac{g}{ml} = \dfrac{4.8 \cancel{kg}}{5.5 \cancel{liters}} \times \dfrac{10^3 \text{ g}}{\cancel{kg}} \times \dfrac{10^{-3} \cancel{liter}}{ml}$

$$= 0.87 \text{ g/ml}$$

Answer column:

grams

50 ml

$\dfrac{1.5 \text{ g}}{ml}$

g,　50 ml,　$\dfrac{1.5 \text{ g}}{ml}$

grams

$50 \times 1.5 \text{ g} = 75 \text{ g}$

How many liters?

liters

2 lb

1.3 g/liter

liters,　2.0 lb

pounds

grams

pounds,　grams

$\dfrac{454 \text{ g}}{lb}$

grams

grams

$\dfrac{\text{liter}}{1.3 \text{ g}}$

liters

Problem 3. $? \text{ liters} = 800 \, g \times \dfrac{ml}{13.6 \, g} \times \dfrac{10^{-3} \text{ liter}}{ml}$

$$= 5.88 \times 10^{-2} \text{ liter}$$

Problem 4. $\text{Volume} = (10)^3 \text{ cm}^3 = 10^3 \text{ ml}$

$$? \, \dfrac{g}{ml} = \dfrac{21 \, kg}{10^3 \text{ ml}} \times \dfrac{10^3 \text{ g}}{kg} = 21 \text{ g/ml}$$

SECTION B Specific Gravity

Sometimes another measurement of the mass-volume characteristics of a substance is used in preference to density. Such a measure is the specific gravity (sp. gr.), which is the ratio of the weight of the substance to the weight of water that occupies the same volume. Since a density is weight per milliliter, then you can see that specific gravity also equals the ratio of the densities.

So we can write

$$\text{Specific gravity} = \frac{\text{weight of sample}}{\text{weight of water occupying same volume}}$$

or

$$\text{Specific gravity} = \frac{\text{density of sample}}{\text{density of water}}.$$

Water is generally the reference for solids and liquids. For gases the reference is usually to air.

It is often said that specific gravity is a dimensionless quantity since it is g/g. However, this is not exactly true, since the dimensions are not simply grams but are grams substance and grams water. It is the old story of pounds of potatoes and pounds of onions in the same problem.

As you probably know, the density of all substances varies with temperature. When you wish to convert from the density of a substance to its specific gravity, you must know at what temperature you have measured it and also what the density of water is at that temperature. The densities of water at various temperatures are tabulated below to three decimal places. Greater precision will require your getting values from a handbook.

T (°C)	Density (g/ml)	T (°C)	Density (g/ml)
0°–12°	1.000	27°–30°	0.996
13°–18°	0.999	31°–33°	0.995
19°–23°	0.998	33°–36°	0.994
24°–26°	0.997	37°–39°	0.993

As you see from the table, the density of water is 1.0 g/ml expressed to two significant figures from a temperature of 0°C to 40°C. If the table continued to 100°C, the value would still be 1.0 g/ml. Thus the specific gravity would equal the density to two significant figures. If you require a greater number of significant figures, then you must perform the calculation.

Problem: what is the density of a sample if its specific gravity is 1.13 at 25°C? This is so easy to solve that we need not use dimensional analysis. The relation is

$$\text{Specific gravity} = \frac{\text{density of sample}}{\underline{\hspace{2cm}}}$$

density of water

At 25°C the _____ is 0.997 g/ml. So you may write

density of water

$$1.13 = \frac{\text{density of sample}}{0.997},$$

$$1.12661 = \text{density of sample},$$

$$1.13 = \text{density of sample to three significant figures.}$$

SECTION C   Percentage

Like density, percent (%) is a conversion factor. Recall the Latin prefixes used in naming the metric units. "Cent" means the

number ____, so when you say that wine is 14% alcohol, this is a

100

shorthand way of saying that there are 14 parts of alcohol per ____

100

parts of wine. As a fraction we write

$$\frac{14 \text{ parts alcohol}}{100 \text{ parts wine}}.$$

The percentage is the number of parts of a component in 100 parts of the total substance which contains the component.

The dimensions used to express the number of parts may vary, and you must be careful to note this. Thus a weight percent in the metric system would be

$$\frac{\text{grams component}}{100 \text{ g total substance}}.$$

Similarly, a volume percent would be

$$\frac{\text{liters components}}{\underline{\hspace{3cm}}},$$

100 liters total substance

or also

$$\frac{\text{milliliters component}}{\underline{\hspace{3cm}}}.$$

100 ml total substance

For weight percent you could also have

$$\frac{\underline{\hspace{3cm}}}{100 \text{ kg total substance}}.$$

kg component

The unit in the numerator must be the same as the unit in the denominator. (There are some mixed dimension percentages, but these are so unscientific and confusing that you should never use them.)

Here is a simple problem. What is the weight percent of salt in sea water if there are 5.0 g salt in 125 g sea water? You must

grams

reword the question to ask how many _____ of salt there are per

100 g

_____ sea water. The dimensions of the answer will be

grams salt, 100 g sea water

_____, and the information given is _____. There

$$\frac{5.0 \text{ g salt}}{125 \text{ g sea water}}$$

is one conversion factor in the problem, _____. So you set the problem up in the usual way,

grams salt, 100 g sea water

$$? \underline{\hspace{1.5cm}} = \underline{\hspace{3cm}}.$$

Multiplying this by the conversion factor given in the problem

grams sea water

cancels the unwanted dimension, _____, and leaves only the dimensions of the answer, which must be

$$? \text{ g salt} = 100 \cancel{\text{ g sea water}} \times \frac{5.0 \text{ g salt}}{125 \cancel{\text{ g sea water}}}$$

$$= 4.0 \text{ g salt}.$$

You now know that there are 4.0 g salt per 100 g sea water, so the sea water must be 4.0% salt.

In doing percentage problems the unit used in the numerator is always the same as that in the denominator. Consequently, you must be very careful to specify what you are talking about. Thus, you must write grams salt and grams sea water; it is not enough simply to write grams.

One way that the problems can become a little more complicated is when the dimensions of the conversion factor are not those of the question asked and the information given. This shouldn't be too much of a chore to convert. Consider this problem: what is the volume percent of fat in milk if 25 gal of milk contains 11 liters of fat? First you must reword the problem to ask how many

100 ml

milliliters of fat per _____ milk. The question asked is

How many milliliters fat?, 100 ml milk

"_____" and the information now given is_____.

$$\frac{11 \text{ liters fat}}{25 \text{ gal milk}}$$

The conversion factor given in the problem is _____. Starting in the usual way, you write

ml fat, 100 ml milk

$$? \underline{\hspace{1.5cm}} = \underline{\hspace{2cm}}.$$

However, the volume dimension given in the conversion factor for milk is _____, so you must first convert the dimensions of the information given, 100 ml milk, to gallons. It will take two conversion factors to do this. You have

> ? ml fat = 100 ml milk
>
> $$\times \frac{\text{qt milk}}{946 \text{ ml milk}} \times \frac{\text{gal milk}}{4 \text{ qt milk}}.$$

Now you can multiply by the conversion factor given in the problem,

> ? ml fat = 100 ml milk
>
> $$\times \frac{\text{qt milk}}{946 \text{ ml milk}} \times \frac{\text{gal milk}}{4 \text{ qt milk}}$$
>
> $$\times \frac{11 \text{ liters fat}}{25 \text{ gal milk}}$$

The unwanted dimension now is _____, and the dimensions of

the answer will be _____. This is a very simple conversion.

You multiply by the conversion factor _____.

> ? ml fat = 100 ml milk
>
> $$\times \frac{\text{qt milk}}{946 \text{ ml milk}} \times \frac{\text{gal milk}}{4 \text{ qt milk}}$$
>
> $$\times \frac{11 \text{ liters fat}}{25 \text{ gal milk}} \times \frac{\text{ml fat}}{10^{-3} \text{ liter fat}}$$

> ? ml fat = 12 ml fat = 12%

Although the last problem looks a little messy, it really isn't. You just keep multiplying by conversion factors to get from the dimensions of the information given, through the conversion factor given in the problem (the "bridge"), to the dimensions of the answer.

Look at Problem 8 in the pretest. A solution of HCl in water has a density of 1.1 g/ml. What is the weight percent (wt %) of HCl in the solution if 150 ml of the solution contains 30 g HCl? First you

must reword the question to ask how many _____ are in

_____.

There are two conversion factors given in the problem. The first is the density of the solution. The conversion factor is (writing it exactly, so that you don't confuse grams solution with grams HCl)

_____. The second conversion factor is "the bridge;" it is

_____.

Now you start the problem in the usual way.

> ? _____ = _____

Right margin answers:

gallons

liters fat

milliliters fat
$$\frac{\text{ml fat}}{10^{-3} \text{ liter fat}}$$

grams HCl

100 g solution

$$\frac{1.1 \text{ g soln.}}{\text{ml soln.}}$$
$$\frac{30 \text{ g HCl}}{150 \text{ ml soln.}}$$

g HCl,   100 g soln.

grams soln.

ml soln.
_____
1.1 g soln.

milliliters soln.

30 g HCl
_____
150 ml soln.

The unwanted dimension is _____, so you look at your conversion factors and find one which has this dimension in the denominator. The one which does this is _____. Multiply. The unwanted dimension now is _____, so you can multiply by the conversion factor_____. The only dimension remaining now is grams HCl, which is the dimension of the answer. Therefore, the answer must be

$$? \text{ g HCl} = 100 \cancel{\text{ g soln.}} \times \frac{\cancel{\text{ml soln.}}}{1.1 \cancel{\text{ g soln.}}} \times \frac{30 \text{ g HCl}}{150 \cancel{\text{ ml soln.}}}$$

$$= 18 \text{ g HCl}$$
$$\text{or } 18\%.$$

Notice that once again you must be very careful to write not only the unit (grams or milliliters) but also what it is a unit of. The density is grams/milliliter of *solution*, so you must write grams solution per milliliter solution. If you don't, in this case you will end up confusing grams HCl with grams solution.

Percent Purity

One percentage that is very often of interest to chemists is the percent purity of a substance. It is almost always expressed as a weight percent since volumes vary with temperature. Therefore, if you wish to know the percent purity of an impure sample, you must determine how many grams pure component are in 100 g impure sample. There is a danger of confusing the grams of pure substance with grams sample, because of the way the problem is often worded. For example, "What is the percent purity of an iron ore sample if . . . ?" In such a case you must carefully distinguish between impure sample and pure iron. The best way to do this is always to refer to the total sample as impure (imp.); you can reword the question to ask how many grams iron are in 100 g imp.

Here is an example. What is the percent purity of a sample of gold ore if it is found that 2.0 kg ore yields 90 g gold metal? The

grams gold, 100 g imp.

90 g gold
_____
2.0 kg imp.

grams gold, 100 g imp.

reworded question is: how many _____ are in _____? The conversion factor ("bridge") given in the problem is

grams impure

kilograms impure

grams impure

kilograms impure

10⁻³ kg imp.
_____
g imp.

_____. Starting in the usual way, you write

? _____ = _____.

The unwanted dimension, which must be cancelled, is

_____. The dimension of weight of impure sample in the "bridge" is _____. Consequently, in order to be able to use the bridge, you must convert _____ to

_____. This is easily done using the conversion factor _____. Now you can multiply by the conversion factor, and this leaves as the only dimension _____, which is the dimension of the answer. Therefore, the final answer will be

$$? \text{ g gold} = 100 \cancel{\text{ g imp.}} \times \frac{10^{-3} \cancel{\text{ kg imp.}}}{\cancel{\text{ g imp.}}} \times \frac{90 \text{ g gold}}{2.0 \cancel{\text{ kg imp.}}}$$

$$= 4.5 \text{ g gold} = 4.5\%.$$

You must have noticed that the conversion factor called the "bridge" has been mentioned more and more often. This will become increasingly important as you move on to problems really involving chemistry. This "bridge" is a conversion factor which connects the substance about which the question is asked with the substance in the information given. In solving a problem you always go from the information given across the "bridge" to the substance in the answer.

Problem 9 in the pretest is a good example of a chemical reaction "bridge." What is the percent purity of a sample of potassium permanganate if 21.0 g of the sample react with exactly 35.0 ml of a sodium oxalate solution ? It is known that 10.0 ml of the sodium oxalate solution react with 5.70 g pure potassium permanganate.

You are trying to determine how many grams potassium permanganate are in 100 g impure sample. There are two conversion factors given in the problem. First, 21.0 g impure per 35.0 ml sod. ox. soln., which, written as a fraction, is

_____. The second is a "bridge" since it relates sodium oxalate solution to pure potassium permanganate. Written as a fraction it is _____.

Starting in the usual way, you write

? _____ = _____.

The unwanted dimension, which must be cancelled, is

_____; as you see, the conversion factor

_____ contains that dimension. Multiplying by this conversion factor leaves the unwanted dimension

_____. However, this appears in the "bridge," so

next you multiply by _____.

The only dimension left is _____, which is the dimension of the answer. Therefore, the answer must be

$$? \text{ g pot. perm.} = 100 \cancel{\text{ g imp.}} \times \frac{35.0 \cancel{\text{ ml sod. ox. soln.}}}{21.0 \cancel{\text{ g imp.}}}$$

$$\times \frac{5.70 \text{ g pot. perm.}}{10.0 \cancel{\text{ ml sod. ox. soln.}}}$$

$$= 95.0 \text{ g pot. perm.}$$

$$= 95.0\% \text{ pure.}$$

grams gold

$$\frac{21.0 \text{ g imp.}}{35.0 \text{ ml sod. ox. soln.}}$$

$$\frac{10.0 \text{ ml sod. ox. soln.}}{5.70 \text{ g pot. perm.}}$$

g pot. perm.,   100 g imp.

$$\frac{\text{grams impure}}{35.0 \text{ ml sod. ox. soln.}}$$

$$\frac{21.0 \text{ g imp.}}{}$$

ml sod. ox. sol.

$$\frac{5.70 \text{ g pot. perm.}}{10.0 \text{ ml sod. ox. soln.}}$$

g. pot. perm.

Using Percentage as a Conversion Factor

In problems where the percentage is given, you can use it as the conversion factor given in the problem. It is the "bridge" between pure component and impure substance.

Look at Problem 6 in the pretest. A sample of calcite is 87.0% $CaCO_3$ by weight. How many grams of calcite are required to yield

How many grams of calcite?

125 g $CaCO_3$? The question asked is "_____."

The information given is _____. Starting the problem in the usual way, then, you write

? g calcite, 125 g $CaCO_3$

_____ = _____.

The percentage is the conversion factor given in the problem.

$$\frac{100 \text{ g calcite}}{87.0 \text{ g } CaCO_3}$$

Writing this as a fraction gives _____. (In this problem the impure substance has a name, calcite.) Multiplying by this factor,

grams calcite

all dimensions cancel except _____, and so the answer must be

$$? \text{ g calcite} = 125 \text{ g } \cancel{CaCO_3} \times \frac{100 \text{ g calcite}}{87.0 \text{ g } \cancel{CaCO_3}}$$

$$= 144 \text{ g calcite.}$$

Do Problem 7 in the pretest. The setup is shown below, but try it before you check this setup.

$$? \text{ g iron} = 600 \text{ g } \cancel{ore} \times \frac{35.0 \text{ g iron}}{100 \text{ g } \cancel{ore}}$$

$$= 210 \text{ g iron}$$

Look at Problem 10 in the pretest, which is a perfect example of how carefully you must label all the dimensions. The problem asks how many grams of NaCl are in 500 ml of a solution which is 26.0 wt % NaCl and has a density of 1.20 g/ml. The dimensions of

grams NaCl

the answer will be _____. The information given is

500 ml soln.

_____. The conversion factors given in the problem are

$$\frac{26.0 \text{ g NaCl}}{100 \text{ g soln.}}, \quad \frac{1.20 \text{ g soln.}}{\text{ml soln.}}$$

_____ and _____. The density is of the *solution*.

Now to set up the problem.

$$? \text{ g NaCl} = 500 \text{ ml soln.} \times \frac{1.20 \text{ g soln.}}{\text{ml soln.}} \times \frac{26.0 \text{ g NaCl}}{100 \text{ g soln.}}$$

$$= 500 \times 1.20 \times \frac{26.0}{100} \text{ g NaCl}$$

$$= 156 \text{ g NaCl}$$

The Sum of the Percentages of All Components Equals 100

Since the percentage of a component equals the number of parts of the component in 100 parts of the total substance, it is obvious that if you add up all the percentages of all the components they must equal 100. So if you have a solution made up of only NaCl in water and it is 15% by weight of NaCl, the weight percent of

water must be ___.

85%

Consequently, if you have a substance made up of two components and the percentage of one is given, you really have two different conversion factors at your disposal. In the example given above, 15 wt % NaCl in water, the two conversion factors

would be _____ and _____.

$$\frac{15 \text{ g NaCl}}{100 \text{ g soln.}}, \quad \frac{85 \text{ g water}}{100 \text{ g soln.}}$$

If you invert one of these conversion factors and multiply it by the other, you get a combined conversion factor which relates

grams NaCl to grams water. This would be _____.

$$\frac{15 \text{ g NaCl}}{85 \text{ g water}}$$

Try one. How many grams of benzene are there in 5.0 g of a 40 wt % solution of naphthalene in benzene? The dimensions of

the answer will be _____. The information given is

grams benzene

_____. The conversion factor given in the problem as it is

$$5.0 \text{ g soln.}$$
$$\frac{40 \text{ g naphthalene}}{100 \text{ g soln.}}$$

written is _____.

You see that the conversion factor is about one component, naphthalene, and the dimensions of the answer are about the other component, benzene. You must therefore work out the equivalent

conversion factor for benzene. It will be _____, which you

$$\frac{60 \text{ g benzene}}{100 \text{ g soln.}}$$

get by subtracting 40 from ___.

100

Now the problem is very simple. Starting in the usual way, you write

$$? \text{_____} = \text{_____}.$$

g benzene, 5.0 g soln.

Then you multiply by the conversion factor _____, and all the unwanted dimensions cancel except the dimensions of the answer, which must be

$$\frac{60 \text{ g benzene}}{100 \text{ g soln.}}$$

$$? \text{ g benzene} = 5.0 \, \cancel{\text{g soln.}} \times \frac{60 \text{ g benzene}}{100 \, \cancel{\text{g soln.}}}$$

$$= 3.0 \text{ g benzene.}$$

PROBLEM SET

After doing each problem, check the answer pages where you will find the setup as well as the final answer. If you miss a problem, go over the setup carefully to see just where you went wrong. The problems are divided up according to the section in the text which applies to them, and even subdivided into types of problems. In each such division the problems get more and more complicated as you go along. If you don't get a problem, don't go on to the next one until you understand how to do it.

Section A

1. What is the density of gasoline if 400 ml weigh 280 g?

2. What is the density of hydrogen gas if 100 liters weigh 8.93×10^{-3} kg?

3. What is the density of milk in g/ml if 1.00 qt weighs 1.00 kg?

4. What is the density of earth if 1.00 cu yd weighs 22.0 lb? Express your answer in British units, lb/ft³.

5. It is believed that dwarf stars may be made up of tightly packed hydrogen nuclei. Calculate the density of a hydrogen nucleus in tons/cm³ from the fact that the diameter of the spherical nucleus is 1.0×10^{-5} Å and 6×10^{23} nuclei weigh 1.0 g. The volume of a sphere is $1/6\pi d^3$, where $d =$ diameter. (π equals 3.1412.)

6. The density of chloroform is 1.49 g/ml. A 100 g sample of chloroform will occupy how many milliliters?

7. How many ounces does a 250 ml sample of sulfuric acid (density = 1.30 g/ml) weigh?

8. The density of brass is 8.0 g/ml. A cube of brass 2.0 in. on an edge will weigh how many grams?

9. The price of gold is $135 per Troy ounce. (This is the weight unit for precious metals and 1.0 Troy oz equals 1.1 avoirdupois (normal) oz.) If you had $1000, how large a cube of gold (in inches cubed) could you buy? The density of gold is 19 g/ml.

Section B

10. What is the specific gravity of a liquid whose density is 0.943 g/ml at 25°C? The density of water at this temperature is 0.997 g/ml.

11. What is the density of a metal if its specific gravity is 6.85 at 18°C? The density of water at this temperature is 0.999 g/ml.

12. A chunk of ore weighing 3.05 g when submerged in water displaces 0.642 ml of the water. The density of water at this temperature is 0.998 g/ml. What is the specific gravity of the ore?

Section C

13. What is the weight percent of gold in a sample of gold ore if it was found that 1.00 g of the ore contains 45 mg of gold?

14. What is the volume percent of oxygen in air if it is found that 150 liters of air contain 48 g of oxygen. The density of oxygen at this temperature is 1.4 g/liter.

15. What is the weight percent of sugar in a solution whose density is 1.10 g/ml if it was found that 1.0 gal of the solution contains 2.0 lb of sugar?

16. What is the percent purity of an alcohol solution whose density is 0.84 g/ml if 8.0 ml of the solution contain 2.0 ml of alcohol? The density of pure alcohol is 0.79 g/ml.

17. What is the percent purity of a sample of galena (lead sulfide ore) if it was found that 2.5 g liberate 220 ml of H_2S gas on reaction with HCl? It was found that 4.8 g pure lead sulfide liberate 0.45 liter H_2S gas under the same conditions.

18. How many grams of tin are contained in a 2.0 lb sample of ore which is 85% pure tin?

19. How many cubic meters of gold-bearing rock which is 3.5% gold must be processed to produce $5 worth of gold? The specific gravity at 10°C of the rock is 11. The price of gold is $180 per avoirdupois (normal) ounce.

20. A mining machine can dig out 15 cu yd of coal per minute and costs $3.50 per hour to operate. The density of the coal is 4.8 g/ml and the coal is 98% pure carbon. How much will it cost to dig out 5.0 tons of carbon? (*Hint:* before you start, write down all the conversion factors in the problem.)

PROBLEM SET ANSWERS

1. $? \dfrac{g}{ml} = \dfrac{280 \text{ g}}{400 \text{ ml}} = 0.700 \dfrac{g}{ml}$

2. $? \dfrac{g}{liter} = \dfrac{8.93 \times 10^{-3} \cancel{kg}}{100 \text{ liters}} \times \dfrac{10^3 \text{ g}}{\cancel{kg}}$

   $= 8.93 \times 10^{-2} \dfrac{g}{liter}$ (per liter because it is a gas)

3. $? \dfrac{g}{ml} = \dfrac{1.00 \cancel{kg}}{1.00 \cancel{qt}} \times \dfrac{10^3 \text{ g}}{\cancel{kg}} \times \dfrac{\cancel{qt}}{946 \text{ ml}}$

   $= 1.06 \dfrac{g}{ml}$

4. $? \dfrac{lb}{ft^3} = \dfrac{22.0 \text{ lb}}{1.00 \cancel{yd^3}} \times \dfrac{\cancel{yd^3}}{(3)^3 \text{ ft}^3}$

   $= 0.815 \dfrac{lb}{ft^3}$

5. volume $= \dfrac{\pi}{6}(1.0 \times 10^{-5}\ \text{Å})^3 = 5.2 \times 10^{-16}\ \text{Å}^3$ per nucleus

$$? \frac{\text{tons}}{\text{cm}^3} = \frac{1.0\ \cancel{g}}{6 \times 10^{23}\ \cancel{\text{nuclei}}} \times \frac{\cancel{\text{nucleus}}}{5.2 \times 10^{-16}\ \cancel{\text{Å}^3}} \times \frac{(10^{10})^3\ \cancel{\text{Å}^3}}{\cancel{\text{m}^3}}$$

$$\times \frac{(10^{-2})^3\ \cancel{\text{m}^3}}{\text{cm}^3} \times \frac{\cancel{\text{lb}}}{454\ \cancel{g}} \times \frac{\text{ton}}{2000\ \cancel{\text{lb}}}$$

$$= \frac{1}{6 \times 10^{23}} \times \frac{1}{5.2 \times 10^{-16}} \times 10^{30} \times 10^{-6}$$

$$\times \frac{1}{454} \times \frac{1}{2000} \times \frac{\text{tons}}{\text{cm}^3}$$

(It doesn't matter if you do not have these in the same order.)

$$= 3.5 \times 10^9\ \frac{\text{tons}}{\text{cm}^3}$$

(If you are able to do this problem, you are really doing well!)

6. $? \text{ml} = 100\ \cancel{g} \times \dfrac{\text{ml}}{1.49\ \cancel{g}} = 67.1\ \text{ml}$

7. $? \text{oz} = 250\ \cancel{\text{ml}} \times \dfrac{1.30\ \cancel{g}}{\cancel{\text{ml}}} \times \dfrac{\cancel{\text{lb}}}{454\ \cancel{g}} \times \dfrac{16\ \text{oz}}{\cancel{\text{lb}}}$

$\qquad = 11.5\ \text{oz}$

8. Volume $= (2.0)^3\ \text{in.}^3 = 8.0\ \text{in.}^3$

$$? \text{g} = 8.0\ \cancel{\text{in.}^3} \times \frac{(2.54)^3\ \cancel{\text{cm}^3}}{\cancel{\text{in.}^3}} \times \frac{\cancel{\text{ml}}}{\cancel{\text{cm}^3}} \times \frac{8.0\ \text{g}}{\cancel{\text{ml}}}$$

$\qquad = 1.0 \times 10^3\ \text{g}$

9. $? \text{in.}^3 = \$1000 \times \dfrac{\cancel{\text{oz Troy}}}{\$135} \times \dfrac{1.1\ \cancel{\text{oz}}}{\cancel{\text{oz Troy}}} \times \dfrac{\cancel{\text{lb}}}{16\ \cancel{\text{oz}}}$

$$\times \frac{454\ \cancel{g}}{\cancel{\text{lb}}} \times \frac{\cancel{\text{ml}}}{19\ \cancel{g}} \times \frac{\cancel{\text{cm}^3}}{\cancel{\text{ml}}} \times \frac{\text{in.}^3}{(2.54)^3\ \cancel{\text{cm}^3}}$$

$\qquad = 0.74\ \text{in.}^3$

(If you were able to do this one, you are a hot shot!)

10. Specific gravity $= \dfrac{0.943\ \text{g/ml}}{0.997\ \text{g/ml}} = 0.946$

11. $6.85 = \dfrac{\text{density}}{0.999}; \quad 6.84 = \text{g/ml} = \text{density}$

12. Density ore $\dfrac{g}{ml} = \dfrac{3.05\ g}{0.642\ ml} = 4.75\ g/ml$

(The volume of displaced water equals the volume of the submerged sample.)

sp. gr. $= \dfrac{4.75\ g/ml}{0.998\ g/ml} = 4.76$

13. ? g gold $= 100\ \cancel{g\ imp.} \times \dfrac{45\ \cancel{mg\ gold}}{1.00\ \cancel{g\ imp.}} \times \dfrac{10^{-3}\ g\ gold}{\cancel{mg\ gold}}$

$= 4.5\ g\ gold = 4.5\%$

14. ? liters oxygen $= 100\ \cancel{liters\ air} \times \dfrac{48\ \cancel{g\ oxygen}}{150\ \cancel{liters\ air}}$

$\times \dfrac{liter\ oxygen}{1.4\ \cancel{g\ oxygen}}$

$= 23\ liters\ oxygen = 23\%$

15. ? g sugar $= 100\ \cancel{g\ soln.} \times \dfrac{\cancel{ml\ soln.}}{1.10\ \cancel{g\ soln.}} \times \dfrac{\cancel{qt\ soln.}}{946\ \cancel{ml\ soln.}}$

$\times \dfrac{\cancel{gal\ soln.}}{4\ \cancel{qt\ soln.}} \times \dfrac{2.0\ \cancel{lb\ sugar}}{\cancel{gal\ soln.}} \times \dfrac{454\ g\ sugar}{\cancel{lb\ sugar}}$

$= 100 \times \dfrac{1}{1.10} \times \dfrac{1}{946} \times \dfrac{1}{4} \times 2.0 \times 454\ g\ sugar$

$= 22\ g\ sugar = 22\%$

16. ? g alcohol $= 100\ \cancel{g\ soln.} \times \dfrac{\cancel{ml\ soln.}}{0.84\ \cancel{g\ soln.}}$

$\times \dfrac{2.0\ \cancel{ml\ alcohol}}{8.0\ \cancel{ml\ soln.}} \times \dfrac{0.79\ g\ alcohol}{\cancel{ml\ alcohol}}$

$= 23.5\ g\ alcohol = 24\%$

(Be sure to watch your significant figures.)

17. ? g lead sulfide $= 100\ \cancel{g\ ore} \times \dfrac{220\ \cancel{ml\ H_2S}}{2.5\ \cancel{g\ ore}} \times \dfrac{10^{-3}\ \cancel{liter\ H_2S}}{\cancel{ml\ H_2S}}$

$\times \dfrac{4.8\ g\ lead\ sulfide}{0.45\ \cancel{liter\ H_2S}}$

$= 94\ g\ lead\ sulfide = 94\%$

18. ? g tin $= 2.0\ \cancel{lb\ ore} \times \dfrac{454\ \cancel{g\ ore}}{\cancel{lb\ ore}} \times \dfrac{85\ g\ tin}{100\ \cancel{g\ ore}}$

$= 772\ g\ tin$

$= 7.7 \times 10^2\ g\ tin$

19. $? \text{ m}^3 = \$5 \times \dfrac{\text{oz gold}}{\$180} \times \dfrac{\text{lb gold}}{16 \text{ oz gold}} \times \dfrac{454 \text{ g gold}}{\text{lb gold}}$

$\times \dfrac{100 \text{ g ore}}{3.5 \text{ g gold}} \times \dfrac{\text{ml ore}}{11 \text{ g ore}} \times \dfrac{\text{cm}^3 \text{ ore}}{\text{ml ore}} \times \dfrac{(10^{-2})^3 \text{ m}^3 \text{ ore}}{\text{cm}^3 \text{ ore}}$

$= 5 \times \dfrac{1}{180} \times \dfrac{1}{16} \times 454 \times \dfrac{100}{3.5} \times \dfrac{1}{11} \times 1 \times 10^{-6} \text{ m}^3 \text{ ore}$

$= 2.04 \times 10^{-6} \text{ m}^3 \text{ ore}$

(Notice that the specific gravity is equal to the density to two significant figures.)

20. $? \$ = 5 \text{ tons carbon} \times \dfrac{100 \text{ tons coal}}{98 \text{ tons carbon}} \times \dfrac{2000 \text{ lb coal}}{\text{ton coal}}$

$\times \dfrac{454 \text{ g coal}}{\text{lb coal}} \times \dfrac{\text{ml coal}}{4.8 \text{ g coal}} \times \dfrac{\text{cm}^3 \text{ coal}}{\text{ml coal}}$

$\times \dfrac{\text{in.}^3}{(2.54)^3 \text{ cm}^3} \times \dfrac{\text{yd}^3}{(36)^3 \text{ in.}^3} \times \dfrac{\text{min}}{15 \text{ yd}^3}$

$\times \dfrac{\text{hr}}{60 \text{ min}} \times \dfrac{\$3.50}{\text{hr}}$

$= 5 \times \dfrac{100}{98} \times 2000 \times 454 \times \dfrac{1}{4.8} \times 1 \times \dfrac{1}{16.4} \times \dfrac{1}{4.65} \times 10^4$

$\times \dfrac{1}{15} \times \dfrac{1}{60} \times \$3.50$

$= \$4.9 \times 10^{-3} = \$0.005$

(Hats off to you if you were able to drag your way through this last problem. If you didn't quite make it, that's okay since you will never really face any problem with 10 conversion factors.)

Chapter 5
Mole : A Chemical
Conversion Factor

There will be no pretest for this short chapter. All that is contained in it is the introduction of two new conversion factors. One of these will be absolutely vital to you and will soon become second nature.

SECTION A   Atomic Weight

The size and mass of atoms are so small that it isn't really possible to experimentally determine an absolute mass. However, it is extremely simple to determine masses relative to each other. Consequently, what has been done is to assign an arbitrary number to one specific atom (Most recently, the ^{12}C isotope was chosen and assigned the number 12.00000.) and use this as a measuring stick for all other atoms. Thus, if a magnesium atom weighs twice as much as the C^{12} isotope, the atomic weight of magnesium will

be _____, that is, __ times _____.   24.00000, 2, 12.00000

Since it was found that a titanium atom was four times as heavy as the C^{12} isotope, the atomic weight (at. wt) of titanium is

_____. It is just as simple as that.   48.00000

You will find the atomic weights of the common elements in a table on the front cover of this text. The values are only given to the first place after the decimal, since you will not require more precision than that for the work here. However, much more precise values are known for most elements and can be found either in other textbooks or in a handbook of chemistry and physics.

You may also find in some texts a unit of mass for the atomic weights. It is called atomic mass unit (a.m.u.). This term will not be useful in this text, and so we won't mention it again.

SECTION B   Molecular Weight

As you know, molecules are made up of combinations of atoms chemically bound together. Since the mass of the molecule will equal the sum of the masses of the atoms which make up the molecule, all you have to do is add up all the atomic weights. You will then have the molecular weight (M.W.). This, too, has no units, unless you have labeled the atomic weights in a.m.u.

Chemists have a shorthand method of writing molecules. These are called chemical formulas. The symbols of the atoms which make up the molecule are written, and a subscript (below the line) number is put after the symbol of each atom to tell you how many of that atom is present in the molecule. If there is only one of a given atom in the molecule, you don't bother writing the 1. It is understood.

Consider the molecule H_2SO_4, sulfuric acid. It is made up of

2, 1, 4

__ H atoms, __ S atom, and __ O atoms. (The Atomic Weights Table at the beginning of the text lists the symbols as well as the names of the elements.) Thus, to get the molecular weight of sulfuric acid you add the following.

$$H = 2 \times 1.0 = 2.0$$
$$S = 1 \times 32.1 = 32.1$$
$$\underline{O = 4 \times 16.0 = 64.0}$$
$$98.1$$

To be certain that you have grasped this calculation, determine the molecular weight of $Na_4P_2O_7$, sodium pyrophosphate.

4, 92.0

$$Na = \_ \times 23.0 = \underline{\quad}$$

2, 62.0

$$P = \_ \times 31.0 = \underline{\quad}$$

16.0, 112.0

$$O = 7 \times \underline{\quad} = \underline{\quad}$$

$$\underline{\quad}$$

266.0

$$\underline{\quad}$$

When a group of atoms (called a radical) is present in a molecule more than once, it is customary to enclose the radical in parentheses and with a subscript number to show how many of the radicals are present. Thus, in the molecule $Ca(NO_3)_2$ there are 2 (NO_3^-)

1, 2

radicals. Consequently, the molecule will have a total of __ Ca, __

6

N and __ O atoms.

3

Consider the molecule $Al_2(SO_4)_3$. It will have a total of __ S

12

atoms and __ O atoms.

When one molecule is attached to another in a complex molecule, rather than use parentheses and a subscript number, a dot is put in the middle of the line between the two parts, and a normal number is put in front of the complexed molecule. Thus $Na_2P_4O_7 \cdot 10H_2O$ (sodium pyrophosphate decahydrate) has 10 water molecules attached. The total molecule has 20 H atoms and

17

__ O atoms.

SECTION C The Mole

Since the atomic weights and the molecular weights do not have any dimension, two units called the gram atomic weight and the gram molecular weight were set up so that they would have the same number of grams as the number of the atomic weight and the molecular weight respectively. Thus, if the atomic weight of O is 16.0, then its gram atomic weight is 16.0 g. If the molecular weight

of H_2O is 18.0, then its gram molecular weight is _____. (These expressions seem a little stilted now, and we won't use them.)

It has been determined that it requires 6.02×10^{23} atoms of any element to weigh the atomic weight in grams. It also will require

6.02×10^{23} molecules to weigh the _____ in grams. This number is called Avogadro's Number.

What we are now going to do is define a new unit called the *mole*. In the same way that a million equals 10^6 or a billion equals 10^9, a mole equals 6.02×10^{23}.

We know that a given number of atoms weighs the atomic weight in grams. This same number of molecules weighs the

_____ in grams. Thus, we immediately have four conversion factors.

$$\frac{6.02 \times 10^{23} \text{ atoms}}{\text{mole}} \qquad \frac{6.02 \times 10^{23} \text{ molecules}}{\text{mole}}$$

$$\frac{\text{Atomic weight in g}}{\text{mole}} \qquad \frac{\text{Molecular weight in g}}{\text{mole}}$$

As a chemist, you will have very few opportunities to count numbers of atoms or molecules. The first two conversion factors have very limited uses. However, the second two are going to become absolutely second nature to you. When you specify molecular weight you will give it the dimensions of grams per

mole. Thus, H_2O has a molecular weight of 18.0 _____, and O has

an atomic weight of 16.0 _____.

Here are a few examples to practice on. There are 23.0 g of Na

per _____ of Na. You can write this as a conversion factor,

_____. There are 6.02×10^{23} atoms of Na per _____ Na, and

this can be written as a conversion factor, _____.

There are 16.0 g of oxygen in a mole of O atoms, but there are 32.0 g of oxygen in a mole of O_2 molecules. This is going to be a problem, and you are going to have to be especially careful to write down the complete dimensions. Thus you will write

$$\frac{16.0 \text{ g O}}{\text{mole O}} \qquad \text{and} \qquad \frac{32.0 \text{ g } O_2}{\rule{2cm}{0.4pt}}$$

Formerly a mole of atoms was called a *gram-atom* to eliminate the problem as to whether you were referring to atoms or

---

Margin answers:

18.0 g

molecular weight

molecular weight

$\dfrac{g}{mole}$

$\dfrac{g}{mole}$

mole

$\dfrac{23.0 \text{ g Na}}{\text{mole Na}}$, mole

$\dfrac{6.02 \times 10^{23} \text{ atoms Na}}{\text{mole Na}}$

mole O_2

molecules. We will call everything *mole* but be very certain to specify what comprises the mole.

Some workers even limit the term *mole* only to compounds that exist as molecules; for example, NaCl, which in solution or even in the solid or liquid state exists as Na+ and Cl− ions, could not be expressed in moles. The term *gram-formula weights* is used to make this distinction. We will not be so fussy and will call everything a mole.

There are two types of possible problems concerning moles. The first doesn't concern itself with weight, either in the question asked or in the information given, and only numbers of atoms or molecules are mentioned. For these problems you will use the conversion factors

$$\frac{6.02 \times 10^{23} \text{ atoms}}{\text{mole atoms}} \quad \text{and} \quad \frac{6.02 \times 10^{23} \text{ molecules}}{\rule{3cm}{0.4pt}}.$$

mole molecules

Look at some typical problems.

How many S atoms in 4.000 moles of S? The dimensions of the answer will be _____. The information given is _____.
Starting in the usual way, you write

atoms S, 4.000 moles S

atoms S, 4.000 moles S

$$? \underline{\hspace{2cm}} = \underline{\hspace{2cm}}.$$

moles

The conversion factor having _____ in the denominator and moving toward the dimensions of the answer is

$$\frac{6.02 \times 10^{23} \text{ atoms S}}{\text{moles S}}$$

_____. Multiplying by this factor cancels the unwanted dimension.

$$? \text{ atoms S} = 4.000 \;\cancel{\text{moles S}} \times \frac{6.02 \times 10^{23} \text{ atom S}}{\cancel{\text{mole S}}}$$

$$= 24.08 \times 10^{23} \text{ atoms S}$$

$$= 2.41 \times 10^{24} \text{ atoms S}$$

Here is another example. If you have 2.107×10^{24} molecules of H_2O, how many moles of H_2O do you have?
Starting in the usual way, you write

mole H_2O,
2.107×10^{24} molecules H_2O

$$? \underline{\hspace{2cm}} = \underline{\hspace{3cm}}.$$

If you then multiply by the conversion factor

$$\frac{\text{mole } H_2O}{6.02 \times 10^{23} \text{ molecules } H_2O}$$

_____, the unwanted dimension will cancel leaving

$$? \text{ mole } H_2O = 2.107 \times 10^{24} \;\cancel{\text{molecules } H_2O}$$

$$\times \frac{\text{mole } H_2O}{6.02 \times 10^{23} \;\cancel{\text{molecules } H_2O}}$$

3.50

$$= \underline{\hspace{1.5cm}} \text{ moles } H_2O.$$

Here is a slightly different problem. How many atoms of N are there in 5.4 moles of N_2? There is a conversion factor that you can get from the formula of N_2. It relates the number of N atoms to N_2 molecules. You can therefore write

$$\frac{\text{2 N atoms}}{\rule{2cm}{0.4pt}}$$

1 molecule N_2

Now you start the problem in the usual way.

$$? \underline{\hspace{2cm}} = \underline{\hspace{2cm}}$$

N atoms,   5.4 mole N_2

You can convert to molecules of N_2 by multiplying by the

conversion factor $\underline{\hspace{4cm}}$. The unwanted

dimension left is $\underline{\hspace{2cm}}$, and you can convert this to the dimensions of the question asked by multiplying by the conversion

factor you found in the formula, $\underline{\hspace{2cm}}$. The final setup will then be

$$\frac{6.02 \times 10^{23} \text{ molecules } N_2}{\text{mole } N_2}$$

molecules N_2

$$\frac{\text{2 N atoms}}{\text{1 molecule } N_2}$$

$$? \text{ N atoms} = 5.4 \text{ moles } N_2 \times \frac{6.02 \times 10^{23} \text{ molecules } N_2}{\text{mole } N_2}$$

$$\times \frac{\text{2 N atoms}}{\text{molecule } N_2}$$

$$= \underline{\hspace{3cm}}$$

6.5×10^{24} N atoms

Try this problem. How many H atoms are in 4.2 moles of $(NH_4)_2SO_4$? The conversion factor that you can get from the formula relating H atoms to the molecule is

$\underline{\hspace{0.5cm}}$ H atoms/molecule $(NH_4)_2SO_4$.

8

Now work the problem through. The answer you get is

$\underline{\hspace{4cm}}$.

2.0×10^{25} atoms H

Notice in the above problems that there is no mention of weights. Consequently, you did not use a conversion factor concerning weight, either the atomic or molecular weight. Now here is a problem with weight in it. If you are asked how many grams 6.0 moles of H atoms weigh, the dimensions of the answer

will be $\underline{\hspace{2cm}}$, so now you must use the conversion factor which includes atomic weight. The information given is

grams H

$\underline{\hspace{2cm}}$. Setting up the problem, you write

6.0 moles H

$$? \underline{\hspace{0.5cm}} = \underline{\hspace{2cm}}$$

g H,   6.0 moles H

The conversion factor which will cancel the unwanted

dimension and give you the dimension of the answer is $\underline{\hspace{1cm}}$, the atomic weight of H. Your answer will therefore be

$$\frac{\text{1.0 g H}}{\text{mole H}}$$

$$? \text{ g H} = 6.0 \text{ moles H} \times \frac{\text{1.0 g H}}{\text{mole H}},$$

$$= 6.0 \text{ g H}.$$

If you are asked how many moles of H_2SO_4 are in 49.05 g H_2SO_4, you begin in the same way,

mole H_2SO_4, 49.05 g H_2SO_4

? _____ = _____ .

The conversion factor by which you must multiply is

$\dfrac{\text{mole } H_2SO_4}{98.1 \text{ g } H_2SO_4}$

_____ , the inverted form of the molecular weight of H_2SO_4. This operation cancels the unwanted dimension of the information given and leaves only the dimensions of the answer.

$$? \text{ moles } H_2SO_4 = 49.05 \cancel{\text{ g } H_2SO_4} \times \frac{\text{mole } H_2SO_4}{98.1 \cancel{\text{ g } H_2SO_4}}$$

5.00×10^{-1}

$$= \underline{\hspace{2cm}} \text{ mole } H_2SO_4$$

The conversion factor which relates the molecular or atomic weight to moles of a substance is going to be one of your most frequent tools. Consequently, it will be especially useful to do more problems of this type.

How many grams of H_3AsO_4 will you need to have 0.50 mole of H_3AsO_4? Since the problem deals with weight, you will need

molecular

the conversion factor you get from the _____ weight. It is

mole H_3AsO_4

$\dfrac{141.9 \text{ g } H_3AsO_4}{\underline{\hspace{1.5cm}}}$.

Nothing in the problem mentions the number of atoms or molecules, so you will have no use for a conversion factor containing

Avogadro's Number

_____ , 6.02×10^{23}. Don't try to use it; it will only confuse the picture. Starting in the usual way, then, you write

g H_3AsO_4, 0.50 mole H_3AsO_4

? _____ = _____ .

You can get to the dimensions of the question asked by multiplying by the molecular weight

$\dfrac{141.9 \text{ g } H_3AsO_4}{\text{mole } H_3AsO_4}$

$? \text{ g } H_3AsO_4 = 0.50 \text{ mole } H_3AsO_4 \times \underline{\hspace{2.5cm}}$

71

$= \underline{\hspace{1cm}} \text{ g } H_3AsO_4$

Here is a more complicated problem. How many moles of acetic acid, $C_2H_4O_2$, are there in 75.0 ml of acetic acid? The density of acetic acid is 1.05 g/ml. This problem doesn't deal directly with weight. It gives information about the volume. However, if you have the density of a compound, you can easily convert volume to weight. So there are two conversion factors that you will need. First the density,

g acetic acid

$\dfrac{1.05 \underline{\hspace{2cm}}}{\text{ml acetic acid}}$,

molecular

and then the _____ weight,

60.0

$\dfrac{\underline{\hspace{1cm}} \text{ g acetic acid}}{\underline{\hspace{2.5cm}}}$.

mole acetic acid

Starting in the usual way, then, you write

? _____ = _____ .

Now you simply choose a conversion factor which has

_____ in the denominator in order to cancel the unwanted dimension in the information given. This would be

_____ . Multiplying by this factor gives you

$$? \text{ moles acetic acid} = 75 \;\cancel{\text{ml acetic acid}} \times \frac{1.05 \text{ g acetic acid}}{\cancel{\text{ml acetic acid}}} \;.$$

If you now multiply by the conversion factor which is the molecular

_____ in its reciprocal form, the only dimension left is the dimension of the question asked.

$$? \text{ moles acetic acid} = 75 \;\cancel{\text{ml acetic acid}} \times \frac{1.05 \;\cancel{\text{g acetic acid}}}{\cancel{\text{ml acetic acid}}}$$

$$\times \frac{\text{mole acetic acid}}{60.0 \;\cancel{\text{g acetic acid}}}$$

$$= 1.3 \text{ moles acetic acid}$$

Here is a problem that you can try on your own; there will be just a few hints to get you started. You want to find how many grams of a 7.50 wt % solution of NaCl you need in order to have 0.85 mole of NaCl. The conversion factor given in the problem is found in the statement "7.50 wt % solution of NaCl"; it is

_____ . (If you have forgotten this, you can check it in Section C of Chapter 4.) Since this problem deals with the weight of NaCl (indirectly), you will need the conversion factor which is

the _____ ; it is

$$\frac{58.5 \text{ g NaCl}}{\rule{2cm}{0.4pt}} \;.$$

The tricky part of this problem is asking the right question. It is

"How many grams of _____?" You want to watch this very carefully when you solve problems yourself.

Now you are ready to set up the problem and work it. If you got 663. g soln. as your answer, you're doing just fine. If you found 1.1 or 3.7 g as your answer, you have not labeled your dimensions correctly so that they will cancel. The setup should be

$$? \text{ g soln.} = 0.85 \;\cancel{\text{mole NaCl}} \times \frac{58.5 \;\cancel{\text{g NaCl}}}{\cancel{\text{mole NaCl}}} \times \frac{100 \text{ g soln.}}{7.50 \;\cancel{\text{g NaCl}}}$$

This type of problem concerning the number of moles and some property that can be related to the weight of the compound can

---

(right margin answers)

moles acetic acid,   75 ml acetic acid

milliliters acetic acid

$$\frac{1.05 \text{ g acetic acid}}{\text{ml acetic acid}}$$

weight

$$\frac{7.50 \text{ g NaCl}}{100 \text{ g soln.}}$$

molecular weight

mole NaCl

solution

only become more complex if the weight must be found through various conversion factors. However, you can always use as your

molecular

"bridge" the conversion factor which is the _____ weight

grams

expressed in its correct units, _____/mole. You may, of course, use it in its reciprocal (inverted form).

It is possible to have a problem which involves both the weight of a compound and the number of atoms or molecules. In these cases you must use both conversion factors, the molecular weight and the one containing Avogadro's Number. If you are asked, for example, how many molecules of CO_2 are present in 3.2 g of CO_2, you start the problem in the usual way,

molecules CO_2, 3.2 g CO_2

$$? \underline{\hspace{3cm}} = \underline{\hspace{2cm}}.$$

The only conversion factor you know which contains grams CO_2 is

molecular

$\dfrac{\text{mole } CO_2}{44.0 \text{ g } CO_2}$

the _____ weight. Multiplying by this you have

$$? \text{ molecules } CO_2 = 3.2 \text{ g } CO_2 \times \underline{\hspace{2.5cm}}$$

cancel

Notice that you used it in its inverted form in order to _____ the unwanted dimension in the information given. The unwanted

mole CO_2

dimension left is _____, and this can be converted to the dimension of the question asked simply by multiplying by

$\dfrac{6.02 \times 10^{23} \text{ molecules } CO_2}{\text{mole } CO_2}$

_____. The final setup will then be

$$? \text{ molecules } CO_2 = 3.2 \text{ g } \cancel{CO_2} \times \frac{\text{mole } \cancel{CO_2}}{44.0 \text{ g } \cancel{CO_2}}$$

$$\times \frac{6.02 \times 10^{23} \text{ molecules } CO_2}{\text{mole } \cancel{CO_2}}$$

$$= 4.4 \times 10^{22} \text{ molecules } CO_2.$$

Here is a more involved example of this type of problem. If the density of iron is 7.8 g/ml, how many iron atoms are contained in a cube of pure iron 2.0 cm on an edge? First you must calculate the volume of Fe (the symbol for iron).

$$\text{Volume} = (2.0 \text{ cm})^3 = 8.0 \text{ cm}^3$$

Now the question can be reworded as, "How many Fe atoms are in 8.0 cm³?" Starting in the usual way, you have

atoms Fe, 8.0 cm³

$$? \underline{\hspace{2cm}} = \underline{\hspace{1.5cm}}.$$

You know the conversion factor, 1.0 ml/cm³, and also the conversion factor given in the problem, 7.8 g Fe/ml Fe, which is the density. Multiplying by these two conversion factors now

grams Fe

$\dfrac{\text{mole Fe}}{55.8 \text{ g Fe}}$

leaves the unwanted dimension _____. You can cancel this unwanted dimension by multiplying by the conversion factor _____.

This now leaves the unwanted dimension _____, which can be converted to the dimensions of the answer by multiplying by the conversion factor _____. Now the only dimension left is the dimension of the answer, so the answer must be

$$? \text{ atoms Fe} = 8.0 \cancel{cm^3} \times \frac{\cancel{ml}}{\cancel{cm^3}} \times \frac{7.8 \text{ g } \cancel{Fe}}{\cancel{ml}}$$

$$\times \frac{\cancel{\text{mole Fe}}}{55.8 \text{ g } \cancel{Fe}} \times \frac{6.02 \times 10^{23} \text{ atoms Fe}}{\cancel{\text{mole Fe}}}$$

$$= 6.7 \times 10^{23} \text{ atoms Fe.}$$

(Remember that you should be writing the conversion factors down in your own setup as you go along.)

The only way in which problems can become still more complicated is to have more conversion factors given in the problem. Here is such an example.

The price of an HCl solution is $6.00 for 2.0 kg. How much would 6.0 moles of HCl cost if the solution contains 310 g HCl per liter and its density is 1.1 g/ml? The first thing to do is to write down all the conversion factors given in the problem.

$$\frac{\$6.00}{\underline{\hspace{2cm}}} \qquad \frac{310 \text{ g HCl}}{\underline{\hspace{2cm}}} \qquad \frac{1.1 \underline{\hspace{1cm}}}{\text{ml} \underline{\hspace{1cm}}}$$

It is absolutely vital that you write the dimensions in such a way as to distinguish between grams solution and grams HCl. Since the problem is concerned only with weight and moles, you will need the conversion factor relating weight of HCl to moles HCl. This is

the molecular weight, _____.

Now start in the usual way by writing

? _____ = _____.

The problem now is very simple. You just keep multiplying by conversion factors, cancelling unwanted dimensions until you finally arrive at the dimensions of the answer. Do it yourself and check your results against the answer below.

$$? \text{ dollars} = 6.0 \cancel{\text{ moles HCl}} \times \frac{36.5 \text{ g } \cancel{HCl}}{\cancel{\text{mole HCl}}} \times \frac{\cancel{\text{liter soln.}}}{310 \text{ g } \cancel{HCl}}$$

$$\times \frac{\cancel{\text{ml soln.}}}{10^{-3} \cancel{\text{liter soln.}}} \times \frac{1.1 \text{ g } \cancel{\text{soln.}}}{\cancel{\text{ml soln.}}} \times \frac{\cancel{\text{kg soln.}}}{10^3 \text{ g } \cancel{\text{soln.}}}$$

$$\times \frac{\$6.00}{2 \cancel{\text{ kg soln.}}}$$

$$= \$2.3 \text{ (to the correct number of significant figures)}$$

You are now ready to do the problem set.

---

mole Fe

$$\frac{6.02 \times 10^{23} \text{ atoms Fe}}{\text{mole Fe}}$$

g soln.

2 kg soln., liter soln., soln.

$$\frac{36.5 \text{ g HCl}}{\text{mole HCl}}$$

dollars, 6.0 moles HCl

PROBLEM SET

1. Using the Table of Atomic Weights at the beginning of the text, calculate the molecular weight for the following compounds.

   (a) HBr (c) H_3PO_4 (e) $(NH_4)_3AsO_3$

   (b) $CaCl_2$ (d) Na_2SO_4 (f) $KF \cdot 2H_2O$

Moles and Numbers of Atoms or Molecules

2. How many atoms of C are there in 8.0 moles of C?

3. A sample containing 3.01×10^{24} molecules of water contains how many moles of water?

4. How many O atoms are there in 3.0 moles of P_2O_5?

5. How many molecules of sugar are there in 150 ml of a solution which contains 3.5 moles of sugar in 12.0 liters of solution?

6. A solution of alcohol in water was prepared by mixing 3.61×10^{22} molecules of alcohol with enough water to have 750 ml of solution. The density of the solution was 0.92 g/ml. How many moles of alcohol are present in 160 g of the solution?

Moles and a Weight Related Property

7. How many moles of HCl are present in a sample of HCl weighing 272 g?

8. How many grams does a sample of $KHSO_3$ weigh if it contains 1.7 moles of $KHSO_3$?

9. How many grams of water must you take to have 2.5 moles of water?

10. How many ml of ethyl alcohol, C_2H_5OH, must you take to have 0.35 mole of ethyl alcohol? The density of ethyl alcohol is 0.79 g/ml.

11. What is the price per mole for NaCl, if 100 g NaCl cost $.05?

12. How many moles of $CaCO_3$ are present in a 450 g sample of limestone which is 94% $CaCO_3$?

13. A solution was prepared by dissolving 0.25 moles of $K_2Cr_2O_7$ in sufficient water to prepare 3.0 liters of solution. How many grams of $K_2Cr_2O_7$ are there in 125 ml of the solution?

14. A solution was prepared by dissolving 33.5 g of $Na_2C_2O_4$ in enough water to give you 1.50 liters of solution. The density of the solution was 1.08 g/ml. How many grams of the solution must you take to have 0.125 mole of $Na_2C_2O_4$?

15. What will be the edge length of a cube of Co which costs $3.50, if the price of Co is $7.00 per mole? The density of Co is 8.9 g/ml.

Weight and Number of Molecules and Atoms

16. How many atoms of Na are there in 92.0 g Na?

17. How many grams will 3.01×10^{24} molecules of H_2SO_4 weigh?

18. How many milliliters of Br_2 must you take to have 2.11×10^{22} molecules of Br_2? The density of Br_2 is 3.10 g/ml.

19. How many H atoms are there in 45.0 g $(NH_4)_3AsO_3$?

20. How many atoms, regardless of what type, are there in 196 g H_2SO_4? (Be careful. There are 7 atoms of all types per molecule.)

21. What is the price for 1.0×10^6 atoms of Pt if a cube 1.0 inches on an edge costs $550? The density of Pt is 22.0 g/ml.

22. How many atoms of O are there in a cube of calcite ore which is 92% $CaCO_3$ and which measures 0.51 cm on an edge? The density of the calcite is 4.26 g/ml.

PROBLEM SET ANSWERS

You should have done all parts of Problem 1 correctly. If you didn't, check your addition and if the error was not there, check the value for the atomic weights you used. If you still cannot find your error, go back and redo Section B in the chapter.

The rest of the problems were divided into groups depending on whether they concerned numbers of molecules or atoms, weight of compound or both. The problems become progressively more complicated as you go through each group. You should be able to do a minimum of three in the first group, seven in the second group, and five in the third group.

1. (a) 80.9   (c) 98.0   (e) 176.9

    (b) 111.1   (d) 142.1   (f) 94.1

2. ? atoms C = $8.0 \text{ moles C} \times \dfrac{6.02 \times 10^{23} \text{ atoms C}}{\text{mole C}}$

    $= 4.8 \times 10^{24}$ atoms C

3. ? moles $H_2O = 3.01 \times 10^{24} \text{ molecules } H_2O \times \dfrac{\text{mole } H_2O}{6.02 \times 10^{23} \text{ molecules } H_2O}$

    $= 5.00$ moles H_2O

4. ? atoms O = $3.0 \text{ moles } P_2O_5 \times \dfrac{6.02 \times 10^{23} \text{ molecules } P_2O_5}{\text{mole } P_2O_5}$

    $\times \dfrac{5 \text{ atoms O}}{\text{molecule } P_2O_5}$

    $= 9.0 \times 10^{24}$ atoms O

5. ? molecules sugar = 150 ml soln. $\times \dfrac{10^{-3} \text{ liter soln.}}{\text{ml soln.}} \times \dfrac{3.5 \text{ moles sugar}}{12.0 \text{ liters soln.}}$

$$\times \dfrac{6.02 \times 10^{23} \text{ molecules sugar}}{\text{mole sugar}}$$

$$= 2.6 \times 10^{22} \text{ molecules sugar}$$

6. ? moles alcohol = 160 g soln. $\times \dfrac{\text{ml soln.}}{0.92 \text{ g soln.}}$

$$\times \dfrac{3.61 \times 10^{22} \text{ molecules alcohol}}{750 \text{ ml soln.}}$$

$$\times \dfrac{\text{mole alcohol}}{6.02 \times 10^{23} \text{ molecules alcohol}}$$

$$= 1.39 \times 10^{-2} \text{ mole alcohol}$$

7. ? moles HCl = 272 g HCl $\times \dfrac{\text{mole HCl}}{36.5 \text{ g HCl}}$

$$= 7.45 \text{ moles HCl}$$

8. ? g $KHSO_3$ = 1.7 moles $KHSO_3$ $\times \dfrac{120.2 \text{ g } KHSO_3}{\text{mole } KHSO_3}$

$$= 2.0 \times 10^2 \text{ g } KHSO_3$$

9. ? g H_2O = 2.5 moles H_2O $\times \dfrac{18.0 \text{ g } H_2O}{\text{mole } H_2O}$

$$= 4.5 \times 10^1 \text{ g } H_2O$$

(Notice in the last two problems the use of scientific notation to express the answers to the correct number of significant figures.)

10. ? ml ethyl alcohol = 0.35 mole ethyl alcohol $\times \dfrac{46.0 \text{ g ethyl alcohol}}{\text{mole ethyl alcohol}}$

$$\times \dfrac{\text{ml ethyl alcohol}}{0.79 \text{ g ethyl alcohol}}$$

$$= 2.0 \times 10^1 \text{ ml ethyl alcohol}$$

11. ? dollars = 1.00 moles NaCl $\times \dfrac{58.5 \text{ g NaCl}}{\text{mole NaCl}} \times \dfrac{\$.05}{100 \text{ g NaCl}}$

$$= \$.0292$$

12. ? mole $CaCO_3$ = 450 g imp. $\times \dfrac{94 \text{ g } CaCO_3}{100 \text{ g imp.}} \times \dfrac{\text{mole } CaCO_3}{100.1 \text{ g } CaCO_3}$

$$= 4.23 \text{ moles } CaCO_3$$

(If you have forgotten how to handle percent purity, check back to Section C of Chapter 4.)

13. $? \text{ g } K_2Cr_2O_7 = 125 \text{ ml soln.} \times \dfrac{10^{-3} \text{ liter soln.}}{\text{ml soln.}}$

$\times \dfrac{0.25 \text{ mole } K_2Cr_2O_7}{3.0 \text{ liters soln.}} \times \dfrac{294.2 \text{ g } K_2Cr_2O_7}{\text{mole } K_2Cr_2O_7}$

$= 3.1 \text{ g } K_2Cr_2O_7$

14. $? \text{ g soln.} = 0.125 \text{ mole } Na_2C_2O_4$

$\times \dfrac{134.0 \text{ g } Na_2C_2O_4}{\text{mole } Na_2C_2O_4} \times \dfrac{1.5 \text{ liters soln.}}{33.5 \text{ g } Na_2C_2O_4}$

$\times \dfrac{\text{ml soln.}}{10^{-3} \text{ liter soln.}} \times \dfrac{1.08 \text{ g soln.}}{\text{ml soln.}}$

$= 8.1 \times 10^2 \text{ g soln.}$

15. First you must determine the volume of the cube in cubic centimeters

$? \text{ cm}^3 = \$3.50 \times \dfrac{\text{mole Co}}{\$7.00} \times \dfrac{58.9 \text{ g Co}}{\text{mole Co}}$

$\times \dfrac{\text{ml Co}}{8.9 \text{ g Co}} \times \dfrac{\text{cm}^3}{\text{ml}}$

$= 3.3 \text{ cm}^3$

The edge length can then be found by taking the cube root of 3.3 cm³.

Edge length cm $= (3.3 \text{ cm}^3)^{1/3} = 1.5 \text{ cm}$

(If your calculator does not find roots, refer to Section B in Chapter 1 to see how to get them using logarithms.)

16. $? \text{ atoms Na} = 92.0 \text{ g Na} \times \dfrac{\text{mole Na}}{23.0 \text{ g Na}} \times \dfrac{6.02 \times 10^{23} \text{ atoms Na}}{\text{mole Na}}$

$= 2.41 \times 10^{24} \text{ atoms Na}$

17. $? \text{ g } H_2SO_4 = 3.01 \times 10^{24} \text{ molecules } H_2SO_4$

$\times \dfrac{\text{mole } H_2SO_4}{6.02 \times 10^{23} \text{ molecules } H_2SO_4} \times \dfrac{98.1 \text{ g } H_2SO_4}{\text{mole } H_2SO_4}$

$= 4.91 \times 10^2 \text{ g } H_2SO_4$

18. $? \text{ ml } Br_2 = 2.11 \times 10^{22} \text{ molecules } Br_2 \times \dfrac{\text{mole } Br_2}{6.02 \times 10^{23} \text{ molecules } Br_2}$

$\times \dfrac{159.8 \text{ g } Br_2}{\text{mole } Br_2} \times \dfrac{\text{ml } Br_2}{3.10 \text{ g } Br_2}$

$= 1.81 \text{ ml } Br_2$

19. $? \text{ atoms H} = 45.0 \text{ g } (NH_4)_3AsO_3 \times \dfrac{\text{mole } (NH_4)_3AsO_3}{176.9 \text{ g } (NH_4)_3AsO_3}$

$\times \dfrac{6.02 \times 10^{23} \text{ molecules } (NH_4)_3AsO_3}{\text{mole } (NH_4)_3AsO_3} \times \dfrac{12 \text{ atoms H}}{\text{molecule } (NH_4)_3AsO_3}$

$= 1.84 \times 10^{24} \text{ atoms H}$

20. ? atoms $= 196 \text{ g H}_2\text{SO}_4 \times \dfrac{\text{mole H}_2\text{SO}_4}{98.1 \text{ g H}_2\text{SO}_4}$

    $\times \dfrac{6.02 \times 10^{23} \text{ molecules H}_2\text{SO}_4}{\text{mole H}_2\text{SO}_4} \times \dfrac{7 \text{ atoms}}{\text{molecule H}_2\text{SO}_4}$

    $= 8.42 \times 10^{24}$ atoms

21. The first thing you must do is to get the volume of the cube in cm^3.

    $\text{cm}^3 = (1.0)^3 \text{ in.}^3 \times \dfrac{(2.54)^3 \text{ cm}^3}{\text{in.}^3} = 16.4 \text{ cm}^3$

    The rest of the problem is straightforward.

    ? dollars $= 1.0 \times 10^6 \text{ Pt atoms} \times \dfrac{\text{mole Pt}}{6.02 \times 10^{23} \text{ Pt atoms}} \times \dfrac{195.1 \text{ g Pt}}{\text{mole Pt}}$

    $\times \dfrac{\text{ml Pt}}{22.0 \text{ g Pt}} \times \dfrac{\text{cm}^3 \text{ Pt}}{\text{ml Pt}} \times \dfrac{\$550}{16.4 \text{ cm}^3}$

    $= \$4.9 \times 10^{-16}$ (In decimal form this is \$.00000000000000049, which seems to be a good bargain. Calculate how much this number, 1.0×10^6, of Pt atoms will weigh.)

    ? g Pt $= 1.0 \times 10^6 \text{ atoms Pt} \times \dfrac{\text{mole Pt}}{6.02 \times 10^{23} \text{ atoms Pt}} \times \dfrac{195.1 \text{ g Pt}}{\text{mole Pt}}$

    $= 3.2 \times 10^{-16}$ g Pt

    (This is roughly 1,000,000,000,000th as much as the dot on the i weighs. Maybe it wasn't such a big bargain.)

22. Once again you must calculate the volume of the sample.

    $\text{cm}^3 = (0.51)^3 \text{ cm}^3 = 0.13 \text{ cm}^3$

    Since $1 \text{ cm}^3 = 1 \text{ ml}$ you can write vol $= 0.13$ ml

    ? atoms O $= 0.13 \text{ ml imp.} \times \dfrac{4.26 \text{ g imp.}}{\text{ml imp.}} \times \dfrac{92 \text{ g CaCO}_3}{100 \text{ g imp.}} \times \dfrac{\text{mole CaCO}_3}{101.1 \text{ g CaCO}_3}$

    $\times \dfrac{6.02 \times 10^{23} \text{ molecules CaCO}_3}{\text{mole CaCO}_3} \times \dfrac{3 \text{ atoms O}}{\text{molecule CaCO}_3}$

    $= 9.1 \times 10^{21}$ atoms O

Chapter 6
Percentage Composition
of Molecules

PRETEST

You should be able to complete this test in 15 minutes.

1. What is the weight percent N in NO?

2. What is the weight percent N in N_2O?

3. What is the weight percent N in NO_2?

4. What is the weight percent C in $Ca(CN)_2$?

5. What is the weight percent C in $C_{16}H_{26}O_4N_2S$ (penicillin)?

6. What is the weight percent S in $Na_2S_2O_3 \cdot 5H_2O$?

7. What is the weight percent N in $(NH_4)_2SO_4$?

8. What is the weight percent water (H_2O) in $Na_2SO_4 \cdot 10H_2O$?

PRETEST ANSWERS

You probably got all these right unless you were not careful in adding up the atomic weights to get the molecular weights. Consequently, the molecular weight is shown along with the answers in order to see if this is the problem.

If you got all of the answers correct except perhaps for one careless mistake in the molecular weights, go directly to the problem set at the end of the chapter. If you really missed several, you had better work through the chapter.

1. 46.7% N   (M.W. = 30.0 g/mole)

2. 63.6% N   (M.W. = 44.0 g/mole)

3. 30.4% N   (M.W. = 46.0 g/mole)

4. 26.1% C   (M.W. = 92.1 g/mole)

5. 56.1% C   (M.W. = 342.1 g/mole)

6. 25.9% S   (M.W. = 248.2 g/mole)

7. 21.2% N   (M.W. = 132.1 g/mole)

8. 55.9% H_2O (M.W. = 322.1 g/mole)

SECTION A   Percentage Composition by Dimensional Analysis

Frequently chemists will want to know how much of a given element is present in a molecule. Generally, they will express this as the weight percent of the element in the compound, which makes it very simple to calculate how much of the element is in a sample of any size of that particular compound. Once you know the formula of the molecule, calculating the percentage composition is very simple. We will start by looking at how you can solve these problems by dimensional analysis, and then you will see a shortcut method for doing the same thing.

First we will find the weight percent of N in NO. Rewording the problem so as to express the question in usable dimensions, you

can ask how many grams of N there are in ____ grams of NO. (If you have forgotten this method for asking questions about percentage, check back in Section C of Chapter 4.)

100

Starting in the usual way, you write

$$? \text{\_\_\_} = \text{_____.}$$

g N, 100 g NO

The dimension that must be cancelled is _____, and the only

grams NO

conversion factor available is the _____ weight of NO. So next

molecular

you multiply by this conversion factor written in its _____

inverted (or reciprocal)

form so that the unwanted dimension cancels.

$$? \text{ g N} = 100 \text{ g NO} \times \text{_____}$$

$$\frac{\text{mole NO}}{\text{30.0 g NO}}$$

The unwanted dimension now is mole ____, but the question

NO

asked concerns grams __. You will need a "bridge" conversion

N

factor which will relate N to ____.

NO

The formula of the molecule, NO, tells you that there is ____ atom of N in one molecule of NO. There will always be the same

one

number of N _____ as there are NO molecules. Consequently, if

atoms

you had 6.02×10^{23} (which is called a _____)NO molecules, you

mole

would have _____ N atoms (which we call a mole of __ atoms). This is a conversion factor and can be written

6.02×10^{23}, N

$$\frac{1 \text{ mole \_\_ atoms}}{1 \text{ \_\_\_\_\_ NO molecules}}.$$

N

mole

You now have your "bridge" which relates N to NO.

Continuing the problem, you cancel,

$$? \text{ g N} = 100 \text{ g NO} \times \frac{\text{mole NO}}{30.0 \text{ g NO}},$$

and then you multiply by the "bridge," inverted so as to cancel the unwanted dimension.

$$? g\ N = 100\ g\ \cancel{NO} \times \frac{mole\ NO}{30.0\ g\ \cancel{NO}} \times \underline{\hspace{3cm}}$$

$$\frac{mole\ N\ atoms}{mole\ NO}$$

The unwanted dimension now is _____, from which you can get directly to the dimensions of the question asked by

mole N atoms

multiplying by the _____ weight of N.

atomic

$$? g\ N = 100\ g\ \cancel{NO} \times \frac{mole\ \cancel{NO}}{30.0\ g\ NO}$$

$$\frac{14.0\ g\ N}{mole\ N\ atoms}$$

$$\times \frac{mole\ N\ atoms}{\cancel{mole\ NO}} \times \underline{\hspace{2cm}}$$

46.7

$$= \underline{\hspace{1cm}} g\ N = 46.7\%\ N$$

(We must be very careful to write "mole N atoms." This is because when you are talking you say "nitrogen," which could be either N or N_2. Be sure to write "mole N" and "mole N_2," or "mole O" and "mole O_2," or "mole H" and "mole H_2," etc.)

Let's try another example. What is the percent N in $(NH_4)_2SO_4$?

molecular

You will need three conversion factors: the _____ weight of $(NH_4)_2SO_4$,

$$\frac{132.1\ g\ (NH_4)_2SO_4}{\underline{\hspace{1.5cm}}\ (NH_4)_2SO_4};$$

mole

atomic

the _____ weight of N,

g N

$$\frac{14.0\ \underline{\hspace{1cm}}}{mole\ \underline{\hspace{0.5cm}}};$$

N

and the "bridge," which will be

2

$$\frac{\underline{\hspace{0.5cm}}\ moles\ N}{mole\ \underline{\hspace{1.5cm}}}.$$

$(NH_4)_2SO_4$

If you didn't realize that there were two N atoms in each $(NH_4)_2SO_4$ molecule, recall that the subscript 2 after the parenthesis means that the entire NH_4 radical is present two times in the molecule. Check back to Section B in Chapter 5 if you are still a little vague about this.

You can now solve the problem of what the weight percent N is in $(NH_4)_2SO_4$. You start by writing

100

$$? g\ N = \underline{\hspace{1cm}} g\ (NH_4)_2SO_4,$$

and then you multiply by the three conversion factors so that all unwanted dimensions cancel, giving as your final setup

$$? \text{ g N} = 100 \text{ g } \cancel{(NH_4)_2SO_4} \times \frac{\text{mole } \cancel{(NH_4)_2SO_4}}{132.1 \text{ g } \cancel{(NH_4)_2SO_4}}$$

$$\times \frac{2 \text{ moles } \cancel{N}}{\text{mole } \cancel{(NH_4)_2SO_4}} \times \frac{14.0 \text{ g N}}{\cancel{\text{mole N}}}$$

$$= 21.2 \text{ g N} = 21.2\% \text{ N.}$$

SECTION B   A Shortcut Method

These problems are actually so simple that it is hardly worth the effort to use a dimensional analysis setup to get the answer. The answer will always result from simply multiplying the ratio of the atomic weight of the atom of interest to the molecular weight of the molecule times the number of atoms of interest in the molecule times 100.

$$\frac{\text{Atomic weight of atom of interest}}{\text{molecular weight of molecule}} \times \text{number of atoms of interest}$$

$$\times 100 = \text{weight percent of atom of interest}$$

Use the shortcut method to determine the weight percent S in $Na_2S_2O_3$. The ratio of the atomic weight of S to the molecular

weight of the compound is 32.1/_____. The number of S atoms   158.2

per molecule is __. Then the weight percent S is   2

$$\frac{32.1}{158.2} \times 2 \times \underline{\quad} = \underline{\qquad}.$$ 100,   40.6%

It's a real snap!

SECTION C   Percentage of a
Complexed Molecule
in a Compound

As was mentioned in Chapter 5, there is a type of molecule which is made up not only of atoms bonded together, but also of molecules bonded to each other. In general, these are called complexes and are represented in their formulas by a dot joining the molecules which are bonded. Thus $Na_2S_2O_3 \cdot 5H_2O$, sodium thiosulfate pentahydrate, is five water molecules bonded to a sodium thiosulfate molecule. Or $CoCl_2 \cdot 2NH_3$ is cobaltous chloride

with ____ ammonia molecules bonded to it. Sometimes you may be   two
asked to find the percentage of the bonded molecule in these compounds.

g H_2O,   100 g $Na_2S_2O_3 \cdot 5H_2O$

5 moles H_2O

gram H_2O

molecular

For example, what is the percent H_2O in $Na_2S_2O_3 \cdot 5H_2O$? First you reword the question to ask how many grams of H_2O are in 100 g $Na_2S_2O_3 \cdot 5H_2O$.

? _____ = _____

Once again, you know the molecular weight of the $Na_2S_2O_3 \cdot 5H_2O$, and, since the "bridge" must relate moles of H_2O to moles of $Na_2S_2O_3 \cdot 5H_2O$, this factor will be

$$\frac{\rule{3cm}{0.4pt}}{\text{mole } Na_2S_2O_3 \cdot 5H_2O} \cdot$$

Since the answer's dimensions are _____, you must have a conversion factor relating moles H_2O to g H_2O. This will be the _____ weight of H_2O. Your setup will therefore be

$$? \text{ g } H_2O = 100 \text{ g } Na_2S_2O_3 \cdot 5H_2O \times \frac{\text{mole } Na_2S_2O_3 \cdot 5H_2O}{248.2 \text{ g } Na_2S_2O_3 \cdot 5H_2O}$$

$$\times \frac{5 \text{ moles } H_2O}{\text{mole } Na_2S_2O_3 \cdot 5H_2O} \times \frac{18.0 \text{ g } H_2O}{\text{mole } H_2O}$$

$$= 100 \times \frac{1}{248.2} \times 5 \times 18.0 \text{ g } H_2O$$

$$= 36.2 \text{ g } H_2O \text{ or } 36.2\% \text{ } H_2O.$$

The shortcut method for setting up this type of problem would simply be to multiply the ratio of the molecular weight of the complexed molecule of interest to the molecular weight of the whole molecule times the number of complexed molecules times 100.

$$\frac{\text{Molecular weight of complexed molecule}}{\text{molecular weight of the whole molecule}}$$

$$\times \text{ number complexed molecules}$$

$$\times 100 = \text{weight percent complexed molecule}$$

PROBLEM SET

1. What is the weight percent O in MgO?

2. What is the weight percent Fe in Fe_2O_3?

3. What is the weight percent Cl in $Mg(ClO_3)_2$?

4. What is the weight percent As in $Ca_3(AsO_4)_2$?

5. What is the weight percent O in N_2O_5?

6. What is the weight percent C in CH_3COONa?

7. What is the weight percent C in C_2H_5OH?

8. What is the weight percent K in K_3PO_4?

9. What is the weight percent S in H_2S?

10. What is the weight percent H in H_2S?

11. What relationship is there between the answers to 9 and 10?

12. What is the weight percent NH_3 in $Cu(OH)_2 \cdot 4NH_3$?

PROBLEM SET ANSWERS

If you answered ten out of the twelve problems correctly, you are ready to proceed to Chapter 7. If you missed more than two, first check to see that you were using the correct molecular weight. If your errors were not in the molecular weight, you had better go over the chapter again.

The molecular weights of the molecules are shown after each answer.

1. 39.7% O   (M.W. =   40.3 g/mole)

2. 69.9% Fe   (M.W. = 159.6 g/mole)

3. 37.1% Cl   (M.W. = 191.3 g/mole)

4. 37.6% As   (M.W. = 398.1 g/mole)

5. 74.1% O   (M.W. = 108.0 g/mole)

6. 29.3% C   (M.W. =   82.0 g/mole)

7. 52.2% C   (M.W. =   46.0 g/mole)

8. 55.3% K   (M.W. = 212.3 g/mole)

9. 94.1% S   (M.W. =   34.1 g/mole)

10.   5.9% H   (M.W. =   34.1 g/mole)

11. The sum of the weight percents is 100.

12. 41.1% NH_3 (M.W. = 165.5 g/mole)

Chapter 7
The Simplest, or
Empirical, Formula

PRETEST

You should be able to finish this test in 20 minutes. Do all the problems and then check your answers.

Determine the empirical formulas for the following.

1. A compound containing 5.9% H and 94.1% O.

2. A compound containing 75% C and 25% H.

3. A compound containing 69.9% Fe and the remainder is O.

4. A compound containing 32.4% Na, 22.6% S and 45.0% O.

5. A compound containing 28.2% N, 8.1% H, 20.8% P, and 42.9% O. It is known to contain the (NH_4) radical. Write your formula accordingly.

6. A compound containing 92.3% C and the remainder H.

7. If by another technique the molecular weight of the compound in question 6 was found to be 78.0 g/mole, what is the molecular formula?

PRETEST ANSWERS

If you got all seven correct, you can go directly to the problem set at the end of the chapter. If you missed number 7 only, read through Section B at the end of the chapter. If you missed more than one out of questions 1–6, you had better go through the entire chapter.

1. HO

2. CH_4

3. Fe_2O_3

4. Na_2SO_4

5. $(NH_4)_3PO_4$

6. CH

7. C_6H_6

*The Percentage
Composition Turned the
Other Way Around*

In the preceding chapter you saw how it is possible to calculate the weight percent of all the atoms if you know the formula for the compound. In this chapter we are going to turn the whole process around. You will determine the formula of a compound if you know the weight percentage of the atoms which make it up.

SECTION A The Empirical Formula

The formula of a compound shows the number of each type of atom in a molecule. However, the ratio of the numbers of atoms will be the same whether you consider 1 molecule, 10 molecules, a mole of molecules, or any sample of the compound.

If you have a percentage composition of a substance, what you really have is the weight of each type of atom in a 100 g sample of the substance. Consequently, all you have to do is to determine from these weights the number of moles of each type of atom. You can then determine their ratio in the molecule. As we will see a little later on, you will get a formula for a molecule which has the correct ratio of the atoms, but not necessarily the correct number. This is why this is referred to as the *empirical* or simplest formula.

For example, what is the empirical formula for a compound which is 52.9% Cl and 47.1% Cu? You know that the compound must have a formula in the form

$$Cu_x Cl_y,$$

where $x =$ the number of Cu atoms per molecule and $y =$ the number of Cl atoms per molecule.

From the percentage composition you can calculate the number of moles of each atom in a 100 g sample, and you know that the ratio in the 100 g sample will be the same as the ratio in the molecule. Let us make that calculation.

$$x = \text{mole Cu} = 47.1 \text{ g Cu} \times \frac{\text{mole Cu}}{63.5 \text{ g Cu}}$$

$$= 0.742 \text{ mole Cu}$$

$$y = \text{mole Cl} = 52.9 \text{ g Cl} \times \frac{\text{mole Cl}}{\underline{\quad} \text{ g Cl}}$$

35.5

1.49

$$= \underline{\quad} \text{ mole Cl}$$

You now know the ratio of atoms in the formula:

1.49

$$Cu_{0.742}Cl\underline{\quad}.$$

We never use anything but whole numbers of atoms in a formula, so you must reduce these values to whole numbers. The best way

to do this is to start by dividing all the values by the smallest
number since this will give you "1" at least for this number. Thus,

$$Cu_{0.742/0.742}Cl_{1.49/0.742},$$

or

$$Cu_1Cl_{\_}.$$

	2

You know that the ratio of Cu atoms to Cl atoms in the molecule
is 1 to 2. However, you do not know that the real formula
(molecular formula) is $CuCl_2$. It might be Cu_2Cl_4, or Cu_3Cl_6, or
any combination with a 1-to-2 ratio.

We will try another one. What is the empirical formula for a
compound which is 18.4% Al, 32.6% S, and the remainder O?
First you know that the sum of the percentages will always be

____. Therefore, it is possible to determine the percent O simply

by subtracting the sum of all the other percentages from ____. This,

then, will give you the percent O as ____%.

100
100
49.0

The formula for the compound must be in the form

$$Al_x S_y O_z$$

and you must now determine the number of moles of each element
in your 100 g sample represented by the percentage. Therefore,

$$x = \text{moles Al} = 18.4 \text{ g Al} \times \frac{\text{mole Al}}{27.0 \text{ g Al}}$$

$$= 0.681 \text{ mole Al}$$

$$y = \text{moles S} = 32.6 \text{ g S} \times _____$$

$$= _____$$

$\dfrac{\text{mole S}}{32.1 \text{ g S}}$
1.02 moles S

$$z = \text{moles O} = _____ \times \frac{\text{mole O}}{16.0 \text{ g O}}$$

$$= _____.$$

49.0 g O
3.06 moles O

The formula can now be written as

$$Al\_\_\_S\_\_\_O\_\_\_.$$

To get a whole number ratio you can divide each of these by

____, and you will now have

$$Al_1 S\_\_\_O\_\_\_.$$

But these are not whole numbers. However, if you multiply each of

them by ____, you will end up with

$$Al_2S_3O_9,$$

which is the simplest, or empirical, formula.

Sometimes you may have to look very hard for a number to
multiply by to get whole numbers for all the atoms. However, in
90% of the formulas that you will run across, it will be 2, 3, 4, or 5.

0.681, 1.02, 3.06
0.681
1.50, 4.50
2

SECTION B   The Molecular Formula

As you have seen in the section above, it is possible to get the simplest or empirical formula simply from the weights of the various atoms in a sample of substance. In order to get the true or molecular formula, you need one more piece of information, the molecular weight of the molecule.

Let's look at question 7 in the pretest. You have determined the empirical formula in question 6 through the following calculation:

$$\dfrac{mole\ C}{12.0\ g\ C}$$

$$? \text{ moles C} = 92.3 \text{ g C} \times \text{_____}$$

7.70 moles C

$$= \text{_____} ,$$

7.7 g H

$$? \text{ moles H} = \text{_____} \times \dfrac{mole\ H}{1.0\ g\ H}$$

7.7 moles H

$$= \text{_____} .$$

The empirical formula is

CH

____.

In question 7 you are told that the molecular weight is 78.0 g/mole. If the molecular formula were the same as the empirical formula,

13.0   CH, then the molecular weight would be _____, which you get by

one   simply adding the atomic weight of ____ C to the atomic weight of

one   ____ H. But it isn't. Therefore, the molecular formula must be some other combination of C's and H's in a one-to-one ratio which will have a molecular weight of 78.0 g/mole.

How many CH's are needed to have a molecular weight of 78.0 g/mole? You divide the true molecular weight by the

13.0   molecular weight of the empirical formula, _____ g/mole.

$$\dfrac{78.0}{13.0} = 6$$

This means that the true molecule must be made up of six empirical

C_6H_6 formulas. So the true molecular formula is _____.

Notice that the formula was written C_6H_6 and not 6CH. By using the subscripts, you indicate that you have one molecule with six C's and six H's bonded together. Writing 6CH would mean that you have six separate molecules, each of which has one C bonded to one H.

You are now ready for the problem set.

PROBLEM SET

Determine the empirical formula for each compound whose percentage composition is shown below.

1. 77.7% Fe and 22.3% O

2. 43% C and 57% O

3. 40.3% K, 26.7% Cr, and 33.0% O

4. 32.0% C, 42.6% O, 18.7% N, and the remainder H

5. 31.9% K, 28.9% Cl, and the remainder O

6. 42.1% C, 51.5% O, and the remainder H

7. 52.8% Sn, 12.4% Fe, 16.0% C, and 18.8% N

Determine the true molecular formulas for the following compounds.

8. 94.1% O and the remainder H with a molecular weight of 34.0 g/mole.

9. 40.0% C, 53.4% O, and the remainder H with a molecular weight of 90.0 g/mole.

10. 37.7% Ce, 28.4% Cl, and 33.9% water (H_2O). The molecular weight is 372.6 and the formula will be written $Ce_xCl_y \cdot zH_2O$.

PROBLEM SET ANSWERS

If you got the correct answers to eight of the ten problems, you may go on to Chapter 8. If you missed Problems 8, 9, and 10, redo Section B of the chapter. If you missed two or more of Problems 1–7, you had better redo the entire chapter. Problem 6 is particularly difficult.

The answers are shown below along with the first formula prior to adjusting the ratio to whole numbers.

1. FeO ($Fe_{1.39}O_{1.39}$)

2. CO ($C_{3.6}O_{3.6}$)

3. K_2CrO_4 ($K_{1.03}Cr_{0.514}O_{2.06}$)

4. $C_2O_2NH_5$ ($C_{2.67}O_{2.67}N_{1.34}H_{6.70}$)

5. $KClO_3$ ($K_{0.816}Cl_{0.814}O_{2.45}$)

6. $C_{12}H_{22}O_{11}$ ($C_{3.51}H_{6.40}O_{3.22}$) (First $\div$ by 3.22; then $\times 11$)

7. $Sn_2FeC_6N_6$ ($Sn_{0.444}Fe_{0.222}C_{1.33}N_{1.34}$)

8. $H_2O_2 \left[H_{5.9}O_{5.9} = HO \text{ (M.W.} = 17.0) \dfrac{34.0}{17.0} = 2 \right]$

9. $C_3H_6O_3 \left[C_{3.33}H_{6.6}O_{3.33} = CH_2O \text{ (M.W.} = 30.0) \dfrac{90.0}{30.0} = 3 \right]$

10. $CeCl_3 \cdot 7H_2O$ [$Ce_{0.269}Cl_{0.800}(H_2O)_{1.88}$]

Chapter 8
Stoichiometry

Weight Relationships From Chemical Reactions

The pretest for this chapter will appear after a few concepts have been introduced.

You are now going to learn how to do some really basic chemical problems, involving such questions as, "How much of one substance will react with how much of another substance to yield how much of a product?" This is called stoichiometry, and it is harder to spell the word than to do the problems. We will start with the "how much" expressed as weight, though in later chapters the "how much" also may be expressed as volumes of gases or solutions.

When two or more substances react to form other molecules, we say that a chemical reaction has occurred. It is possible to express this reaction in words, but it is much simpler to express the compounds as formulas. Thus, the fact that when hydrogen reacts with chlorine it forms hydrogen chloride is best expressed by the chemical equation

$$H_2 + Cl_2 = 2HCl.$$

This equation is said to be balanced. Before you can solve any stoichiometric problems, you must have a balanced equation. Consequently, the next section will cover this topic.

SECTION A Balancing Chemical Equations

A chemical equation is a real equality; the left side must equal the right side. By this we mean that the number of each type of atom on one side will equal the number on the other (and in the case of an equation with charged particles (ions) the charges on both sides must be equal). The way that this is done is to adjust the number of molecules on both sides.

In the example shown above we had H_2 and Cl_2 reacting to produce HCl.

$$H_2 + Cl_2 \neq HCl$$

This equation is not balanced since there are two H atoms on the left and only one H atom on the right. The Cl atoms also do not

balance. There are ＿＿ Cl atoms on the left, and only ＿＿ Cl atom on the right. It is very easy to balance this equation by simply writing that two HCl molecules are formed.

two,   one

$$H_2 + Cl_2 = 2HCl$$

Now the equation is balanced, for there are two H atoms on both

sides and ＿＿＿＿ atoms on both sides. In words the equation now reads, "One hydrogen molecule reacts with one chlorine molecule to yield two hydrogen chloride molecules." Notice that you put the 2 in front of the HCl. This is the way that you show that you have two HCl molecules. If you had written H_2Cl_2, you would have changed the molecule into a new molecule containing two H atoms and two Cl atoms all bonded together instead of the correct molecule containing only one H and one Cl bonded together. These numbers written in front of the formula of the molecule are called coefficients, and they indicate the number of molecules you have.

two Cl

Most equations can be balanced by just looking at them and adjusting the coefficients (number in front) of each molecule. If you have an equation that has one molecule with a complicated formula, it is best to start with that molecule, counting the number of each type atom in it and inserting coefficients on the other side to balance these atoms. Then you work back to balance the number of atoms you now have.

This sounds complex, but another example will clarify the whole picture. Balance the equation

$$Sn(OH)_4 + H_3PO_4 \neq Sn_3(PO_4)_4 + H_2O.$$

The most complicated formula is ＿＿＿＿＿＿＿; it contains

$Sn_3(PO_4)_4$

＿＿＿＿ Sn atoms. To get three Sn atoms on the left, you will have to

three

put a coefficient of ＿＿ in front of the $Sn(OH)_4$ molecule.

3

$$3Sn(OH)_4 + H_3PO_4 \neq Sn_3(PO_4)_4 + H_2O$$

The most complicated molecule also contains ＿＿ P atoms. To get this many P atoms on the left, you will have to put a coefficient of

four

＿＿ in front of the ＿＿＿＿＿ molecule.

4, H_3PO_4

$$3Sn(OH)_4 + 4H_3PO_4 \neq Sn_3(PO_4)_4 + H_2O$$

Now you count up the number of the unbalanced atoms (H and

O) on both sides. On the left side there are ＿＿ H atoms and ＿＿ O

24,   28

atoms. On the right side, in the complicated molecule, there are ＿＿ O

16

atoms. This means that you must have ＿＿ H atoms and the extra ＿＿

24,   12

O atoms in H_2O molecules. Consequently, you put a coefficient

12

of ___ in front of the H_2O. The balanced equation, therefore, is

$$3Sn(OH)_4 + 4H_3PO_4 = Sn_3(PO_4)_4 + 12H_2O.$$

You can always check whether an equation is really balanced by

3

counting atoms on both sides. There are ___ Sn atoms on the left

3, 28

and ___ Sn atoms on the right. There are a total of ___ O atoms on

28, 24

the left and ___ O atoms on the right. There are ___ H atoms on

24, 4

the left and ___ H atoms on the right. There are ___ P atoms on the

4

left and ___ P atoms on the right. There are no charges on either
side of the equation. Therefore, it *must be balanced*.

Balancing equations by inspection takes a little practice. After a
while you will teach yourself some shortcuts (as in the above
example, where the PO_4 radical appears on both sides and can be
counted as a unit) ; but it mainly will take some practice on
your part.

There is one type of equation, called a redox equation, that can
be very, very difficult to balance by inspection. There are several
systematic methods for handling this type of equation which can
be found in any chemistry textbook. However, don't worry about
them for the moment.

SECTION B Molar Stoichiometry

Now that we know what a balanced equation is, let's consider
what it means. Going back to the first example, we have

$$H_2 + Cl_2 = 2HCl.$$

First this equation says that 1 molecule of H_2 will react with 1
molecule of Cl_2 to produce 2 molecules of HCl. However, if you

10

had 10 molecules of H_2 they would react with ___ molecules of

20

Cl_2 to form ___ molecules of HCl; or if you had 6.02×10^{23}

6.02×10^{23}

molecules of H_2, they would react with _____ molecules of

1.204×10^{24}

Cl_2 to form _____ molecules of HCl. Of course,

mole

6.02×10^{23} molecules is called a ___, so that you can say that

1 mole, 2 moles

1 mole of H_2 reacts with ___ of Cl_2 to form ___ of HCl.
Many chemists find this the most practical way to express an
equation since from it you can get some very useful "bridge" type

1

conversion factors. Since there is 1 mole of H_2 per ___ mole of Cl_2,
you can write the conversion factor

$$\frac{1 \text{ mole } H_2}{1 \text{ mole } Cl_2}.$$

And since there is 1 mole of H_2 per _____ of HCl, you can

write the conversion factor _____. The conversion factor that

you could write to relate Cl_2 to HCl would then be _____.

For practice consider the reaction between HCl and Na_2CO_3
which will yield NaCl, CO_2, and H_2O. If you want to do any

stoichiometric calculations, you must first _____ the equation

$$HCl + Na_2CO_3 \neq H_2O + NaCl + CO_2.$$

When you balance the equation you get a coefficient of ____ in front

of the HCl and a coefficient of ____ in front of the NaCl. All of the

other coefficients are ____. From the balanced equation you can get
10 conversion factors (not counting inverted forms). What is the
conversion factor relating moles of HCl to moles of CO_2?

_____. The conversion factor relating NaCl to Na_2CO_3 is

_____.

When you do a problem, you will have to select the appropriate
conversion factor as your "bridge." The reason these conversion
factors are considered "bridges" is that they relate one substance
to a different substance. Stoichiometric problems will always ask a
question about one substance and give some information about a
different substance. In order to solve them, you must go from the
information given, across the "bridge" to the substance asked for in
the question.

Another example will show this clearly. According to the balanced
equation shown below, how many moles of Br_2 are produced when
0.8 mole of KBr react?

$$KClO_3 + 6KBr + 3H_2SO_4 = KCl + 3Br_2 + 3H_2O + 3K_2SO_4$$

The question asked is, "How many moles ____?" The information

given is _____. You will need a "bridge" relating ____ to

____, since these are the substances in the question asked and in
the information given. Such a "bridge" would be the conversion
factor from the balanced equation

$$\frac{\rule{3cm}{0.4pt}}{6 \text{ moles KBr}}.$$

Starting the problem in the usual way, you have

$$? \rule{2cm}{0.4pt} = \rule{3cm}{0.4pt}.$$

If you now multiply by the "bridge," the unwanted dimension of

the information given, _____, cancels leaving as the only

---

2 moles

$\dfrac{1 \text{ mole } H_2}{2 \text{ moles HCl}}$

$\dfrac{1 \text{ mole } Cl_2}{2 \text{ moles HCl}}$

balance

2

2

1

$\dfrac{2 \text{ moles HCl}}{1 \text{ mole } CO_2}$

$\dfrac{2 \text{ moles NaCl}}{1 \text{ mole } Na_2CO_3}$

Br_2

0.8 mole KBr, Br_2

KBr

3 moles Br_2

moles Br_2, 0.8 mole KBr

mole KBr

mole Br$_2$

dimension _____. This is the dimension of the answer, so the answer must be .

$$? \text{ moles Br}_2 = 0.8 \text{ mole } \cancel{\text{KBr}} \times \frac{3 \text{ moles Br}_2}{6 \text{ moles } \cancel{\text{KBr}}}$$

$$= 0.4 \text{ mole Br}_2.$$

Try another. How many moles of H$_2$SO$_4$ are needed to react with 3.7 moles of KClO$_3$ according to the reaction shown above?

moles H$_2$SO$_4$

3.7 moles KClO$_3$

3.7 moles KClO$_3$

The dimensions of the answer will be _____. The

information given is _____. Therefore,

$$? \text{ moles H}_2\text{SO}_4 = \underline{\hspace{3cm}}.$$

H$_2$SO$_4$, KClO$_3$, $\dfrac{3 \text{ moles H}_2\text{SO}_4}{\text{mole KClO}_3}$

mole KClO$_3$

mole H$_2$SO$_4$

The conversion factor by which you must multiply is the one

which relates _____ to _____, and it is _____. The

unwanted dimension, _____, will now cancel leaving as the

only dimension _____, which is the dimension of the answer. Therefore,

$$? \text{ moles H}_2\text{SO}_4 = 3.7 \text{ } \cancel{\text{moles KClO}_3} \times \frac{3 \text{ moles H}_2\text{SO}_4}{1 \text{ } \cancel{\text{mole KClO}_3}}$$

$$= 11.1 \text{ moles H}_2\text{SO}_4$$

Now you should be ready for the Pretest

PRETEST

You should be able to complete this test in about 20 minutes. Part of the material has been covered already; part remains to come in the rest of the chapter.

Section A

Balance the following equations. Do not leave any fractional coefficients.

1. $Ca(OH)_2 + HCl \neq CaCl_2 + H_2O$

2. $Fe + O_2 \neq Fe_2O_3$

3. $Na_2O_2 + H_2O \neq NaOH + O_2$

4. $PbS + O_2 \neq PbO + SO_2$

Section B

(All equations are balanced already.)

5. How many moles of $KClO_3$ are needed to prepare 7.1 moles O_2 according to the reaction

   $2KClO_3 = 2KCl + 3O_2$?

6. How many moles of NaOH are required to react with 0.50 mole H_3PO_4 according to the reaction

   $H_3PO_4 + 3NaOH = Na_3PO_4 + 3H_2O$?

Section C

(All equations are balanced already.)

7. How many grams of $KMnO_4$ are needed to react completely with 100 g Fe according to

   $3KMnO_4 + 5Fe + 24HCl = 5FeCl_3 + 3MnCl_2 + 3KCl + 12H_2O$?

8. How many grams of $Al_2(SO_4)_3$ are produced by the complete reaction of 27.0 g $Al(OH)_3$ from the reaction

   $2Al(OH)_3 + 3H_2SO_4 = Al_2(SO_4)_3 + 6H_2O$?

9. What weight of Cu metal can be obtained by the most efficient method from 454 g $CuSO_4 \cdot 5H_2O$? (You must figure out the "bridge" without an equation.)

Section D

10. How many moles of Al_2O_3 can be produced from a mixture of 2.0 moles Fe_2O_3 and 2.0 moles Al according to the reaction

    $Fe_2O_3 + 2Al = Al_2O_3 + 2Fe$?

PRETEST ANSWERS

If you were able to do all the problems correctly, go straight to the problem set at the end of the chapter. If you had difficulties with questions 1–4, you might look at the somewhat more difficult equations in the problem set at the end of the chapter. Practice is essential. If you still have real troubles, check in your chemistry textbook and ask your fellow students or your instructors.

Questions 5 and 6 cover the material already presented in Chapter 8, and if you did not answer them correctly, redo the first part of the chapter. Questions 7–9 cover the material in Section C, which follows this pretest. If you missed them, read the section; if you got them, skip it. Question 10 is covered in Section D. If you missed it, work through the section carefully.

1. $Ca(OH)_2 + 2HCl = CaCl_2 + 2H_2O$

2. $4Fe + 3O_2 = 2Fe_2O_3$

3. $2Na_2O_2 + 2H_2O = 4NaOH + O_2$

4. $2PbS + 3O_2 = 2PbO + 2SO_2$

5. 4.7 moles $KClO_3$

6. 1.5 moles NaOH

7. 170 g $KMnO_4$

8. 59.2 g $Al_2(SO_4)_3$

9. 116 g Cu (You may assume any reaction with $CuSO_4 \cdot 5H_2O = Cu$.)

10. 1.0 mole Al_2O_3

SECTION C   Weight Stoichiometry

So far we have only considered the number of moles of one substance that will be involved with a number of moles of another. However, in a laboratory you usually determine the exact amount of a substance on a balance and the information is obtained as weight. In metric units this means that you will have grams as your dimensions. Therefore, you will have to convert our stoichiometric calculations from moles to grams. This is very easy since you will always have at your disposal a conversion factor which will do this, that is, the _____ weight.

molecular

Let us look at an example. How many grams of HCl will react with 25.0 g $Ca(OH)_2$ according to the balanced equation

$$2HCl + Ca(OH)_2 = CaCl_2 + 2H_2O?$$

The question asked is "_____." The

How many grams of HCl?

information given is _____. The "bridge" relating HCl

25.0 g $Ca(OH)_2$

to $Ca(OH)_2$ is _____. Starting in the usual way, then, you have

$$\frac{2 \text{ moles HCl}}{\text{mole } Ca(OH)_2}$$

$$? \underline{} = \underline{} .$$

g HCl,   25.0 g $Ca(OH)_2$

(Be sure to write all of this down in your own setup.)

Now you must cancel the unwanted dimension and move toward a dimension in the "bridge." Since the dimensions in the "bridge" are always moles, you must convert grams $Ca(OH)_2$ to

_____ $Ca(OH)_2$. The reciprocal of the molecular weight does

$$\frac{\text{mole}}{\frac{\text{mole } Ca(OH)_2}{74.1 \text{ g } Ca(OH)_2}}$$

exactly this. You therefore multiply by _____, and you now have

$$? \text{ g HCl} = 25.0 \text{ g } \cancel{Ca(OH)_2} \times \frac{\text{mole } Ca(OH)_2}{74.1 \text{ g } \cancel{Ca(OH)_2}} .$$

Now you can use your "bridge" and get from $Ca(OH)_2$ to HCl. Multiplying will give

$$? \text{ g HCl} = 25.0 \text{ g } \cancel{Ca(OH)_2} \times \frac{\text{mole } Ca(OH)_2}{74.1 \text{ g } \cancel{Ca(OH)_2}}$$

$$\times \underline{} .$$

$$\frac{2 \text{ moles HCl}}{\text{mole } Ca(OH)_2}$$

mole HCl

The unwanted dimension now is _____, which can easily be converted to the dimension in the question asked by multiplying by

molecular

the _____ weight of HCl.

$$? \text{ g HCl} = 25.0 \text{ g } \cancel{Ca(OH)_2} \times \frac{\text{mole } \cancel{Ca(OH)_2}}{74.1 \text{ g } \cancel{Ca(OH)_2}}$$

$$\times \frac{2 \text{ moles HCl}}{\cancel{\text{mole } Ca(OH)_2}} \times \underline{}$$

$$\frac{36.5 \text{ g HCl}}{\text{mole HCl}}$$

The only dimension remaining is the dimension of the question asked. Therefore, your answer must be

$$? \text{ g HCl} = 25.0 \text{ g } \cancel{\text{Ca(OH)}_2} \times \frac{\text{mole } \cancel{\text{Ca(OH)}_2}}{74.1 \text{ g } \cancel{\text{Ca(OH)}_2}}$$

$$\times \frac{2 \text{ moles } \cancel{\text{HCl}}}{\text{mole } \cancel{\text{Ca(OH)}_2}} \times \frac{36.5 \text{ g HCl}}{\text{mole } \cancel{\text{HCl}}}$$

$$= 24.6 \text{ g HCl.}$$

As you see, these problems are not much more difficult than molar stoichiometry problems. You simply use the molecular weights as conversion factors to get from weight in grams to moles. You must always go to a dimension in moles, since the "bridge" from the balanced equation is always expressed in moles.

Try another problem. How many grams of NaOH are needed when 2.00 kg H_3PO_4 are used in the reaction

$$H_3PO_4 + NaOH \neq Na_3PO_4 + H_2O?$$

The first thing you must do in order to solve any stoichiometry

balance	problem is to _____ the equation. This can be done by putting
3, 3	a _ in front of the NaOH and a _ in front of the H_2O. Since the
H_3PO_4	problem asks about NaOH and gives information about _____, the "bridge" conversion factor will contain both of these
$\dfrac{3 \text{ moles NaOH}}{\text{mole } H_3PO_4}$	substances. The "bridge" will be _____. Starting the problem in the usual way, you have
g NaOH, 2.00 kg H_3PO_4	$? \underline{\qquad} = \underline{\qquad\qquad}.$

The unwanted dimension in the information given is

_____, but molecular weights are expressed in grams.

kilograms H_3PO_4	The conversion is simple; you multiply by _____. You can
$\dfrac{10^3 \text{ g } H_3PO_4}{\text{kg } H_3PO_4}$	get to the dimensions in the "bridge," moles H_3PO_4, by multiplying by the conversion factor from the molecular weight.

$$? \text{ g NaOH} = 2.00 \text{ kg } \cancel{H_3PO_4} \times \frac{10^3 \text{ g } H_3PO_4}{\cancel{\text{kg } H_3PO_4}} \times \underline{\qquad\qquad}$$

$\dfrac{\text{mole } H_3PO_4}{98.0 \text{ g } H_3PO_4}$	Using the "bridge" you can move out of the H_3PO_4 units and into the NaOH units. So you multiply by _____.
$\dfrac{3 \text{ moles NaOH}}{\text{mole } H_3PO_4}$	

$$? \text{ g NaOH} = 2.00 \text{ kg } \cancel{H_3PO_4} \times \frac{10^3 \text{ g } \cancel{H_3PO_4}}{\cancel{\text{kg } H_3PO_4}}$$

$$\times \frac{\text{mole } \cancel{H_3PO_4}}{98.0 \text{ g } \cancel{H_3PO_4}} \times \frac{3 \text{ moles NaOH}}{\text{mole } \cancel{H_3PO_4}}$$

moles NaOH	The unwanted dimension left is _____, but the dimensions of

the answer are _____. It is simple to convert from moles to

weight by using the conversion factor from the _____.

You therefore multiply by _____.
   The only dimension left is the dimension of the answer, which must be

$$? \text{ g NaOH} = 2.00 \, \cancel{\text{kg } H_3PO_4} \times \frac{10^3 \, \cancel{\text{g } H_3PO_4}}{\cancel{\text{kg } H_3PO_4}} \times \frac{\cancel{\text{mole } H_3PO_4}}{98.0 \, \cancel{\text{g } H_3PO_4}}$$

$$\times \frac{3 \, \cancel{\text{moles NaOH}}}{\cancel{\text{mole } H_3PO_4}} \times \frac{40.0 \text{ g NaOH}}{\cancel{\text{mole NaOH}}}$$

$$= 2.45 \times 10^3 \text{ g NaOH}$$

grams NaOH

molecular weight
40.0 g NaOH

$$\frac{40.0 \text{ g NaOH}}{\text{mole NaOH}}$$

SECTION D   Reaction Controlling Component

So far we have only considered those cases in which one reacting component is present in a specified amount and any others are assumed to be present in sufficient amount to give a complete reaction. What would happen if you were given specified amounts of several reactants? Let's take a look at our first example,

$$H_2 + Cl_2 = 2HCl.$$

This equation says that 1 mole of H_2 will react with 1 mole of Cl_2. What would happen if we started with 1 mole of H_2 but only 0.75

mole of Cl_2? The 0.75 mole of Cl_2 could only react with ____ mole of H_2. The excess H_2 would simply remain unchanged. We say that the Cl_2 controls the reaction because it is the limiting (smaller) amount, which is like saying that a chain is only as strong as its weakest link. It controls the amount of HCl that will be formed. Any stoichiometric calculations that you make concerning the HCl must therefore be based on the amount of Cl_2.

0.75

Consequently, in the above example, you could only form ____ moles HCl.
   Consider another reaction:

$$2Na + Cl_2 = 2NaCl.$$

This balanced equation says that ____ moles of Na will react with

____ mole of Cl_2 to form ____ moles of NaCl. Thus, if you had only 1 mole of Na, you would need 1/2 mole of Cl. If you are going to compare moles Cl_2 to moles Na to determine the limiting amount that controls the amount of NaCl formed, you must divide the moles of Na given in the problem by 2 before you compare it with the number of moles of Cl_2 given in the problem. The smaller of the two is the limiting amount.
   Consider the reaction

$$4Fe + 3O_2 = 2Fe_2O_3.$$

1.5

2

1, 2

If you are given amounts of Fe and O_2 in the problem and you want to determine which amount is limiting and therefore controls the yield of Fe_2O_3, you must divide the number of moles of Fe by

3

4 and the number of moles of O_2 by ___ before you compare them. What you have done is simply to divide the number of moles for each reactant given in the problem by the coefficient which appears before that component in the balanced equation. Then you can compare and see which component is limiting. In any calculations you will only use this "controlling" component, as in the following example.

How many moles of $BaSO_4$ will be produced from a mixture of 3.5 moles H_2SO_4 and 2.5 moles $BaCl_2$ using the reaction

$$H_2SO_4 + BaCl_2 = BaSO_4 + 2HCl?$$

First you must determine which is the limiting reactant, H_2SO_4 or $BaCl_2$. The coefficients in the balanced equation are both 1, so nothing more than a direct comparison of numbers of moles is necessary to see which is limiting.

Number moles H_2SO_4 = 3.5

Number moles $BaCl_2$ = 2.5

$BaCl_2$

Since there are fewer moles of _____, it is the controlling reactant. You now solve the problem in the usual manner, using the $BaCl_2$ and forgetting about the amount of H_2SO_4 since it is in excess.

2.5 moles $BaCl_2$

? moles $BaSO_4$ = _____

Multiplying by the "bridge" then gives you the answer.

$$? \text{ moles } BaSO_4 = 2.5 \text{ moles } BaCl_2 \times \frac{\text{mole } BaSO_4}{\text{mole } BaCl_2}$$

$$= 2.5 \text{ moles } BaSO_4$$

Here is another example where the coefficients are not all 1 and the amounts are given as weights and not moles. How many grams of $Al_2(SO_4)_3$ will be produced by the reaction of 225 g $Al(OH)_3$ with 784 g H_2SO_4 following the reaction

$$2Al(OH)_3 + 3H_2SO_4 = Al_2(SO_4)_3 + 6H_2O?$$

First you must determine the number of moles of each reactant.

g $Al(OH)_3$

$$? \text{ moles } Al(OH)_3 = 225 \text{ g } Al(OH)_3 \times \frac{\text{mole } Al(OH)_3}{78.0 \text{ _____}}$$

$$= 2.88 \text{ moles } Al(OH)_3$$

$\dfrac{\text{mole } H_2SO_4}{98.1 \text{ g } H_2SO_4}$

7.99

$$? \text{ moles } H_2SO_4 = 784 \text{ g } H_2SO_4 \times \text{_____}$$

$$= \text{_____ moles } H_2SO_4$$

Since the balanced equation has a coefficient of 2 in front of the $Al(OH)_3$, in order to make a comparison between the two

reactants you must divide the number of moles of $Al(OH)_3$ by __.
In the same way, there is a coefficient of 3 before the H_2SO_4, so

you must _____ the number of moles of H_2SO_4 by __.

$$\frac{2.88 \text{ moles } Al(OH)_3}{2} = 1.44$$

$$\frac{7.99 \text{ moles } H_2SO_4}{3} = 2.66$$

You can now see that the _____ will be the limiting
component since its number is smaller. You can now solve the
problem as you did in Section C using only the data concerning

the _____. You disregard the H_2SO_4 because it is in _____.
Do the problem yourself. Did you get an answer of 494 g
$Al_2(SO_4)_3$? If you didn't, check your setup against the one below.

$$? \text{ g } Al_2(SO_4)_3 = 225 \text{ g } \cancel{Al(OH)_3} \times \frac{\text{mole } \cancel{Al(OH)_3}}{78.0 \text{ g } \cancel{Al(OH)_3}}$$

$$\times \frac{1 \text{ mole } \cancel{Al_2(SO_4)_3}}{2 \text{ moles } \cancel{Al(OH)_3}} \times \frac{342.3 \text{ g } Al_2(SO_4)_3}{\cancel{\text{mole } Al_2(SO_4)_3}}$$

Try one more, but this time we will simply decide which of the
reactants is limiting. Given 15.8 g $KMnO_4$, 6.80 g H_2O_2, and an
excess of HCl reacting according to the following equation

$$2KMnO_4 + 5H_2O_2 + 6HCl = 2MnCl_2 + 5O_2 + 8H_2O + 2KCl.$$

Which of the reactants is controlling (limiting)?

The first thing you must do is to determine the number of _____
of each reactant.

$$? \text{ moles } KMnO_4 = \underline{\hspace{3cm}} \times \frac{\text{mole } KMnO_4}{158 \text{ g } KMnO_4}$$

$$= 0.100 \text{ mole } KMnO_4$$

$$? \text{ moles } \underline{\hspace{1.5cm}} = 6.80 \text{ g } H_2O_2 \times \frac{\text{mole } H_2O_2}{34.0 \underline{\hspace{1cm}}}$$

$$= 0.200 \text{ mole } H_2O_2$$

In order to compare the two reactants to determine which is

limiting, you must _____ the number of moles of $KMnO_4$ by 2 and

the number of moles of H_2O_2 by __. These numbers come from the

balanced _____.

$$\frac{0.100 \text{ mole } KMnO_4}{2} = 0.0500$$

$$\frac{0.200 \text{ mole } \underline{\hspace{1cm}}}{5} = 0.0400$$

Consequently, the limiting reactant is _____.

2

divide,   3

$Al(OH)_3$

$Al(OH)_3$,   excess

moles

15.8 g $KMnO_4$

H_2O_2,
g H_2O_2

divide

5

equation

H_2O_2

H_2O_2

PROBLEM SET

Check the answers to your problems on the following pages immediately on completing each problem. The setups as well as the answers are shown so that you can locate your error.

Section A

Balance the following equations using no fractional coefficients.

1. $CaH_2 + H_2O \neq Ca(OH)_2 + H_2$

2. $NH_3 + O_2 \neq N_2 + H_2O$

3. $Al + Fe_3O_4 \neq Al_2O_3 + Fe$

4. $SiO_2 + C \neq SiC + CO$

5. $Cu + H_2SO_4 \neq CuSO_4 + SO_2 + H_2O$

6. $C_2H_5OH + O_2 \neq CO_2 + H_2O$

7. $HNO_3 + I_2 \neq HIO_3 + NO_2 + H_2O$

You will use the following balanced equations in the rest of the problems. The letter of the equation will be noted in the questions.

(a) $MgCO_3 = MgO + CO_2$

(b) $2Na + Cl_2 = 2NaCl$

(c) $Fe_2O_3 + 2Al = Al_2O_3 + 2Fe$

(d) $2Al(OH)_3 + 3H_2SO_4 = Al_2(SO_4)_3 + 6H_2O$

(e) $2KMnO_4 + 5H_2C_2O_4 + 6HCl$
$\quad = 2MnCl_2 + 10CO_2 + 2KCl + 8H_2O$

Section B

8. How many moles of CO_2 are produced according to equation (a) by the reaction of 6.0 moles of $MgCO_3$?

9. If 1.6 moles $Al_2(SO_4)_3$ are produced by reaction (d), how many moles of H_2O are also produced?

10. According to equation (e), 1.5 moles of $KMnO_4$ will react completely with how many moles of $H_2C_2O_4$?

Section C

11. How many grams of Al_2O_3 are produced by the complete reaction of 0.20 mole of Al according to equation (c)?

12. When 0.45 mole CO_2 is produced by equation (e), how many grams of H_2O are also produced?

13. The complete reaction of 4.6 g Na according to equation (b) will yield how many grams of NaCl?

14. How many grams of H_2SO_4 will be required for the complete reaction of 65.0 g of $Al(OH)_3$ according to equation (d)?

15. According to equation (a), 4.0 kg $MgCO_3$ will yield how many grams of CO_2?

16. How many grams of HCl will be required for the complete reaction of 316 g $KMnO_4$ according to equation (e)?

Section D

17. How many moles of Al_2O_3 can be produced by the reaction of 7.0 moles Fe_2O_3 and 3.0 moles Al according to equation (c)?

18. How many moles of $MnCl_2$ can be produced by the reaction of 5.0 moles $KMnO_4$, 3.0 moles $H_2C_2O_4$, and 22 moles HCl according to equation (e)?

19. How many grams of NaCl can be produced by reacting 100 g Na with 100 g Cl_2 according to equation (b)?

20. How many grams of Fe are produced by reacting 2.00 kg Al with 300 g Fe_2O_3 following the reaction of equation (c)?

21. How many grams of which reactant are left over in Problem 20?

PROBLEM SET ANSWERS

These problems have been divided up into sections corresponding to the sections in the chapter You should be able to balance at least six out of the seven equations in Problems 1–7. If you couldn't, redo Section A. You should be able to do Problems 8, 9, and 10. If you couldn't, redo Section B in the chapter. If you missed only one problem in Section C (Problems 11–16), you are doing fine. If you missed more, you had better redo the section. Section D is such a nuisance that if you got only three correct, that will have to do. If you had trouble with more than three in the group Problems 17–21, you had better redo Section D.

1. $CaH_2 + 2H_2O = Ca(OH)_2 + 2H_2$

2. $4NH_3 + 3O_2 = 2N_2 + 6H_2O$

3. $8Al + 3Fe_3O_4 = 4Al_2O_3 + 9Fe$

4. $SiO_2 + 3C = SiC + 2CO$

5. $Cu + 2H_2SO_4 = CuSO_4 + SO_2 + 2H_2O$

6. $C_2H_5OH + 3O_2 = 2CO_2 + 3H_2O$

7. $I_2 + 10HNO_3 = 2HIO_3 + 10NO_2 + 4H_2O$

8. 6.0 moles CO_2 $\left(\text{? moles } CO_2 = 6.0 \text{ moles } \cancel{MgCO_3} \times \dfrac{\text{mole } CO_2}{\text{mole } \cancel{MgCO_3}} \right)$

9. 9.6 moles H_2O $\left(\text{? moles } H_2O = 1.6 \text{ moles } \cancel{Al_2(SO_4)_3} \times \dfrac{6 \text{ moles } H_2O}{\text{mole } \cancel{Al_2(SO_4)_3}} \right)$

10. 3.8 moles $H_2C_2O_4$ $\left(\text{? moles } H_2C_2O_4 = \underline{1.5 \text{ moles } KMnO_4} \right.$

$$\left. \times \frac{5 \text{ moles } H_2C_2O_4}{2 \text{ moles } KMnO_4} \right)$$

11. 10 g Al_2O_3 $\left(\text{? g } Al_2O_3 = 0.20 \text{ mole } Al \times \frac{\text{mole } Al_2O_3}{2 \text{ moles } Al} \times \frac{102 \text{ g } Al_2O_3}{\text{mole } Al_2O_3} \right)$

12. 6.5 g H_2O $\left(\text{? g } H_2O = 0.45 \text{ mole } CO_2 \times \frac{8 \text{ moles } H_2O}{10 \text{ moles } CO_2} \times \frac{18.0 \text{ g } H_2O}{\text{mole } H_2O} \right)$

13. 12 g NaCl $\left(\text{? g NaCl} = 4.6 \text{ g Na} \times \frac{\text{mole Na}}{23.0 \text{ g Na}} \times \frac{2 \text{ moles NaCl}}{2 \text{ moles Na}} \right.$

$$\left. \times \frac{58.5 \text{ g NaCl}}{\text{mole NaCl}} \right)$$

14. 122 g H_2SO_4 $\left(\text{? g } H_2SO_4 = 65.0 \text{ g } Al(OH)_3 \times \frac{\text{mole } Al(OH)_3}{78.0 \text{ g } Al(OH)_3} \right.$

$$\left. \times \frac{3 \text{ moles } H_2SO_4}{2 \text{ moles } Al(OH)_3} \times \frac{98.1 \text{ g } H_2SO_4}{\text{mole } H_2SO_4} \right)$$

15. 2.1×10^3 g CO_2 $\left(\text{? g } CO_2 = 4.0 \text{ kg } MgCO_3 \times \frac{10^3 \text{ g } MgCO_3}{\text{kg } MgCO_3} \right.$

$$\times \frac{\text{mole } MgCO_3}{84.3 \text{ g } MgCO_3} \times \frac{\text{mole } CO_2}{\text{mole } MgCO_3}$$

$$\left. \times \frac{44.0 \text{ g } CO_2}{\text{mole } CO_2} \right)$$

16. 219 g HCl $\left(\text{? g HCl} = 316 \text{ g } KMnO_4 \times \frac{\text{mole } KMnO_4}{158 \text{ g } KMnO_4} \right.$

$$\left. \times \frac{6 \text{ moles HCl}}{2 \text{ moles } KMnO_4} \times \frac{36.5 \text{ g HCl}}{\text{mole HCl}} \right)$$

17. 1.5 moles Al_2O_3 $\left(\text{First determine which is limiting.} \right.$

    7.0 moles Fe_2O_3

$$\frac{3.0 \text{ moles Al}}{2} = 1.5 \quad \text{(limiting)}$$

Then solve

$$\left. \text{? moles } Al_2O_3 = 3 \text{ moles Al} \times \frac{\text{mole } Al_2O_3}{2 \text{ moles Al}} \cdot \right)$$

18. 1.2 moles $MnCl_2$ (First determine which is limiting.

$$\frac{5.0 \text{ moles } KMnO_4}{2} = 2.5$$

$$\frac{3.0 \text{ moles } H_2C_2O_4}{5} = 0.66 \quad (\text{limiting})$$

$$\frac{22 \text{ moles } HCl}{6} = 3.7$$

Then solve

$$? \text{ moles } MnCl_2 = 3.0 \, \cancel{\text{moles } H_2C_2O_4} \times \frac{2 \text{ moles } MnCl_2}{5 \, \cancel{\text{moles } H_2C_2O_4}} \; .)$$

19. 165 g NaCl (First determine which is limiting.

$$\text{mole Na} = 100 \, \cancel{\text{g Na}} \times \frac{\text{mole Na}}{23.0 \, \cancel{\text{g Na}}}$$

$$= 4.35 \text{ moles Na} \, ; \; \frac{4.35 \text{ moles Na}}{2} = 2.17$$

$$\text{mole } Cl_2 = 100 \, \cancel{\text{g } Cl_2} \times \frac{\text{mole } Cl_2}{71.0 \, \cancel{\text{g } Cl_2}} = 1.41 \text{ mole } Cl_2 \quad (\text{limiting})$$

Then solve

$$? \text{ g NaCl} = 100 \, \cancel{\text{g } Cl_2} \times \frac{\cancel{\text{mole } Cl_2}}{71.0 \, \cancel{\text{g } Cl_2}}$$

$$\times \frac{2 \, \cancel{\text{moles NaCl}}}{\cancel{\text{mole } Cl_2}} \times \frac{58.5 \text{ g NaCl}}{\cancel{\text{mole NaCl}}} \; .)$$

20. 210 g Fe (First determine which is limiting.

$$\text{mole Al} = 2.00 \, \cancel{\text{kg Al}} \times \frac{10^3 \, \cancel{\text{g Al}}}{\cancel{\text{kg Al}}} \times \frac{\text{mole Al}}{27.0 \, \cancel{\text{g Al}}} = 74.0 \text{ mole Al} \, ;$$

$$\frac{74.0 \text{ mole Al}}{2} = 37.0$$

$$\text{mole } Fe_2O_3 = 300 \, \cancel{\text{g } Fe_2O_3} \times \frac{\text{mole } Fe_2O_3}{159.6 \, \cancel{\text{g } Fe_2O_3}}$$

$$= 1.89 \text{ mole } Fe_2O_3 \quad (\text{limiting})$$

Then solve

$$? \text{ g Fe} = 300 \text{ g } Fe_2O_3 \times \frac{\text{mole } Fe_2O_3}{159.6 \text{ g } Fe_2O_3}$$

$$\times \frac{2 \text{ moles Fe}}{\text{mole } Fe_2O_3} \times \frac{55.8 \text{ g Fe}}{\text{mole Fe}} \cdot \Bigg)$$

21. 1.90 kg Al   (Since you have determined in Problem 20 that the Fe_2O_3 is limiting and will therefore be completely used up, all you must do is to find out how much Al will be used by the Fe_2O_3 and subtract this amount from the amount of Al you started with.

$$? \text{ g Al} = 300 \text{ g } Fe_2O_3 \times \frac{\text{mole } Fe_2O_3}{159.6 \text{ g } Fe_2O_3}$$

$$\times \frac{2 \text{ moles Al}}{\text{mole } Fe_2O_3} \times \frac{27.0 \text{ g Al}}{\text{mole Al}}$$

$$= 102 \text{ g Al will be used.}$$

Since 2000 g Al were present, then 1898 g Al must remain. Rounded off to the correct number of significant figures, this is 1.90 kg.)

Chapter 9
Complicated Stoichiometry

PRETEST

These questions may take you as long as 30 minutes to do. As you will see, they are fairly complicated.

1. How many pounds of NaOH will be required to produce 500 g Na_2SO_4 according to the reaction

   $H_2SO_4 + 2NaOH = Na_2SO_4 + 2H_2O$?

2. How many tons of water (H_2O) will be produced if 18.0 tons of Na_2SO_4 are produced according to the equation in question 1?

3. It was found that a 35.0 g sample of iron ore contains 19.0 g Fe. How many grams of H_2 will be produced by 25.0 g of ore according to the reaction

   $2Fe + 6HCl = 2FeCl_3 + 3H_2$?

4. What is the percent purity of a sample of iron ore if a 50.0 g sample of the impure ore will produce 2.00 g H_2 according to the reaction in question 3?

5. What is the percent purity of a sample of iron ore if 2.00 pounds of the ore require 5.15 liters of an HCl solution (density = 0.990 g/ml), which contains 12.4 g HCl in 50.0 g of solution? The reaction is the same as in question 3.

6. How many avoirdupois ounces (16 oz/lb) of Ag can be prepared by the reaction of $20 worth of Ag_2S ore? The ore is 54.0 weight percent Ag_2S and costs $5 per kg. The reaction used is

   $Ag_2S + Zn + 2H_2O = 2Ag + H_2S + Zn(OH)_2$.

7. How many liters of H_2 gas at 1.0 atmosphere pressure and 0°C will be needed to react with 6.40 g O_2 according to the equation shown below? The density of H_2 gas at this temperature and pressure is 0.0893 g/liter.

   $2H_2 + O_2 = 2H_2O$

8. How many grams of $CaCO_3$ will be produced if 2.90 g $Fe_2(CO_3)_3$ is reacted with an excess of HCl to produce $FeCl_3$ and CO_2, as is shown in the first equation below, and then the CO_2 gas is bubbled into a solution of $Ca(OH)_2$ to produce $CaCO_3$, as shown in the second equation?

   $Fe_2(CO_3)_3 + 6HCl = 2FeCl_3 + 3CO_2 + 3H_2O$

   $CO_2 + Ca(OH)_2 = CaCO_3 + H_2O$

PRETEST ANSWERS

If you did all eight questions and got the correct answers you are really doing beautifully. Each involves not only one new type of conversion factor, but also may have in it the conversion factors of the preceding questions. Consequently, you should start in the body of the chapter at the section where you first bogged down. If you were able to do questions 1 and 2, you can skip Section A. If you got questions 1 through 4, you can skip Sections A and B. If you got questions 1–5 and 7, you can skip Sections A, B, and C. If you got questions 1 through 7 correctly, you can skip Sections A, B, C, and D. And if you only missed question 8, all you need to do is Section E.

Incidentally, all the questions, except 5, are used as examples in the text, so you will see how they are worked out.

1. 0.620 lb NaOH

2. 4.56 tons H_2O

3. 0.735 g H_2

4. 73.9% Fe

5. 71.0% Fe

6. 66.3 oz Ag

7. 9.0 liters H_2

8. 2.99 g $CaCO_3$

The principle for solving the problems in this chapter will be exactly the same as that used for the simpler problems in Chapter 8. The difference is only that you will be using more conversion factors in solving each problem. All stoichiometry problems are basically the same. You are asked a question about a molecule (or atom) and are given information about another. They are related to each other through a balanced equation from which you can set up a "bridge" conversion factor. The "bridge" contains the two substances expressed in moles.

To solve a problem, first you note the dimensions of the answer. Then you write down the dimensions of the information given and, using conversion factors, convert this to moles of that substance so that you can multiply by the "bridge." You now have moles of the substance in the answer and you convert this to the desired units of the question asked.

In the following sections we will separate the types of conversions you must use to get to moles so that you can use the "bridge."

SECTION A   Nonmetric Units

We can start with question 1 of the pretest. "How many pounds of NaOH will be required to produce 500 g of Na_2SO_4 according to the balanced equation

$$H_2SO_4 + 2NaOH = Na_2SO_4 + 2H_2O?"$$

The question asked is "_____."

How many lb NaOH

The information given is "_____."

500 g Na_2SO_4

The "bridge" conversion factor from the balanced equation which relates the substance in the question asked to the substance

in the information given is _____.

$$\frac{2 \text{ moles NaOH}}{\text{mole } Na_2SO_4}$$

So starting in the usual way

$$? \text{ lb NaOH} = 500 \text{ g } Na_2SO_4$$

Now you want to convert g Na_2SO_4 to _____ Na_2SO_4 so that you can use the "bridge." The conversion factor which will do

moles

this is _____, the reciprocal of the molecular weight.

$$\frac{\text{mole } Na_2SO_4}{142.1 \text{ g } Na_2SO_4}$$

Now, if you multiply by the "bridge" conversion factor, the

unwanted dimension is _____. You must now convert this to

mole NaOH

the dimensions of the answer, _____.

lb NaOH

You can get from moles NaOH to grams NaOH using the

conversion factor _____, and then you can get from grams NaOH to the dimensions of the answer by using the

$$\frac{40.0 \text{ g NaOH}}{\text{mole NaOH}}$$

conversion factor _____. The only dimension left which does not cancel is pounds NaOH, which is the dimension of the

$$\frac{\text{lb NaOH}}{454 \text{ g NaOH}}$$

answer. So the answer must be

$$? \text{ lb NaOH} = 500 \text{ g } \cancel{Na_2SO_4} \times \frac{\cancel{\text{mole } Na_2SO_4}}{142.1 \text{ g } \cancel{Na_2SO_4}}$$

$$\times \frac{2 \text{ moles } \cancel{NaOH}}{\cancel{\text{mole } Na_2SO_4}} \times \frac{40.0 \text{ g } \cancel{NaOH}}{\cancel{\text{mole NaOH}}}$$

$$\times \frac{\text{lb NaOH}}{454 \text{ g } \cancel{NaOH}}$$

$$= 0.620 \text{ lb NaOH}.$$

This was the first question on the pretest. It is really quite simple.

Here is another example where the question asked is in metric units but the information given is not. How many grams of H_2O are produced by the reaction of 11 oz H_2SO_4 according to the reaction

$$2NaOH + H_2SO_4 = Na_2SO_4 + 2H_2O?$$

Starting in the usual way, you have

11 oz H_2SO_4

moles

$$? \text{ g } H_2O = \underline{\hspace{2cm}}.$$

You must get from ounces H_2SO_4 to _____ H_2SO_4 so that you can use the "bridge." Since the molecular weight is given in g/mole, you must first get to grams H_2SO_4. This will take two

$\dfrac{\text{lb } H_2SO_4}{16 \text{ oz } H_2SO_4}$, $\dfrac{454 \text{ g } H_2SO_4}{\text{lb } H_2SO_4}$

conversion factors $\underline{\hspace{2cm}} \times \underline{\hspace{2cm}}$.

Now that the unwanted dimension is grams H_2SO_4, you can get to moles H_2SO_4 by using the inverted molecular weight,

$\dfrac{\text{mole } H_2SO_4}{98.1 \text{ g } H_2SO_4}$

$\underline{\hspace{2cm}}$. Finish the problem off and check your answer with what is shown below.

$$? \text{ g } H_2O = 11 \cancel{\text{ oz } H_2SO_4} \times \frac{\cancel{\text{lb } H_2SO_4}}{16 \cancel{\text{ oz } H_2SO_4}} \times \frac{454 \text{ g } \cancel{H_2SO_4}}{\cancel{\text{lb } H_2SO_4}}$$

$$\times \frac{\cancel{\text{mole } H_2SO_4}}{98.1 \text{ g } \cancel{H_2SO_4}} \times \frac{2 \text{ moles } \cancel{H_2O}}{\cancel{\text{mole } H_2SO_4}} \times \frac{18.0 \text{ g } H_2O}{\cancel{\text{mole } H_2O}}$$

$$= 1.1 \times 10^2 \text{ g } H_2O \quad \text{(Note significant figures.)}$$

What will happen if you have a problem where neither the question asked nor the information given is in metric units? You can, of course, convert both into and out of metric units, as the example below will show.

How many tons of H_2O will be produced when 18.0 tons of Na_2SO_4 are prepared according to the equation

$$2NaOH + H_2SO_4 = Na_2SO_4 + 2H_2O?$$

(This was question 2 on the pretest.)

Starting in the usual way, you have

tons H_2O, 18.0 tons Na_2SO_4

$\dfrac{2000 \text{ lb } Na_2SO_4}{\text{ton } Na_2SO_4}$

$\dfrac{454 \text{ g } Na_2SO_4}{\text{lb } Na_2SO_4}$

$$? \underline{\hspace{1.5cm}} = \underline{\hspace{2cm}}.$$

You can get from tons Na_2SO_4 to grams Na_2SO_4 if you multiply by the two conversion factors, $\underline{\hspace{2cm}}$ and

$\underline{\hspace{2cm}}$.

Now you use the inverted form of the molecular weight of Na_2SO_4 to get to moles of Na_2SO_4 and then use the "bridge" to get to moles of H_2O. (You should be writing these down as we go along.) Once at moles of H_2O, you can get to grams H_2O by using the conversion factor, _____, the molecular weight.

$$\frac{18.0 \text{ g } H_2O}{\text{mole } H_2O}$$

There remain only the conversions to get grams H_2O to

_____, the dimensions of the question asked. You can do this in two steps by using the conversion factors _____ and

$$\frac{\text{tons } H_2O}{\dfrac{\text{lb } H_2O}{454 \text{ g } H_2O}}$$

$$\frac{\dfrac{\text{ton } H_2O}{2000 \text{ lb } H_2O}}{}$$

_____. You then find the answer through the steps below.

$$? \text{ tons } H_2O = 18.0 \text{ tons } Na_2SO_4 \times \frac{2000 \text{ lb } Na_2SO_4}{\text{ton } Na_2SO_4}$$

$$\times \frac{454 \text{ g } Na_2SO_4}{\text{lb } Na_2SO_4} \times \frac{\text{mole } Na_2SO_4}{142.1 \text{ g } Na_2SO_4}$$

$$\times \frac{2 \text{ moles } H_2O}{\text{mole } Na_2SO_4} \times \frac{18.0 \text{ g } H_2O}{\text{mole } H_2O}$$

$$\times \frac{\text{lb } H_2O}{454 \text{ g } H_2O} \times \frac{\text{ton } H_2O}{2000 \text{ lb } H_2O}$$

$$= 4.56 \text{ tons } H_2O.$$

You may have noticed that when you wrote down the numbers you had 2000×454 and $1/2000 \times 1/454$. These numbers cancel completely, so it is not worth the effort to put them in at all. There is a shortcut that you can make when you have all your units nonmetric.

Recall that when we defined the "mole," we said that it is really called the "gram-mole" and is defined as the number of molecules which weigh the same as the molecular weight in grams. We could have defined a different "mole" as the number of molecules which weigh the molecular weight in pounds, or as the number of molecules which weigh the molecular weight in tons. The first would be called a "pound-mole" and the second a "ton-mole." The terms pound-mole and ton-mole do not mean pound $\times$ mole or ton $\times$ mole, and you cannot cancel the individual parts of these names; they are to be treated as single units. The "bridge" conversion factor which you set up from the balanced equation could be in pound-moles or ton-moles, as you choose.

Thus, in the balanced equation

$$H_2 + Cl_2 = 2HCl$$

the "bridge" relating ton-moles of H_2 to HCl would be

_____, and the "bridge" connecting pound-moles of H_2 and NH_3 in the equation

$$N_2 + 3H_2 = 2NH_3$$

would be _____.

$$\frac{\text{ton-mole } H_2}{2 \text{ ton-moles } HCl}$$

$$\frac{3 \text{ lb-moles } H_2}{2 \text{ lb-moles } NH_3}$$

$$\frac{36.5 \text{ tons HCl}}{\text{ton-mole HCl}}$$

$$\frac{17.0 \text{ lb NH}_3}{\text{lb-mole NH}_3}$$

Of course, the molecular weight conversion factors are now expressed in pounds/pound-mole and tons/ton-mole; so the

molecular weight of HCl in ton-moles would be _____, and the molecular weight of NH_3 in pound-moles would be

_____.

Let's try the preceding problem again, but since both the question asked and the information given are in tons, we will use ton-moles. Starting in the usual way, then, we have

$$? \text{ tons H}_2\text{O} = \underline{\hspace{4cm}}.$$

18.0 tons Na_2SO_4

$$\frac{\text{ton-mole Na}_2\text{SO}_4}{142.1 \text{ tons Na}_2\text{SO}_4}$$

$$\frac{2 \text{ ton-moles H}_2\text{O}}{\text{ton-mole Na}_2\text{SO}_4}$$

$$\frac{18.0 \text{ tons H}_2\text{O}}{\text{ton-mole H}_2\text{O}}$$

The reciprocal of the molecular weight expressed in ton-moles for Na_2SO_4 is _____. It is now possible to use a "bridge" in ton-moles relating Na_2SO_4 to H_2O; it is

_____. Ton-moles H_2O can be related to the dimensions of the answer by using the molecular weight in tons,

_____. You now have left as the only dimension, the dimension of the question asked. The setup for the whole problem would now be

$$? \text{ tons H}_2\text{O} = 18.0 \text{ } \cancel{\text{tons Na}_2\text{SO}_4} \times \frac{\cancel{\text{ton-mole Na}_2\text{SO}_4}}{142.1 \text{ } \cancel{\text{tons Na}_2\text{SO}_4}}$$

$$\times \frac{2 \text{ } \cancel{\text{ton-moles H}_2\text{O}}}{\cancel{\text{ton-mole Na}_2\text{SO}_4}} \times \frac{18.0 \text{ tons H}_2\text{O}}{\cancel{\text{ton-mole H}_2\text{O}}}$$

$$= 18.0 \times \frac{1}{142.1} \times 2 \times 18.0 \text{ tons H}_2\text{O}$$

$$= 4.56 \text{ tons H}_2\text{O}.$$

As you can see, this procedure has shortened the setup quite a bit by eliminating four conversion factors. However, since no more multiplications or divisions would be needed in the long way (all the numbers in the four extra conversion factors simply cancel), if you do not care to use the pound-moles and ton-moles, you really don't have to.

SECTION B Impure Substances

Frequently the substance in the information given in the problem will not be pure, and sometimes the substance in the question asked will not be pure. It is even possible to have problems where both the substance asked for and the given substance are impure. This doesn't present too much difficulty since you will always have conversion factors given in the problem which will relate pure substance to impure substance.

Consider question 3 on the pretest. You are told that 35.0 g iron ore contain 19.0 g Fe. How many grams of H_2, then, can be produced by the reaction of 25.0 g of ore according to the equation

$$2Fe + 6HCl = 2FeCl_3 + 3H_2?$$

The iron ore given in the problem is impure, but the conversion factor given in the problem relating pure Fe to iron ore is

_____. Thus, starting in the usual way, you have

? _____ = _____.

The unwanted dimension is _____, but this is contained in the conversion factor given in the problem. Now you multiply by this factor.

$$? \text{ g } H_2 = 25.0 \text{ g ore} \times \frac{19.0 \text{ g Fe}}{35.0 \text{ g ore}}$$

This operation leaves _____ as the new unwanted dimension, which is the point where you would have started the problem if you had only pure substances. Finish the problem and then compare your answer with the setup shown below.

$$? \text{ g } H_2 = 25.0 \text{ g ore} \times \frac{19.0 \text{ g Fe}}{35.0 \text{ g ore}} \times \frac{\text{mole Fe}}{55.8 \text{ g Fe}}$$

$$\times \frac{3 \text{ moles } H_2}{2 \text{ moles Fe}} \times \frac{2.02 \text{ g } H_2}{\text{mole } H_2}$$

$$= 0.735 \text{ g } H_2$$

That is really all there is to this type of problem. You go from weight of impure substance to grams of pure substance. The reciprocal molecular weight will convert grams to moles. Then, using the "bridge," you cross over to the substance in the question asked in moles. Finally, you convert this to the desired units, such as grams, pounds, or whatever.

One word of caution: you must be extremely careful to label the dimensions completely so that you will not cancel grams impure with grams pure. If you make this mistake, you will mess up the whole thing.

It is very common to express the purity of a substance in percent. This is the number of grams of pure substance in 100 g of impure substance. If you are shaky, refer back to Section C in Chapter 4. In these cases, then, the conversion factor given in the problem will be grams of pure substance per _____ of impure substance. Setting up the problem will then be the same as in the type you have already done.

Here is a problem to show you how this works. How many grams of air, which is 22.5 wt % O_2, will be needed for the complete oxidation of 7.20 g pentane according to the reaction

$$C_5H_{12} + 8O_2 = 5CO_2 + 6H_2O?$$

$\dfrac{19.0 \text{ g Fe}}{35.0 \text{ g ore}}$

g H_2,   25.0 g ore

grams ore

grams Fe

100 g

$$\frac{22.5 \text{ g } O_2}{100 \text{ g air}}$$

$$\frac{O_2}{\text{mole } C_5H_{12}}$$

$$\frac{}{8 \text{ moles } O_2}$$

molecular

$$\frac{\text{g } O_2}{\text{mole } C_5H_{12}}$$

g air, 7.20 g C_5H_{12}

grams C_5H_{12}, moles C_5H_{12}

bridge

molecular weight

$$\frac{8 \text{ moles } O_2}{\text{mole } C_5H_{12}}$$

moles O_2

grams O_2

The conversion factor given in the problem is the weight percent O_2 in air. You can write this as the fraction _____.

You might start your problems by writing the conversion factors that you think you will need. First you will need a "bridge" which will relate pentane, C_5H_{12}, to ____. You get this from the balanced equation, and it will be _____. Since you are dealing with weights of the two reactants, you also will need the conversion factors which relate weights to moles; these will be the _____ weights,

$$\frac{72.0 \text{ g } C_5H_{12}}{\rule{2cm}{0.4pt}} \qquad \frac{32.0 \rule{1.5cm}{0.4pt}}{\text{mole } O_2} .$$

Now you can start in the usual way.

$$? \rule{1cm}{0.4pt} = \rule{3cm}{0.4pt}$$

Incidentally, you should have noticed by now that the most difficult step in setting up any problem is the first one—determining the exact dimensions of the question asked and noting the information given in the problem. Once you are over this hump, everything falls into place.

The unwanted dimension in the information given is _____. You must convert this to _____ so that you can use your "_____." You do this by multiplying by the inverted form of the _____.

$$? \text{ g air} = 7.20 \text{ g } \cancel{C_5H_{12}} \times \frac{\text{mole } C_5H_{12}}{72.0 \text{ g } \cancel{C_5H_{12}}}$$

Now that the unwanted dimension is mole C_5H_{12} you can multiply by your "bridge:"

$$? \text{ g air} = 7.20 \text{ g } \cancel{C_5H_{12}} \times \frac{\text{mole } C_5H_{12}}{72.0 \text{ g } \cancel{C_5H_{12}}} \times \rule{3cm}{0.4pt} .$$

The unwanted dimension now is _____, which you can convert to _____ by multiplying by the molecular weight.

$$? \text{ g air} = 7.20 \text{ g } \cancel{C_5H_{12}} \times \frac{\text{mole } \cancel{C_5H_{12}}}{72.0 \text{ g } \cancel{C_5H_{12}}}$$
$$\times \frac{8 \text{ moles } \cancel{O_2}}{\text{mole } \cancel{C_5H_{12}}} \times \frac{32.0 \text{ g } O_2}{\text{mole } \cancel{O_2}}$$

You now have as the unwanted dimension grams O_2, which can be converted to the dimensions of the question asked by using the conversion factor given in the problem, the weight _____ of O_2 in air.

percent

$$? \text{ g air} = 7.20 \text{ g } C_5H_{12} \times \frac{\text{mole } C_5H_{12}}{72.0 \text{ g } C_5H_{12}} \times \frac{8 \text{ moles } O_2}{\text{mole } C_5H_{12}}$$

$$\times \frac{32.0 \text{ g } O_2}{\text{mole } O_2} \times \frac{100 \text{ g air}}{22.5 \text{ g } O_2}$$

Notice that you have to use the weight percent conversion

factor in its _____ form in order to _____ the unwanted dimension and have only the dimension in the question asked. The numerical answer will then be

inverted, cancel

$$? \text{ g air} = 7.20 \times \frac{1}{72.0} \times 8 \times 32.0 \times \frac{100 \text{ g air}}{22.5}$$

$$= 114 \text{ g air}$$

Often you will be asked for the percent purity of a sample. In such cases you must reword the question to, "How many grams pure substance in 100 g impure sample?" The question asked and the information given are all in this one sentence.

Question 4 on the pretest is an example of this type. You are asked to find the percent purity of a sample of Fe ore, if a 50.0 g sample of the impure ore will produce 2.00 g H_2 according to the reaction

$$2Fe + 6HCl = 2FeCl_3 + 3H_2$$

Rewording the question, you ask how many _____ are in

grams Fe

_____ impure ore? (If you have forgotten this method, go back to Section C in Chapter 4.) There is a conversion factor given in the

100 g

problem; it says that there are 50.0 g impure ore per _____.

2.00 g H_2

$$\frac{50.0 \text{ g imp.}}{2.00 \text{ g } H_2}$$

Writing this as a fraction, you will have _____ (It isn't a bad idea to write down all conversion factors given in the problem before you even start the solution.) The "bridge" you will use from

the balanced equation must relate Fe to ____ (the only other

H_2

2 moles Fe

substance mentioned in the problem). It will be _____.

3 moles H_2

Now you are ready to set up the problem.

$$? \text{ g Fe} = \text{_____}$$

100 g imp.

The unwanted dimension in the information given is _____,

grams impure

and the only conversion factor which has this in it is _____. The unwanted dimension now is grams H_2. Multiplying by the

$$\frac{2.00 \text{ g } H_2}{50.0 \text{ g imp.}}$$

inverted form of the molecular weight, _____, you now have

$$\frac{\text{mole } H_2}{2.0 \text{ g } H_2}$$

moles H_2 and can use the "bridge" to get to moles Fe. Multiplying by the "bridge" will give

$$? \text{ g Fe} = 100 \, \underline{\text{g imp.}} \times \frac{2.00 \, \underline{\text{g } H_2}}{50.0 \, \underline{\text{g imp.}}} \times \frac{\text{mole } H_2}{2.0 \, \underline{\text{g } H_2}} \times \underline{\hspace{3cm}}.$$

$$\frac{2 \text{ moles Fe}}{3 \text{ moles } H_2}$$

weight

Finally, if you multiply by the atomic _____ of Fe, all dimensions will cancel except the dimension of the answer.

$$? \text{ g Fe} = 100 \, \underline{\text{g imp.}} \times \frac{2.00 \, \underline{\text{g } H_2}}{50.0 \, \underline{\text{g imp.}}} \times \frac{\underline{\text{mole } H_2}}{2.0 \, \underline{\text{g } H_2}}$$

$$\times \frac{2 \, \underline{\text{moles Fe}}}{3 \, \underline{\text{moles } H_2}} \times \frac{55.8 \text{ g Fe}}{\underline{\text{mole Fe}}}$$

$$= 74 \text{ g Fe}/100 \text{ g imp.} = 74\% \text{ Fe}$$

We will see more problems of this type in Section C, and you will have a little more practice in working them.

SECTION C Solutions and Density

A solution of a substance in a nonreacting solvent is really just another form of an impure substance. Thus, if you have a 24 wt %

$$\frac{24 \text{ g HCl}}{100 \text{ g soln.}}$$

solution of HCl, you can write the conversion factor _____. The problem is solved in exactly the same way as in Section B. Try an example.

How many grams of a 24 wt % solution of HCl will be required to react with 75 g Fe according to the balanced equation

$$2Fe + 6HCl = 2FeCl_3 + 3H_2?$$

grams soln.

The question asked is, "How many _____?" (Be very careful; you are not asked, "How many grams HCl?") The

$$75 \text{ g Fe}$$
$$\frac{\text{mole Fe}}{55.8 \text{ g Fe}}$$
moles

information given is _____.

The first conversion factor that you will use is _____. You have used this because you always head toward _____ of a compound so that you can use the "bridge." The next conversion factor will be the "bridge" which relates the compound in the question with the compound in the information given; that factor

$$\frac{6 \text{ moles HCl}}{2 \text{ moles Fe}}$$
$$\frac{36.5 \text{ g HCl}}{\text{mole HCl}}$$
$$\frac{100 \text{ g soln.}}{24 \text{ g HCl}}$$
grams solution

is _____. The next conversion factor by which you will multiply is _____, since this gets you to the conversion factor given in the problem. Finally, when you multiply by the conversion factor _____ given in the problem, all the dimensions cancel except _____. Therefore, the answer must be

$$? \text{ g soln.} = 75 \, \cancel{\text{g Fe}} \times \frac{\cancel{\text{mole Fe}}}{55.8 \, \cancel{\text{g Fe}}} \times \frac{6 \, \cancel{\text{moles HCl}}}{2 \, \cancel{\text{moles Fe}}}$$

$$\times \frac{36.5 \, \cancel{\text{g HCl}}}{\cancel{\text{mole HCl}}} \times \frac{100 \text{ g soln.}}{24 \, \cancel{\text{g HCl}}}$$

$$= 6.1 \times 10^2 \text{ g soln.}$$

(You should be writing your own setup as we go through the problem.)

There will also be problems which, instead of having weight units for the information given or the question asked (or even both), will have volume units. In such cases you will always have the density as a conversion factor given in the problem. This will allow you to convert from volume to weight. The units of density, you recall, are ____ or _____ for gases.

$$\frac{\text{g}}{\text{ml}} \quad \frac{\text{g}}{\text{liter}}$$

Question 7 in the pretest is a good example of this type. You are asked how many liters of H_2 gas at 1.0 atm and 0°C will be needed to react with 6.40 g O_2 according to the equation

$$2H_2 + O_2 = 2H_2O.$$

The density of H_2 gas at the temperature and pressure shown is 0.0893 g/liter.

The density is the conversion factor given in the problem. Written as a fraction it is _____. Thus, starting in the usual way, you have

$$\frac{0.0893 \text{ g } H_2}{\text{liter } H_2}$$

$$? \text{ _____} = \text{ _____}.$$

liters H_2,   6.40 g O_2

You want to move toward the "bridge," so the first conversion factor you multiply by is _____. Next you multiply by the "bridge," which is _____. From this you can get the weight of H_2 in grams if you multiply by the molecular weight _____. The unwanted dimension is _____, and the dimensions that the answer will have are _____.

$$\frac{\text{mole } O_2}{32.0 \text{ g } O_2}$$
$$\frac{2 \text{ moles } H_2}{\text{mole } O_2}$$
$$\frac{2.0 \text{ g } H_2}{\text{mole } H_2}, \text{ grams } H_2$$
liters H_2

But the density gives you this conversion, so finally you multiply by _____, and the only dimension left is the dimension of the answer.

$$\frac{\text{liter } H_2}{0.0893 \text{ g } H_2}$$

$$? \text{ liters } H_2 = 6.40 \, \cancel{\text{g } O_2} \times \frac{\cancel{\text{mole } O_2}}{32.0 \, \cancel{\text{g } O_2}} \times \frac{2 \, \cancel{\text{moles } H_2}}{1 \, \cancel{\text{mole } O_2}}$$

$$\times \frac{2.0 \, \cancel{\text{g } H_2}}{\cancel{\text{mole } H_2}} \times \frac{\text{liter } H_2}{0.0893 \, \cancel{\text{g } H_2}}$$

$$= 9.0 \text{ liters } H_2 \quad \text{to the correct number of significant figures.}$$

(If you don't remember why 8.96 was rounded off to 9.0, refer back to Chapter 1, Section C.)

It is common in problems which deal with solutions to have volume dimensions rather than weight units given. You will always have a density of the solution to use as a conversion factor, but, once again, be very careful that you write complete dimensions. For example, if you have a solution with a density of 1.10 g/ml, you must write the conversion factor as 1.10 g soln./ml soln. Simply writing 1.10 g/ml will not do. You will become confused and cancel grams solution with grams something else. The following example will illustrate this.

How many grams of Mg can be consumed by 25.0 ml of a 7.10 wt % solution of HCl whose density is 0.990 g/ml according to the reaction

$$Mg + 2HCl = MgCl_2 + H_2?$$

When a problem starts to get complicated, remember to write down all the conversion factors given in the problem before you start the solution. There are two such factors in this problem,

$\dfrac{7.10 \text{ g HCl}}{100 \text{ g soln.}}$, $\dfrac{0.990 \text{ g soln.}}{\text{ml soln.}}$

_____ and _____. Note how the density conversion factor is exactly written. Writing down the "bridge" conversion factor will give you something to head for. In this problem the

$\dfrac{1 \text{ mole Mg}}{2 \text{ moles HCl}}$

"bridge" connecting Mg and HCl will be _____. You will also probably need the molecular (or atomic) weights of the substances mentioned in the problem. These would be

$\dfrac{24.3 \text{ g Mg}}{\text{mole Mg}}$, $\dfrac{36.5 \text{ g HCl}}{\text{mole HCl}}$

_____ and _____.
Now you are ready to start.

$\dfrac{25.0 \text{ ml soln.}}{\text{ml solution}}$

? g Mg = _____

The dimension you want to cancel out is _____. The only conversion factor you have that has this dimension in it is

$\dfrac{0.990 \text{ g soln.}}{\text{ml soln.}}$,   g soln.

_____. Now the unwanted dimension is _____. The

$\dfrac{7.10 \text{ g HCl}}{100 \text{ g soln.}}$

grams HCl

conversion factor which contains this dimension is _____.

Now the unwanted dimension is _____, but this can be converted to moles HCl (so that you can use the "bridge") by

$\dfrac{\text{mole HCl}}{36.5 \text{ g HCl}}$

multiplying by _____.

Next you multiply by the "bridge," which is the only conversion factor left containing the unwanted dimension, mole HCl. You are now into conversion factors dealing with Mg, and if you multiply

$\dfrac{24.3 \text{ g Mg}}{\text{mole Mg}}$

by _____ the only dimension left will be the dimension of the question asked. Thus, the final answer is

$$? \text{ g Mg} = 25.0 \text{ ml soln.} \times \dfrac{0.990 \text{ g soln.}}{\text{ml soln.}} \times \dfrac{7.10 \text{ g HCl}}{100 \text{ g soln.}}$$

$$\times \dfrac{\text{mole HCl}}{36.5 \text{ g HCl}} \times \dfrac{\text{mole Mg}}{2 \text{ moles HCl}} \times \dfrac{24.3 \text{ g Mg}}{\text{mole Mg}}$$

$$= 0.585 \text{ g Mg.}$$

SECTION D Miscellaneous Complications

There are any number of assorted complications which may appear in stoichiometry problems, such as cost in dollars, time required to consume or produce something, amount of heat required or released, or even such things as distances possible to travel on burning fuel, or number of insects that can be killed by the products of a reaction. Any time that you have a chemical reaction and can write a balanced equation, you can have a stoichiometry problem concerning any aspect of the substances in the equation. However, you will always have given in the problem some conversion factor or factors which will allow you to get to or from this strange complication and to moles of the substance so that you can use your "bridge."

Once again, write down all the conversion factors given in the problem before you even start to set it up. Then you can simply pick and choose from these in order to cancel the unwanted dimensions. Here is a simple example. The oxidation of S to SO_3 consumes 6.00 moles of O_2 per minute. How many minutes will it take to produce 14.0 moles of SO_3 according to the equation

$$2S + 3O_2 = 2SO_3?$$

First you write down the conversion factor given in the problem.

It is _____. Now you start in the usual way.

$$? \underline{\hspace{2cm}} = \underline{\hspace{3cm}}$$

The unwanted dimension is already in _____, so you can use the "bridge" conversion factor which relates this to

_____. You multiply by _____. The unwanted

dimension now is _____, but the conversion factor given in the problem relates this directly to the dimensions that the answer

must have. If you multiply by _____, the dimensions will all cancel, except the dimension in the question asked. The answer must be

$$? \text{ min} = 14.0 \text{ moles } \cancel{SO_3} \times \frac{3 \text{ moles } \cancel{O_2}}{2 \text{ moles } \cancel{SO_3}} \times \frac{\text{minute}}{6.00 \text{ moles } \cancel{O_2}}.$$

$$= 3.50 \text{ min.}$$

You may find it somewhat difficult to pick the correct "bridge" since the substances involved may be slightly hidden. In the problem above it is obvious that the "bridge" must contain moles SO_3, but you must look around a little to realize that the dimensions of the answer, minutes, can be related only to O_2, since this is in the conversion factor given in the problem. If by chance you choose an incorrect "bridge," you will quickly reach a dead end and realize that you will have to go back and find another "bridge."

$\dfrac{6.00 \text{ moles } O_2}{\text{minute}}$

minutes,   14.0 moles SO_3

moles SO_3

moles O_2, $\dfrac{3 \text{ moles } O_2}{2 \text{ moles } SO_3}$

moles O_2

$\dfrac{\text{minute}}{6.00 \text{ moles } O_2}$

The example above is very simple to work because both the information given and the conversion factor in the problem are already in moles. In a more usual case, they will be in weight and you must make the usual conversions to get to moles so that you can use the "bridge." Here is an example.

If the rate of O_2 consumption in the oxidation of S to SO_3 is 6.00 g O_2 per minute, how many seconds will it take to make 1.75 lb of SO_3 according to the reaction

$$2S + 3O_2 = 2SO_3?$$

$\dfrac{6.00 \text{ g } O_2}{\text{minute}}$

The conversion factor given in the problem is _____. Starting in the usual way, then, you have

sec, 1.75 lb SO_3

? ___ = _____.

pounds SO_3

The unwanted dimension is _____, and you must, in order

moles

to use a "bridge," get to _____ SO_3. It will take two conversion

grams

factors to do this. First you must convert pounds SO_3 to _____ SO_3. You can do this by multiplying by the conversion factor

$\dfrac{454 \text{ g } SO_3}{\text{lb } SO_3}$

_____. Once you have grams SO_3, you can get to moles SO_3 by multiplying by the conversion factor, which is the

molecular weight

_____ in its inverted form. The conversion factor is

$\dfrac{\text{mole } SO_3}{80.1 \text{ g } SO_3}$

_____.

$\dfrac{3 \text{ moles } O_2}{2 \text{ moles } SO_3}$

The unwanted dimension is now moles of SO_3, and you can use

the "bridge," _____, to get you to moles O_2. The reason that you choose this particular "bridge" is that the

conversion factor

_____ given in the problem contains the substance O_2

moles O_2

in it. The unwanted dimension now is _____, but the

grams

conversion factor given in the problem has _____ O_2. To get to grams O_2, you must multiply by the molecular weight of O_2,

$\dfrac{32.0 \text{ g } O_2}{\text{mole } O_2}$, grams O_2

_____. The unwanted dimension now is _____, and you can multiply by the conversion factor given in the problem,

$\dfrac{\text{minute}}{6.00 \text{ g } O_2}$

minutes

_____.

seconds

The last unwanted dimension is _____, but the dimensions of

$\dfrac{60 \text{ sec}}{\text{minute}}$

the answer to the question asked are _____. It is very easy to get from minutes to seconds by multiplying by _____. Finally, the only dimension left is the dimensions of the answer, which must be

$$? \text{ sec} = 1.75 \; \cancel{\text{lb SO}_3} \times \frac{454 \; \cancel{\text{g SO}_3}}{\cancel{\text{lb SO}_3}} \times \frac{\cancel{\text{mole SO}_3}}{80.1 \; \cancel{\text{g SO}_3}}$$

$$\times \frac{3 \; \cancel{\text{moles O}_2}}{2 \; \cancel{\text{moles SO}_3}} \times \frac{32.0 \; \cancel{\text{g O}_2}}{\cancel{\text{mole O}_2}}$$

$$\times \frac{\cancel{\text{minute}}}{6.00 \; \cancel{\text{g O}_2}} \times \frac{60 \text{ seconds}}{\cancel{\text{minute}}}$$

$$= 4.76 \times 10^3 \text{ sec.}$$

Industrial chemists are often interested in the cost of chemical reactions. In such specific problems there will always be some conversion factor having either dollars or cents as one of its units. You approach these problems in the same way as you did the "time" problems. Here is an example.

Ammonia, NH_3, can be produced by the direct reaction of N_2 with H_2 according to the equation

$$N_2 + 3H_2 = 2NH_3.$$

If H_2 costs \$0.01/kg, how much will the H_2 cost to produce 6000 g NH_3?

The conversion factor given in the problem is _____. Starting in the usual way, then, you have

$$? \text{ dollars} = \underline{\hspace{2cm}}.$$

In order to use the "bridge" conversion factor, you must convert

grams NH_3 to _____ NH_3. You can do this by multiplying by

_____, the molecular weight of NH_3 in its inverted form.

The "bridge" you are going to use must have _____ in the

denominator and should have _____ in the numerator (on top) because the conversion factor given in the problem has to do with

H_2. The "bridge" will therefore be _____.
The units for H_2 in the conversion factor given in the problem

are _____ H_2. Therefore, you must convert moles H_2 to

_____ H_2, which can be done in two conversions. First you

multiply by 2.0 g H_2/mole H_2, the _____ weight of H_2. Then

you multiply by _____. The unwanted dimension now is kilograms H_2. This appears in the conversion factor given in the

problem, and multiplying by this conversion factor leaves _____

$\dfrac{\$0.01}{\text{kg } H_2}$

6000 g NH_3

moles

$\dfrac{\text{mole } NH_3}{17.0 \text{ g } NH_3}$

moles NH_3

moles H_2

$\dfrac{3 \text{ moles } H_2}{2 \text{ moles } NH_3}$

kilograms

kilograms

molecular

$\dfrac{\text{kg } H_2}{10^3 \text{ g } H_2}$

dollars

as the only dimension. The answer, then, is

$$? \text{ dollars} = 6000 \text{ g NH}_3 \times \frac{\text{mole NH}_3}{17.0 \text{ g NH}_3} \times \frac{3 \text{ moles H}_2}{2 \text{ moles NH}_3}$$

$$\times \frac{2.0 \text{ g H}_2}{\text{mole H}_2} \times \frac{\text{kg H}_2}{10^3 \text{ g H}_2} \times \frac{\$0.01}{\text{kg H}_2}$$

$$= \$0.011.$$

(Notice that there are two significant figures in the answer. The price of the H_2 is not a measured quantity and has no uncertainty.)

Let us take a look now at question 6 in the pretest. This one really has everything in it except the laboratory sink. However, if you start out by writing all the conversion factors in the problem, it won't be that difficult.

You are asked how many avoirdupois ounces (16 oz/lb) of Ag can be prepared by the reaction of $20 worth of Ag_2S ore. The ore is 54.0 wt % Ag_2S and costs $5/kg. The reaction used is

$$Ag_2S + Zn + 2H_2O = 2Ag + H_2S + Zn(OH)_2.$$

First write down all the conversion factors in the problem.

$$\frac{16 \text{ oz Ag}}{\qquad} \qquad \frac{54.0 \text{ g Ag}_2S}{\qquad}$$

 lb Ag, 100 g ore

$$\frac{\$5.00}{\qquad}$$

kg ore

Ag_2S

$$\frac{2 \text{ moles Ag}}{\text{mole Ag}_2S}$$

The two substances mentioned in the problem are Ag and _____, so the "bridge" conversion factor from the balanced equation will

be _____. Since the problem deals with weights of both substances, it is probable that you will also need their molecular weight conversion factors.

247.9

$$\frac{107.9 \text{ g Ag}}{\qquad} \qquad \frac{\qquad \text{ g Ag}_2S}{\text{mole Ag}_2S}$$

mole Ag

Now you are ready to start. The question asked is

How many ounces Ag, $20

"_____?" The information given is ____.

oz Ag, $20

? ____ = ____

dollars

The unwanted dimension that you wish to cancel is _____. In your collection of conversion factors, there is only one which

$$\frac{\text{kg ore}}{\$5}$$

contains this dimension, so you must multiply by _____. The

kilograms ore

unwanted dimension now is _____, and the only

conversion factor left having this substance is _____. As you see, one weight of ore is given in kilograms, and the other is in grams. You must convert from one to the other by multiplying by

_____.

Now you can multiply by the conversion factor given in the

problem, _____, and the new unwanted dimension is

_____. In order to be able to use the "bridge" conversion

factor, you must have _____ Ag_2S, which you can get if you

multiply by the molecular weight conversion factor, _____ (in its inverted form).

Now you can multiply by the "bridge" conversion factor,

_____. The unwanted dimension now is in the substance of the question asked, and all that you must do is get to the weight dimension ounces. First you can get to grams Ag by multiplying by

_____, the atomic weight of Ag. Then you can get from

grams Ag to pounds Ag by multiplying by _____. Finally, you can get from pounds Ag to ounces Ag by multiplying by the

conversion factor _____. The only dimension left now is

_____, and that is the dimension that the answer must have. Therefore, the answer must be

$$? \text{ oz Ag} = \$20 \times \frac{\text{kg ore}}{\$5} \times \frac{10^3 \text{ g ore}}{\text{kg ore}} \times \frac{54.0 \text{ g Ag}_2\text{S}}{100 \text{ g ore}}$$

$$\times \frac{\text{mole Ag}_2\text{S}}{247.9 \text{ g Ag}_2\text{S}} \times \frac{2 \text{ moles Ag}}{\text{mole Ag}_2\text{S}} \times \frac{107.9 \text{ g Ag}}{\text{mole Ag}}$$

$$\times \frac{\text{lb Ag}}{454 \text{ g Ag}} \times \frac{16 \text{ oz Ag}}{\text{lb Ag}}$$

$$= 66.3 \text{ oz Ag}$$

If you were able to work your way through that last example, you can do anything!

Occasionally you may have a problem which asks for the total cost of the reactants in a chemical equation. What this amounts to is more than one problem. You simply determine the cost of each reactant separately and then add them all up. An example of this type appears in the problem set at the end of this chapter.

SECTION E Consecutive Reactions

Sometimes you will have a problem where the chemical reaction occurs in a series of steps, one following the other. This is called a series of consecutive reactions. You may be asked about a product

---

$$\frac{54.0 \text{ g Ag}_2\text{S}}{100 \text{ g ore}}$$

$$\frac{10^3 \text{ g ore}}{\text{kg ore}}$$

$$\frac{54.0 \text{ g Ag}_2\text{S}}{100 \text{ g ore}}$$

grams Ag_2S

mole

$$\frac{\text{mole Ag}_2\text{S}}{247.9 \text{ g Ag}_2\text{S}}$$

$$\frac{2 \text{ moles Ag}}{\text{mole Ag}_2\text{S}}$$

$$\frac{107.9 \text{ g Ag}}{\text{mole Ag}}$$

$$\frac{\text{lb Ag}}{454 \text{ g Ag}}$$

$$\frac{16 \text{ oz Ag}}{\text{lb Ag}}$$

ounces Ag

in the last reaction and be given information about a reactant in the first. In such a case you must write balanced equations for all the steps and use more than a single "bridge." An example will make this very clear.

How many moles of $CaCO_3$ will be produced when 0.10 mole $Fe_2(CO_3)_3$ are reacted with an excess of HCl according to equation I to produce $FeCl_3$, H_2O, and CO_2, and then the CO_2 is reacted with $Ca(OH)_2$ according to equation II to produce $CaCO_3$?

(I) $6HCl + Fe_2(CO_3)_3 = 3CO_2 + 2FeCl_3 + 3H_2O$

(II) $CO_2 + Ca(OH)_2 = CaCO_3 + H_2O$

How many moles $CaCO_3$?	The question asked is, "_____" The
0.10 mole $Fe_2(CO_3)_3$	information given is _____.
0.10 mole $Fe_2(CO_3)_3$	? moles $CaCO_3$ = _____

You are already at moles, so you are ready for a "bridge" conversion factor. You therefore look for the substance which is

CO_2	common to both equations; it is ____. Your "bridge" must contain both moles $Fe_2(CO_3)_3$ and moles CO_2, and you can get such a
$\dfrac{3 \text{ moles } CO_2}{\text{mole } Fe_2(CO_3)_3}$	"bridge" from equation I; it is _____.
mole CO_2	Your unwanted dimension now is _____, and your answer
mole $CaCO_3$	must have the dimensions _____. You must look for a "bridge" which will relate these two. There is one that you can get
$\dfrac{\text{mole } CaCO_3}{\text{mole } CO_2}$	from equation II; it is _____. If you multiply by this second
mole $CaCO_3$	"bridge," the only dimension left is _____, which is the dimension of the answer.

$$? \text{ mole } CaCO_3 = 0.10 \text{ mole } Fe_2CO_3 \times \frac{3 \text{ moles } CO_2}{\text{mole } Fe_2CO_3}$$

$$\times \frac{\text{mole } CaCO_3}{\text{mole } CO_2}$$

$$= 0.30 \text{ mole } CaCO_3$$

This problem was essentially question 8 on the pretest, except that the amounts were changed and were all expressed in moles. In question 8 you will have to convert from grams to moles by

molecular   using the _____ weight conversion factors. Try to do it and see if you come up with the correct answer.

Notice that if you combine the two "bridge" conversion factors,

$$\frac{3 \text{ moles } CO_2}{\text{mole } Fe_2(CO_3)_3} \times \frac{\text{mole } CaCO_3}{\text{mole } CO_2},$$

mole CO_2 cancels. This leaves a new conversion factor,

$$\frac{3 \text{ moles CaCO}_3}{\text{mole Fe}_2(\text{CO}_3)_3}.$$

You could have gotten this combined "bridge" by simply adding the two equations together. However, equation II has to be multiplied through by 3 so that the CO_2's will cancel.

$$6HCl + Fe_2(CO_3)_3 = 3CO_2 + 2FeCl_3 + 3H_2O$$
$$3CO_2 + 3Ca(OH)_2 = 3CaCO_3 + 3H_2O$$
$$\overline{6HCl + Fe_2(CO_3)_3 + 3Ca(OH)_2 = 2FeCl_3 + 3CaCO_3 + 6H_2O}$$

This procedure of adding the equations together is preferable if the substance in which you are interested appears on the same side of two equations. An example of this would be the production of H_2 by the burning of C. The problem, for example, is to find how many grams of H_2 will be formed by burning 100 g C following the two consecutive reactions shown below.

$$C + H_2O = H_2 + CO$$

$$CO + H_2O = H_2 + CO_2$$

Notice that H_2 appears on the right-hand side of both equations. Therefore, you add them together. (No adjustment is necessary since the CO will cancel as the equations stand.) Adding them gives you the combined equation

$$C + \underline{\hspace{2cm}} = \underline{\hspace{2cm}} + CO_2.$$

The "bridge" relating C to H_2 from this combined equation is

$$\underline{\hspace{3cm}}.$$

The problem can now be solved in the normal way using this "bridge."

$$? \text{ g } H_2 = \underline{\hspace{2cm}}.$$

Next you multiply by $\underline{\hspace{2cm}}$. Next you multiply by $\underline{\hspace{2cm}}$.

Next you multiply by $\underline{\hspace{2cm}}$. The only dimension left is

$\underline{\hspace{2cm}}$, and so the answer must be $\underline{\hspace{2cm}}$.

You should be rather good now at doing these, so try the problem set which follows.

$2H_2O, \quad 2H_2$

$$\frac{2 \text{ moles } H_2}{\text{mole C}}$$

$$\frac{100 \text{ g C}}{} \quad \frac{\text{mole C}}{12.0 \text{ g C}}, \quad \frac{2 \text{ moles } H_2}{\text{mole C}}$$

$$\frac{2.0 \text{ g } H_2}{\text{mole } H_2}$$

grams H_2,   33 g H_2

PROBLEM SET

This is a very long problem set. To save yourself some time, don't bother to solve the problems for numerical answers; simply set them up. On the following pages you will find the setups and also the setups repeated without the dimensions which have cancelled.

Some of the problems are rather complicated. If you really think that you are an expert at the setups, you can try them; if they just annoy you, skip them. So that you know which problems they are, they are marked with an asterisk before the number. You don't have to bother to work out all the molecular weights; a table of those you will need is given below.

	g/mole		g/mole		g/mole		g/mole
Ag	107.9	CO	28.0	HNO_3	63.0	NaOH	40.0
Al	27.0	C_2H_5OH	46.0	H_2S	34.1	NO	30.0
$AlCl_3$	133.5	$C_6H_{12}O_6$	180.0	H_2SO_4	98.1	P	31.0
$BiCl_3$	315.5	Cl_2	71.0	KCl	74.6	S	32.1
Bi_2S_3	514.3	$H_2C_2O_4$	90.0	$KMnO_4$	158.0	SO_2	64.1
$Ca_3(PO_4)_2$	310.3	HCl	36.5	Na	23.0	SO_3	80.1
C	12.0						

The following twelve problems will use the balanced equations shown below. You will be told in the problem which equation is referred to.

(a) $2AlCl_3 + 3H_2SO_4 = Al_2(SO_4)_3 + 6HCl$

(b) $2Na + 2H_2O = 2NaOH + H_2$

(c) $2KMnO_4 + 5H_2C_2O_4 + 6HCl = 2MnCl_2 + 10CO_2 + 8H_2O + 2KCl$

(d) $11H_2O + 12CO_2 = C_{12}H_{22}O_{11} + 12O_2$ (photosynthesis)

(e) $2Al + 2NaOH + 2H_2O = 2NaAlO_2 + 3H_2$

(f) $2BiCl_3 + 3H_2S = Bi_2S_3 + 6HCl$

1. How many grams of H_2 can be obtained from the reaction of 4.0 lb Na according to reaction (b)?

2. If 16.0 ounces of Al are reacted with an excess of NaOH as in reaction (e), how many kilograms of H_2 are formed?

3. What is the maximum number of grams of Bi_2S_3 that can be obtained by the reaction of 1.10 lb $BiCl_3$ (f)?

*4. A 5.0 ton sample of Al requires how many kilograms of NaOH for complete reaction (e)?

5. How many tons of H_2SO_4 are required to react completely with 16.0 tons $AlCl_3$ (a)?

6. A sample of $KMnO_4$ weighing 7.20 oz will require how many ounces of $H_2C_2O_4$ (c)?

7. A 75 g sample of bauxite, an aluminum ore, contains 8.0 g Al. How many grams of H_2 can be prepared by reacting 425 g bauxite (e)?

8. How many grams of H_2SO_4 are there in 25.0 g of an H_2SO_4 solution, if 90.0 g of the solution are required to react with 52.0 g $AlCl_3$ (a)?

*9. A 6.0 oz sample of impure $KMnO_4$ contains only 90.0 g $KMnO_4$. How many grams of an 85% pure sample of $H_2C_2O_4$ will be needed to react with 14.0 oz of the impure $KMnO_4$ (c)?

10. How many pounds of an 85.0% pure sample of H_2S will be required to completely react with 4.0 lb $BiCl_3$ (f)?

11. What is the percent purity of an $H_2C_2O_4$ solution if 35.0 g of the solution produces 0.30 g KCl (c)?

*12. What is the percent purity of an Al ore if 4.0 kg of the ore produces 0.80 lb H_2 (e)?

The following sixteen problems will use the balanced equations shown below.

(g) $C_6H_{12}O_6 = 2C_2H_5OH + 2CO_2$

(h) $Ca_3(PO_4)_2 + 3SiO_2 + 5C = 3CaSiO_3 + 5CO + 2P$

(i) $3Ag + 4HNO_3 = 3AgNO_3 + NO + 2H_2O$

(j) $2KMnO_4 + 16HCl = 2MnCl_2 + 5Cl_2 + 8H_2O + 2KCl$

(k) (1) $S + O_2 = SO_2$

    (2) $2SO_2 + O_2 = 2SO_3$

    (3) $SO_3 + H_2O = H_2SO_4$

    (4) $H_2SO_4 + 2NaOH = Na_2SO_4 + 2H_2O$

13. Glucose, $C_6H_{12}O_6$, can be enzymatically reduced to alcohol, C_2H_5OH, according to reaction (g). If it takes 3.0 hours to produce 5.0 kg alcohol, how many days will it take to consume 2.0 tons sugar?

14. How many grams of Cl_2 can be produced by oxidizing 300 ml of an HCl solution which is 25.0 wt % HCl and has a density of 1.13 g/ml (j)?

*15. What is the density of Cl_2 gas in grams/liter if the Cl_2 gas produced when 31.6 g $KMnO_4$ is reacted with an excess of HCl [according to equation (j)] occupies 4.40 liters?

16. What is the percent purity of a sample of silver (Ag) ore if a 5.00 g sample of the ore will react with 11.0 ml of an HNO_3 solution which is 17.00 wt % HNO_3 and whose density is 1.10 g/ml (i)?

*17. If 1.00 mole of SO_2 gas will occupy 22.4 liters at a temperature of 0°C and a pressure of 1.0 atm, what will be the volume of SO_2 gas at this temperature and pressure which will be produced by the oxidation of 1.00 cubic meter of S whose density is 1.8 g/ml (k-1)?

18. How many milliliters of an H_2SO_4 solution whose density is 1.40 g/ml and which is 50 wt % H_2SO_4 will react with 35.0 ml of a 20.0 wt % solution of NaOH whose density is 0.90 g/ml (k-4)?

19. What is the density of the CO (in g/liter) if a 550 g sample of $Ca_3(PO_4)_2$ which is 60.0% pure gives 110 liters of the CO gas (h)?

20. If the price of glucose, $C_6H_{12}O_6$, is $0.15 per lb, what would it cost to prepare 150 kg of alcohol, C_2H_5OH (g)?

21. What is the cost of the coke, C, needed for the preparation of 4.0 tons of phosphorus, P, if the cost of coke is $15 per ton (h)?

22. What is the total raw material cost to produce 10.0 lb of P, if the cost of $Ca_3(PO_4)_2$ is $0.12 per pound, the cost of SiO_2 is $0.05 per kg, and the cost of C is $15 per ton (h)?

*23. How many dollars worth of Ag can be obtained from 3.0 cu yd of silver ore whose density is 8.0 g/ml if 2.0 lb of the ore can produce 2.2 liters of NO? The density of the NO is 1.34 g/liter, and the price of Ag is $6.00 per oz (16 oz/lb) (i)?

24. Sulfur, S, is oxidized to SO_3 according to equations (k-1) and (k-2). How many grams of SO_3 can be produced by the oxidization of 3.0 lb S?

*25. If 128 g of S are oxidized to SO_3 according to equations (k-1) and (k-2) and the SO_3 is reacted with water to form 3.0 liters of solution according to equation (k-3), what will be the wt % H_2SO_4 in the solution you have made? The density of the solution is 1.08 g/ml.

26. How many milliliters of a 30.0 wt % solution of NaOH will be required to react with the H_2SO_4 produced from reactions (k-1), (k-2), and (k-3) if you started with 16.05 g S (k-4)? Density of the NaOH solution is 1.33 g/ml.

*27. What is the percent purity of a sample of S if, when 35.0 g of the impure S is oxidized to SO_3 (k-1) and (k-2), and the SO_3 is dissolved in excess H_2O, this solution requires 500 ml of a solution of NaOH which contains 4.0 moles of NaOH per liter of NaOH solution (k-4) for complete reaction?

*28. How many hours would it take to produce $400 worth of alcohol, C_2H_5OH, if you can produce 5.0 kg alcohol in 3.0 hours and the price of glucose is $0.15/lb? The selling price of alcohol is twice the cost of its raw material, glucose ($C_6H_{12}O_6$) (g)

PROBLEM SET ANSWERS

The setups for the problems are shown below, and the numbers by which you will end up multiplying and dividing are given. Your answer is correct even if you do not have these numbers in exactly the same order as shown below. Simply check and see that all the numbers are present in both the numerators and denominators of your solution. If you can do any fifteen of these very complicated problems, you are doing fine and can proceed to the next chapter.

1. $? g\ H_2 = 4.0\ \cancel{lb\ Na} \times \dfrac{454\ \cancel{g\ Na}}{\cancel{lb\ Na}} \times \dfrac{\cancel{mole\ Na}}{23.0\ \cancel{g\ Na}}$

$\times \dfrac{\cancel{mole\ H_2}}{2\ \cancel{moles\ Na}} \times \dfrac{2.0\ g\ H_2}{\cancel{mole\ H_2}}$

$= 4.0 \times 454 \times \dfrac{1}{23.0} \times \dfrac{1}{2} \times 2.0\ g\ H_2$

2. $? \text{ kg H}_2 = 16.0 \text{ oz Al} \times \dfrac{\text{lb Al}}{16 \text{ oz Al}} \times \dfrac{454 \text{ g Al}}{\text{lb Al}} \times \dfrac{\text{mole Al}}{27.0 \text{ g Al}}$

$\times \dfrac{3 \text{ moles H}_2}{2 \text{ moles Al}} \times \dfrac{2.0 \text{ g H}_2}{\text{mole H}_2} \times \dfrac{\text{kg H}_2}{10^3 \text{ g H}_2}$

$= 16.0 \times \dfrac{1}{16} \times 454 \times \dfrac{1}{27.0} \times \dfrac{3}{2} \times 2.0 \times \dfrac{1}{10^3} \text{ kg H}_2$

3. $? \text{ g Bi}_2\text{S}_3 = 1.10 \text{ lb BiCl}_3 \times \dfrac{454 \text{ g BiCl}_3}{\text{lb BiCl}_3} \times \dfrac{\text{mole BiCl}_3}{315.5 \text{ g BiCl}_3}$

$\times \dfrac{\text{mole Bi}_2\text{S}_3}{2 \text{ moles BiCl}_3} \times \dfrac{514.3 \text{ g Bi}_2\text{S}_3}{\text{mole Bi}_2\text{S}_3}$

$= 1.10 \times 454 \times \dfrac{1}{315.5} \times \dfrac{1}{2} \times 514.3 \text{ g Bi}_2\text{S}_3$

4. $? \text{ kg NaOH} = 5.0 \text{ tons Al} \times \dfrac{2000 \text{ lb Al}}{\text{ton Al}} \times \dfrac{454 \text{ g Al}}{\text{lb Al}} \times \dfrac{\text{mole Al}}{27.0 \text{ g Al}}$

$\times \dfrac{2 \text{ moles NaOH}}{2 \text{ moles Al}} \times \dfrac{40.0 \text{ g NaOH}}{\text{mole NaOH}} \times \dfrac{\text{kg NaOH}}{10^3 \text{ g NaOH}}$

$= 5.0 \times 2000 \times 454 \times \dfrac{1}{27.0} \times \dfrac{2}{2} \times 40.0 \times \dfrac{1}{10^3} \text{ kg NaOH}$

5. $? \text{ tons H}_2\text{SO}_4 = 16.0 \text{ tons AlCl}_3 \times \dfrac{\text{ton-mole AlCl}_3}{133.5 \text{ tons AlCl}_3} \times \dfrac{3 \text{ ton-moles H}_2\text{SO}_4}{2 \text{ ton-moles AlCl}_3}$

$\times \dfrac{98.1 \text{ tons H}_2\text{SO}_4}{\text{ton-mole H}_2\text{SO}_4}$

$= 16.0 \times \dfrac{1}{133.5} \times \dfrac{3}{2} \times 98.1 \text{ tons H}_2\text{SO}_4$

(If you didn't use ton-mole, you will have both 2000 and 454 appearing in the numerator and denominator and cancelling.)

6. $? \text{ oz H}_2\text{C}_2\text{O}_4 = 7.20 \text{ oz KMnO}_4 \times \dfrac{\text{oz-mole KMnO}_4}{158.0 \text{ oz KMnO}_4} \times \dfrac{5 \text{ oz-moles H}_2\text{C}_2\text{O}_4}{2 \text{ oz-moles KMnO}_4}$

$\times \dfrac{90.0 \text{ oz H}_2\text{C}_2\text{O}_4}{\text{oz-mole H}_2\text{C}_2\text{O}_4}$

$= 7.20 \times \dfrac{1}{158.0} \times \dfrac{5}{2} \times 90.0 \text{ oz H}_2\text{C}_2\text{O}_4$

(If you didn't use ounce-moles, you will have 16 and 454 appearing in both the numerator and the denominator and cancelling.)

7. $? \text{ g H}_2 = 425 \text{ g bauxite} \times \dfrac{8.0 \text{ g Al}}{75 \text{ g bauxite}} \times \dfrac{\text{mole Al}}{27.0 \text{ g Al}} \times \dfrac{3 \text{ moles H}_2}{2 \text{ moles Al}}$

$\times \dfrac{2.0 \text{ g H}_2}{\text{mole H}_2}$

$= 425 \times \dfrac{8.0}{75} \times \dfrac{1}{27.0} \times \dfrac{3}{2} \times 2.0 \text{ g H}_2$

8. $? \text{ g } H_2SO_4 = 25.0 \text{ g soln.} \times \dfrac{52.0 \text{ g } AlCl_3}{90.0 \text{ g soln.}} \times \dfrac{\text{mole } AlCl_3}{133.5 \text{ g } AlCl_3}$

$$\times \dfrac{3 \text{ moles } H_2SO_4}{2 \text{ moles } AlCl_3} \times \dfrac{98.1 \text{ g } H_2SO_4}{\text{mole } H_2SO_4}$$

$$= 25.0 \times \dfrac{52.0}{90.0} \times \dfrac{1}{133.5} \times \dfrac{3}{2} \times 98.1 \text{ g } H_2SO_4$$

9. $? \text{ g imp. } H_2C_2O_4 = 14.0 \text{ oz imp. } KMnO_4 \times \dfrac{90.0 \text{ g } KMnO_4}{6.0 \text{ oz imp. } KMnO_4}$

$$\times \dfrac{\text{mole } KMnO_4}{158 \text{ g } KMnO_4} \times \dfrac{5 \text{ moles } H_2C_2O_4}{2 \text{ moles } KMnO_4}$$

$$\times \dfrac{90.0 \text{ g } H_2C_2O_4}{\text{mole } H_2C_2O_4} \times \dfrac{100 \text{ g imp. } H_2C_2O_4}{85 \text{ g } H_2C_2O_4}$$

$$= 14.0 \times \dfrac{90.0}{6.0} \times \dfrac{1}{158} \times \dfrac{5}{2} \times 90.0 \times \dfrac{100}{85} \text{ g imp. } H_2C_2O_4$$

10. $? \text{ lb imp. } H_2S = 4.0 \text{ lb } BiCl_3 \times \dfrac{\text{lb-mole } BiCl_3}{315.5 \text{ lb } BiCl_3} \times \dfrac{3 \text{ lb-moles } H_2S}{2 \text{ lb-moles } BiCl_3}$

$$\times \dfrac{34.1 \text{ lb } H_2S}{\text{lb-mole } H_2S} \times \dfrac{100 \text{ lb imp. } H_2S}{85.0 \text{ lb } H_2S}$$

$$= 4.0 \times \dfrac{1}{315.5} \times \dfrac{3}{2} \times 34.1 \times \dfrac{100}{85.0} \text{ lb imp. } H_2S$$

11. $? \text{ g } H_2C_2O_4 = 100 \text{ g soln.} \times \dfrac{0.30 \text{ g } KCl}{35.0 \text{ g soln.}} \times \dfrac{\text{mole } KCl}{74.6 \text{ g } KCl}$

$$\times \dfrac{5 \text{ moles } H_2C_2O_4}{2 \text{ moles } KCl} \times \dfrac{90.0 \text{ g } H_2C_2O_4}{\text{mole } H_2C_2O_4}$$

$$= 100 \times \dfrac{0.30}{35.0} \times \dfrac{1}{74.6} \times \dfrac{5}{2} \times 90.0\%$$

12. $? \text{ g } Al = 100 \text{ g ore} \times \dfrac{\text{kg ore}}{10^3 \text{ g ore}} \times \dfrac{0.80 \text{ lb } H_2}{4.0 \text{ kg ore}} \times \dfrac{454 \text{ g } H_2}{\text{lb } H_2}$

$$\times \dfrac{\text{mole } H_2}{2.0 \text{ g } H_2} \times \dfrac{2 \text{ moles } Al}{3 \text{ moles } H_2} \times \dfrac{27.0 \text{ g } Al}{\text{mole } Al}$$

$$= 100 \times \dfrac{1}{10^3} \times \dfrac{0.80}{4.0} \times 454 \times \dfrac{1}{2.0} \times \dfrac{2}{3} \times 27.0 \text{ g } Al$$

13. $? \text{ days} = 2.0 \text{ tons } C_6H_{12}O_6 \times \dfrac{2000 \text{ lb } C_6H_{12}O_6}{\text{ton } C_6H_{12}O_6} \times \dfrac{454 \text{ g } C_6H_{12}O_6}{\text{lb } C_6H_{12}O_6}$

$$\times \dfrac{\text{mole } C_6H_{12}O_6}{180.0 \text{ g } C_6H_{12}O_6} \times \dfrac{2 \text{ moles } C_2H_5OH}{\text{mole } C_6H_{12}O_6} \times \dfrac{46.0 \text{ g } C_2H_5OH}{\text{mole } C_2H_5OH}$$

$$\times \dfrac{\text{kg } C_2H_5OH}{10^3 \text{ g } C_2H_5OH} \times \dfrac{3.0 \text{ hr}}{5.0 \text{ kg } C_2H_5OH} \times \dfrac{\text{day}}{24 \text{ hr}}$$

$$= 2.0 \times 2000 \times 454 \times \dfrac{1}{180.0} \times 2 \times 46.0 \times \dfrac{1}{10^3} \times \dfrac{3}{5} \times \dfrac{1}{24} \text{ days}$$

14. $? \text{ g Cl}_2 = 300 \text{ ml soln.} \times \dfrac{1.13 \text{ g soln.}}{\text{ml soln.}} \times \dfrac{25.0 \text{ g HCl}}{100 \text{ g soln.}} \times \dfrac{\text{mole HCl}}{36.5 \text{ g HCl}}$

$\times \dfrac{5 \text{ moles Cl}_2}{16 \text{ moles HCl}} \times \dfrac{71.0 \text{ g Cl}_2}{\text{mole Cl}_2}$

$= 300 \times 1.13 \times \dfrac{25.0}{100} \times \dfrac{1}{36.5} \times \dfrac{5}{16} \times 71.0 \text{ g Cl}_2$

15. $\dfrac{? \text{ g Cl}_2}{\text{liter Cl}_2} = \dfrac{31.6 \text{ g KMnO}_4}{4.40 \text{ liter Cl}_2} \times \dfrac{\text{mole KMnO}_4}{158 \text{ g KMnO}_4} \times \dfrac{5 \text{ moles Cl}_2}{2 \text{ moles KMnO}_4}$

$\times \dfrac{71.0 \text{ g Cl}_2}{\text{mole Cl}_2}$

$= \dfrac{31.6}{4.40} \times \dfrac{1}{158} \times \dfrac{5}{2} \times 71.0 \dfrac{\text{g Cl}_2}{\text{liter Cl}_2}$

16. $? \text{ g Ag} = 100 \text{ g imp.} \times \dfrac{11.0 \text{ ml soln.}}{5.00 \text{ g imp.}} \times \dfrac{1.10 \text{ g soln.}}{\text{ml soln.}} \times \dfrac{17.00 \text{ g HNO}_3}{100 \text{ g soln.}}$

$\times \dfrac{\text{mole HNO}_3}{63.0 \text{ g HNO}_3} \times \dfrac{3 \text{ moles Ag}}{4 \text{ moles HNO}_3} \times \dfrac{107.9 \text{ g Ag}}{\text{mole Ag}}$

$= 100 \times \dfrac{11.0}{5.00} \times 1.10 \times \dfrac{17.00}{100} \times \dfrac{1}{63.0} \times \dfrac{3}{4} \times 107.9 \text{ g Ag}$

17. $? \text{ liters SO}_2 = 1.00 \text{ m}^3 \text{ S} \times \dfrac{\text{cm}^3 \text{ S}}{(10^{-2})^3 \text{ m}^3 \text{ S}} \times \dfrac{\text{ml S}}{\text{cm}^3 \text{ S}} \times \dfrac{1.8 \text{ g S}}{\text{ml S}}$

$\times \dfrac{\text{mole S}}{32.1 \text{ g S}} \times \dfrac{\text{mole SO}_2}{\text{mole S}} \times \dfrac{22.4 \text{ liters SO}_2}{1.00 \text{ mole SO}_2}$

$= 1.00 \times \dfrac{1}{10^{-6}} \times 1 \times 1.8 \times \dfrac{1}{32.1} \times 1 \times \dfrac{22.4}{1.0} \text{ liters SO}_2$

18. $? \text{ ml H}_2\text{SO}_4 \text{ soln.}$

$= 35.0 \text{ ml NaOH soln.} \times \dfrac{0.90 \text{ g NaOH soln.}}{\text{ml NaOH soln.}}$

$\times \dfrac{20.0 \text{ g NaOH}}{100 \text{ g NaOH soln.}} \times \dfrac{\text{mole NaOH}}{40.0 \text{ g NaOH}}$

$\times \dfrac{\text{mole H}_2\text{SO}_4}{2 \text{ moles NaOH}} \times \dfrac{98.1 \text{ g H}_2\text{SO}_4}{\text{mole H}_2\text{SO}_4}$

$\times \dfrac{100 \text{ g H}_2\text{SO}_4 \text{ soln.}}{50 \text{ g H}_2\text{SO}_4} \times \dfrac{\text{ml H}_2\text{SO}_4 \text{ soln.}}{1.40 \text{ g H}_2\text{SO}_4 \text{ soln.}}$

$= 35.0 \times 0.90 \times \dfrac{20.0}{100} \times \dfrac{1}{40.0} \times \dfrac{1}{2} \times 98.1 \times \dfrac{100}{50} \times \dfrac{1 \text{ ml H}_2\text{SO}_4 \text{ soln.}}{1.40}$

(In this type of problem you must be especially careful not to confuse the two different solutions. Write the complete dimension to indicate which solution it is.)

19. $\dfrac{?\ \text{g CO}}{\text{liter CO}} = \dfrac{550\ \text{g imp.}}{110\ \text{liters CO}} \times \dfrac{60.0\ \text{g Ca}_3(\text{PO}_4)_2}{100\ \text{g imp.}} \times \dfrac{\text{mole Ca}_3(\text{PO}_4)_2}{310.3\ \text{g Ca}_3(\text{PO}_4)_2}$

$\times \dfrac{5\ \text{moles CO}}{\text{mole Ca}_3(\text{PO}_4)_2} \times \dfrac{28.0\ \text{g CO}}{\text{mole CO}}$

$= \dfrac{550}{110} \times \dfrac{60.0}{100} \times \dfrac{1}{310.3} \times 5 \times 28.0\ \dfrac{\text{g CO}}{\text{liter CO}}$

20. $?\ \text{dollars} = 150\ \text{kg C}_2\text{H}_5\text{OH} \times \dfrac{10^3\ \text{g C}_2\text{H}_5\text{OH}}{\text{kg C}_2\text{H}_5\text{OH}} \times \dfrac{\text{mole C}_2\text{H}_5\text{OH}}{46.0\ \text{g C}_2\text{H}_5\text{OH}}$

$\times \dfrac{\text{mole C}_6\text{H}_{12}\text{O}_6}{2\ \text{moles C}_2\text{H}_5\text{OH}} \times \dfrac{180.0\ \text{g C}_6\text{H}_{12}\text{O}_6}{\text{mole C}_6\text{H}_{12}\text{O}_6} \times \dfrac{\text{lb C}_6\text{H}_{12}\text{O}_6}{454\ \text{g C}_6\text{H}_{12}\text{O}_6}$

$\times \dfrac{\$0.15}{\text{lb C}_6\text{H}_{12}\text{O}_6}$

$= 150 \times 10^3 \times \dfrac{1}{46.0} \times \dfrac{1}{2} \times 180.0 \times \dfrac{1}{454} \times \$0.15$

21. $?\ \text{dollars} = 4.0\ \text{tons P} \times \dfrac{\text{ton-mole P}}{31.0\ \text{tons P}} \times \dfrac{5\ \text{ton-moles C}}{2\ \text{ton-moles P}}$

$\times \dfrac{12.0\ \text{tons C}}{\text{ton-mole C}} \times \dfrac{\$15}{\text{ton C}}$

$= 4.0 \times \dfrac{1}{31.0} \times \dfrac{5}{2} \times 12.0 \times \$15.$

(This problem is much easier to solve by using ton-moles. If you didn't, you will have 2000 and 454 cancelling.)

22. This is really three separate problems. You must calculate the cost of $\text{Ca}_3(\text{PO}_4)_2$, then the cost of SiO_2, and finally the cost of C. Then, when you have done that, you add all the costs up. Below only the separate cost calculations are shown. Since we are not solving to a final numerical answer, it isn't possible to add them.

$?\ \text{dollars} = 10.0\ \text{lb P} \times \dfrac{\text{lb-mole P}}{31.0\ \text{lb P}} \times \dfrac{\text{lb-mole Ca}_3(\text{PO}_4)_2}{2\ \text{lb-moles P}}$

$\times \dfrac{310.3\ \text{lb Ca}_3(\text{PO}_4)_2}{\text{lb-mole Ca}_3(\text{PO}_4)_2} \times \dfrac{\$0.12}{\text{lb Ca}_3(\text{PO}_4)_2}$

$= 10.0 \times \dfrac{1}{31.0} \times \dfrac{1}{2} \times 310.3 \times \$0.12$

$?\ \text{dollars} = 10.0\ \text{lb P} \times \dfrac{\text{kg P}}{2.2\ \text{lb P}} \times \dfrac{\text{kg-mole P}}{31.0\ \text{kg P}} \times \dfrac{3\ \text{kg-moles SiO}_2}{2\ \text{kg-moles P}}$

$\times \dfrac{60.1\ \text{kg SiO}_2}{\text{kg-mole SiO}_2} \times \dfrac{\$0.05}{\text{kg SiO}_2}$

$= 10.0 \times \dfrac{1}{2.2} \times \dfrac{1}{31.0} \times \dfrac{3}{2} \times 60.1 \times \$0.05$

$$? \text{ dollars} = 10.0 \, \cancel{lb \, P} \times \frac{ton \, P}{2000 \, \cancel{lb \, P}} \times \frac{ton\text{-}mole \, P}{31.0 \, \cancel{tons \, P}} \times \frac{5 \, ton\text{-}moles \, C}{2 \, \cancel{ton\text{-}moles \, P}}$$

$$\times \frac{12.0 \, \cancel{tons \, C}}{\cancel{ton\text{-}mole \, C}} \times \frac{\$15}{\cancel{ton \, C}}$$

$$= 10.0 \times \frac{1}{2000} \times \frac{1}{31.0} \times \frac{5}{2} \times 12.0 \times \$15$$

(Notice that to simplify these, the conversion was immediately from pounds P to whatever weight unit the cost was expressed in, pounds, kilograms, or tons. Then the "bridge" was expressed in pound-moles, kilogram-moles, and ton-moles.

23. This is the type of complex problem where you should always start by writing all the conversion factors given in the problem. Here they are.

$$\frac{8.0 \text{ g ore}}{\text{ml ore}} ; \quad \frac{2.0 \text{ lb ore}}{2.2 \text{ liters NO}} ; \quad \frac{1.8 \text{ g NO}}{\text{liter NO}} ; \quad \frac{\$2.05}{\text{oz Ag}} ; \quad \frac{16 \text{ oz Ag}}{\text{lb Ag}}$$

$$? \text{ dollars} = 3.0 \, \cancel{yd^3 \, ore} \times \frac{(36)^3 \, \cancel{in.^3 \, ore}}{\cancel{yd^3 \, ore}} \times \frac{(2.54)^3 \, \cancel{cm^3 \, ore}}{\cancel{in.^3 \, ore}} \times \frac{\cancel{ml \, ore}}{\cancel{cm^3 \, ore}}$$

$$\times \frac{8.0 \, \cancel{g \, ore}}{\cancel{ml \, ore}} \times \frac{\cancel{lb \, ore}}{454 \, \cancel{g \, ore}} \times \frac{2.2 \, \cancel{liter \, NO}}{2.0 \, \cancel{lb \, ore}} \times \frac{1.34 \, \cancel{g \, NO}}{\cancel{liter \, NO}}$$

$$\times \frac{\cancel{mole \, NO}}{30.0 \, \cancel{g \, NO}} \times \frac{3 \, \cancel{moles \, Ag}}{\cancel{mole \, NO}} \times \frac{107.9 \, \cancel{g \, Ag}}{\cancel{mole \, Ag}} \times \frac{\cancel{lb \, Ag}}{454 \, \cancel{g \, Ag}}$$

$$\times \frac{16 \, \cancel{oz \, Ag}}{\cancel{lb \, Ag}} \times \frac{\$6.00}{\cancel{oz \, Ag}}$$

$$= 3.0 \times (36)^3 \times (2.54)^3 \times 1 \times 8.0 \times \frac{1}{454} \times \frac{2.2}{2.0} \times 1.34 \times \frac{1}{30.0}$$

$$\times 3 \times 107.9 \times \frac{1}{454} \times 16 \times \$6.00$$

24. Since the substances in the problem do not appear on the same side of the equations, we can use two "bridges."

$$? \text{ g } SO_3 = 3.0 \, \cancel{lb \, S} \times \frac{454 \, \cancel{g \, S}}{\cancel{lb \, S}} \times \frac{\cancel{mole \, S}}{32.1 \, \cancel{g \, S}}$$

$$\times \frac{\cancel{mole \, SO_2}}{\cancel{mole \, S}} \times \frac{2 \, \cancel{moles \, SO_3}}{2 \, \cancel{moles \, SO_2}} \times \frac{80.1 \text{ g } SO_3}{\cancel{mole \, SO_3}}$$

$$= 3.0 \times 454 \times \frac{1}{32.1} \times 1 \times \frac{2}{2} \times 80.1 \text{ g } SO_3$$

25. $? \text{ g H}_2\text{SO}_4 = 100 \text{ g soln.} \times \dfrac{\text{ml soln.}}{1.08 \text{ g soln.}} \times \dfrac{10^{-3} \text{ liter soln.}}{\text{ml soln.}}$

$$\times \dfrac{128 \text{ g S}}{3.0 \text{ liter soln.}} \times \dfrac{\text{mole S}}{32.1 \text{ g S}} \times \dfrac{\text{mole SO}_2}{\text{mole S}}$$

$$\times \dfrac{2 \text{ moles SO}_3}{2 \text{ moles SO}_2} \times \dfrac{\text{mole H}_2\text{SO}_4}{\text{mole SO}_3} \times \dfrac{98.1 \text{ g H}_2\text{SO}_4}{\text{mole H}_2\text{SO}_4}$$

$$= 100 \times \dfrac{1}{1.08} \times 10^{-3} \times \dfrac{128}{3.0} \times \dfrac{1}{32.1}$$

$$\times 1 \times \dfrac{2}{2} \times 1 \times 98.1\%$$

26. $? \text{ ml NaOH soln.} = 16.05 \text{ g S} \times \dfrac{\text{mole S}}{32.1 \text{ g S}} \times \dfrac{\text{mole SO}_2}{\text{mole S}} \times \dfrac{2 \text{ moles SO}_3}{2 \text{ moles SO}_2}$

$$\times \dfrac{\text{mole H}_2\text{SO}_4}{\text{mole SO}_3} \times \dfrac{2 \text{ moles NaOH}}{\text{mole H}_2\text{SO}_4} \times \dfrac{40.0 \text{ g NaOH}}{\text{mole NaOH}}$$

$$\times \dfrac{100 \text{ g NaOH soln.}}{30.0 \text{ g NaOH}} \times \dfrac{\text{ml NaOH soln.}}{1.33 \text{ g NaOH soln.}}$$

$$16.05 \times \dfrac{1}{32.1} \times 1 \times \dfrac{2}{2} \times 1 \times 2 \times 40.0 \times \dfrac{100}{30.0}$$

$$\times \dfrac{1}{1.33} \text{ ml soln.}$$

27. $? \text{ g S} = 100 \text{ g imp.} \times \dfrac{500 \text{ ml NaOH soln.}}{35.0 \text{ g imp.}} \times \dfrac{10^{-3} \text{ liter NaOH soln.}}{\text{ml NaOH soln.}}$

$$\times \dfrac{4.0 \text{ moles NaOH}}{\text{liter NaOH soln.}} \times \dfrac{\text{mole H}_2\text{SO}_4}{2 \text{ moles NaOH}} \times \dfrac{\text{mole SO}_3}{\text{mole H}_2\text{SO}_4}$$

$$\times \dfrac{2 \text{ moles SO}_2}{2 \text{ moles SO}_3} \times \dfrac{\text{mole S}}{\text{mole SO}_2} \times \dfrac{32.1 \text{ g S}}{\text{mole S}}$$

$$= 100 \times \dfrac{500}{35.0} \times 10^{-3} \times 4.0 \times \dfrac{1}{2} \times 1 \times \dfrac{2}{2} \times 1 \times 32.1 \text{g S}$$

28. You must start by saying, "If the alcohol is worth $400, and it is twice as expensive as the sugar, then the sugar must have cost $200." Then you can set up the problem.

$$? \text{ hr} = \$200 \times \dfrac{\text{lb C}_6\text{H}_{12}\text{O}_6}{\$0.15} \times \dfrac{454 \text{ g C}_6\text{H}_{12}\text{O}_6}{\text{lb C}_6\text{H}_{12}\text{O}_6} \times \dfrac{\text{mole C}_6\text{H}_{12}\text{O}_6}{180 \text{ g C}_6\text{H}_{12}\text{O}_6}$$

$$\times \dfrac{2 \text{ moles C}_2\text{H}_5\text{OH}}{\text{mole C}_6\text{H}_{12}\text{O}_6} \times \dfrac{46.0 \text{ g C}_2\text{H}_5\text{OH}}{\text{mole C}_2\text{H}_5\text{OH}} \times \dfrac{\text{kg C}_2\text{H}_5\text{OH}}{10^3 \text{ g C}_2\text{H}_5\text{OH}}$$

$$\times \dfrac{3 \text{ hr}}{5.0 \text{ kg C}_2\text{H}_5\text{OH}}$$

$$= 200 \times \dfrac{1}{0.15} \times 454 \times \dfrac{1}{180} \times 2 \times 46.0 \times \dfrac{1}{10^3} \times \dfrac{3}{5.0} \text{ hr}$$

Chapter 10
Gas Laws

PRETEST

If you have had some previous experience with gas law calculations you may want to try this pretest. You should be able to complete it in 40 minutes.

1. How many atmospheres pressure is 380 mm Hg pressure?

2. What is the temperature in °K when the temperature is 30°C?

3. If 2.0 liters of hydrogen gas at 27°C and a pressure of 6.0 atm has its pressure increased to 9.0 atm and its temperature lowered to −23°C, what will be its new volume?

4. If 300 ml of oxygen gas at a temperature of 350°K and a pressure of 780 torr has its volume reduced to 0.150 liter and its pressure raised to 3.0 atm, what will its new temperature be?

5. What must the pressure be in mm Hg to have a gas at 18°C occupy the same volume as it would at STP?

6. How many moles of an ideal gas will occupy 2.0 liters at a temperature of 27°C and a pressure of 0.5 atm?

7. When 16.0 g of He gas is contained in a 2.0 liter vessel at a temperature of 100°K, what will its pressure be in atmospheres?

8. What is the density of O_2 gas in grams per liter at 27°C and 152 cm Hg pressure?

9. What is the molecular weight of an unknown gas if 3.0 liters at a pressure of 1.0 atm and a temperature of 27°C weighs 2.4 g?

10. What is the molecular weight of a gas whose density is 4.50 g/liter at −73°C and a pressure of 380 mm Hg?

11. What will be the pressure in a 4.0 liter container which is at 27°C and contains 2.0 moles of H_2 and 5.0 moles of He gases?

12. What is the molecular weight of a gas if 70.0 ml collected over water at a temperature of 27°C and a total pressure of 756.5 mm Hg weighs 0.080 g when dry? The vapor pressure of water at this temperature is 26.5 mm Hg.

13. A sample of N_2 gas diffuses through a small hole at a rate of 2.6 ml per minute. At what rate would Cl_2 gas diffuse through the same hole under the same conditions?

14. If 20.0 g of an unknown gas diffuses through a pinhole in the same time that it takes 40.0 g of CH_4 to diffuse through the same pinhole under the same conditions, what is the molecular weight of the unknown gas?

PRETEST ANSWERS

All the questions are worked out in the body of the chapter. If you were able to do questions 1 and 2 you can skip Section A. Questions 3, 4, and 5 are covered in Section B. If you got them right you may skip that section, and you may skip Section C if you got numbers 6, 7, and 8. Section D contains the solutions to 9 and 10. If you were able to do numbers 11 and 12 you may skip Section E. Section F contains the subject of questions 13 and 14.

1. 0.500 atm

2. 303°K

3. 1.1 liters

4. 238°C (511°K)

5. 810 mm Hg

6. 0.04 mole

7. 16 atm

8. 2.60 g/liter

9. 20 g/mole

10. 148 g/mole

11. 4.3×10^1 atm

12. 29 g/mole

13. 1.6 ml/min

14. 64 g/mole

Most gases, particularly at high temperature and low pressure, have the same relationship between the temperature, pressure, volume, and number of moles of the gas. This relationship can be expressed mathematically as

$$PV = nRT.$$

This expression is called an equation of state or the *ideal gas equation*. It says that the product of the pressure (P) times volume (V) equals the number of moles (n) times a constant (R) times the absolute temperature (T).

For example, if you have 1 mole of a gas at 1 atmosphere pressure and 273°K (0°C), it will occupy 22.4 liters. Putting these values in the ideal gas equation, you have

$$(1.0 \text{ atm}) (22.4 \text{ liters}) = (1.0 \text{ mole}) (R) (273°\text{K}),$$

and, solving for R by cross-multiplying

$$\frac{(1.0 \text{ atm}) (22.4 \text{ liters})}{(1.0 \text{ mole}) (273°\text{K})} = R$$

$$\frac{0.0820 \text{ (liter) (atm)}}{\text{(mole) (°K)}} = R.$$

R is, of course, a conversion factor, but it has an odd set of dimensions. There are _____ different units. However, density has two different units, _____; and molecular weight has two different units _____. When you get to Chapter 16, you will have the molal boiling and freezing point constants, which are conversion factors with three different units.

The numerical value for a conversion factor depends on which units you use. For example, there are 5280 feet per mile or 63,360 inches per mile. We will try to be consistent and use only the units liters, atmospheres, moles, and °K for R, and then you only have to remember one number, 0.0820.

SECTION A Gas Dimensions

The ideal gas equation only holds true if the temperature is expressed in °K. This is also called the absolute temperature. It is simple to change from °C to °K: you simply add 273.16 to the temperature in °C and you have the temperature in °K.

$$°\text{K} = 273.16 + °\text{C}$$

We will use just 273 (3 significant figures). Thus a temperature of 100°C would be equal to _____°K; a temperature of 533°K would be equal to _____°C; and a temperature of -10°C would be equal to _____°K.

Margin answers:

four

$\dfrac{\text{g}}{\text{ml}}$

$\dfrac{\text{g}}{\text{mole}}$

373

260

263

There are many ways of expressing pressure. When you talk

about the air pressure in the tires on your car, you use _____ per   pounds
inch^2 (p.s.i.). In countries using metric units they talk about
kilograms per meter^2. In most scientific work there also are two
other units used. One is millimeters Hg, or "torrs," which represent
the height of a column of mercury that the pressure can support.
The other unit used is *atmospheres* (atm); an atmosphere is the
average barometric pressure at sea level.

The conversion factors needed to get from one unit to another
are

$$\frac{760 \text{ mm Hg}}{\text{atm}}, \quad \frac{760 \text{ torr}}{\text{atm}}, \quad \frac{14.7 \text{ p.s.i.}}{\text{atm}}, \quad \frac{76 \text{ cm Hg}}{\text{atm}}.$$

Using the numerical value of R as 0.0820, you will always want

to express your pressure in _____. Thus, a pressure of   atmospheres

1520 torr will equal _____ atm, and a pressure of 380 mm Hg will   2.000

equal _____ atm.   0.500

SECTION B Change of Conditions Problems

There are really only two types of gas-law problems. We will examine the simplest first. You have a sample of gas at certain conditions of temperature, pressure, and volume (called its variables or parameters) and you change one or two of the variables and want to know what will be the value of the third variable after the change. These problems are so easy that we won't use dimensional analysis.

The reason that this is so simple is that since you are talking about the same sample of gas, the number of moles (n) will be the same both before and after the change; that is, n is a constant. R is always a constant, and so you can write the ideal gas equation as

$$\frac{PV}{T} = nR = \text{constant.}$$

Now consider the gas under its initial set of conditions. The pressure is P', the volume is V', and the temperature is T'. You can then write

nR $\qquad$
$$\frac{P'V'}{T'} = \underline{\quad}.$$

If you change the variables so that the pressure is P'', the volume is V'', and the temperature is T'', you can write

nR $\qquad$
$$\frac{P''V''}{T''} = \underline{\quad}.$$

Since both of these expressions equal the same nR, they must equal each other.

$$\frac{P'V'}{T'} = \frac{P''V''}{T''}$$

In a change of conditions problem, you will have the initial conditions, P', V', and T', all given. You will also have two of the final variables given and will only have to calculate the third. It is very simple. However, just so that you don't get confused, you will do yourself a big favor if you take the time to write a small table.

$P' =$ $\qquad\qquad$ $P'' =$
$V' =$ $\qquad\qquad$ $V'' =$
$T' =$ $\qquad\qquad$ $T'' =$

Fill in all the values given in the problem, using a question mark for the unknown parameter. Be very careful that the temperature is expressed in °K. You cannot work in any other units, although pressure and volume can be in any units so long as they are the same for the initial as for the final condition. (All conversion factors will cancel that way.) Once you have your table set up, you can then put the values in the expression

$$\frac{P'V'}{T'} = \underline{\hspace{2cm}}$$

| $\dfrac{P''V''}{T''}$ |

and solve for the unknown final parameter.

Question 3 in the pretest is a simple example of this type of problem. If 2.0 liters of hydrogen gas at 27°C and a pressure of 6.0 atm has its pressure increased to 9.0 atm and its temperature lowered to -23°C, what will be its new volume?

First you set up your table.

$P' = 6.0$ atm $P'' = 9.0$ atm

$V' = 2.0$ liters $V'' = ?$ liters

$T' = 300$°K $T'' = \underline{\hspace{1.5cm}}$

| 250°K |

Then you have

$$\frac{P'V'}{T'} = \frac{P''V''}{\underline{\hspace{1cm}}} ,$$

| T'' |

and, putting in the values from the table, you have

$$\frac{(6.0 \text{ atm}) (2.0 \text{ liters})}{300°\text{K}} = \frac{(9.0 \text{ atm}) (? \text{ liters})}{\underline{\hspace{1.5cm}}} .$$

| 250°K |

Cross-multiplying gives you

$$\frac{(6.0 \cancel{\text{ atm}}) (2.0 \text{ liters}) (\underline{\hspace{1cm}})}{(9.0 \cancel{\text{ atm}}) (\underline{\hspace{1cm}})} = ? \text{ liters}$$

| 250°K |
| 300°K |

$$\left(\frac{6.0}{9.0}\right)\left(\frac{250}{300}\right)(2.0 \text{ liters}) = 1.1 \text{ liters} = ? \text{ liters}.$$

Perhaps now is a good time to learn an algebraic trick called "cross-multiplication." If you know this already, you can skip the next short section.

Cross-multiplication

When you have any equality (equation) it is possible to multiply or divide both sides of the equality and still have an equality. An example will prove this to you. Consider

$$\frac{6}{8} = \frac{3}{4} \quad \text{or} \quad 0.75 = 0.75.$$

If you multiply both sides by 2 you have

$$\frac{12}{8} = \frac{6}{4} \quad \text{or} \quad 1.5 = 1.5; \quad \text{(Both sides are still equal.)}$$

Or, if you divide both sides by 3, you have

$$\frac{6}{24} = \frac{3}{12} \quad \text{or} \quad 0.25 = 0.25. \quad \text{(Both sides are still equal.)}$$

Let us do this now with letters rather than numbers. Consider

$$ax = by.$$

We will divide both sides by a,

$$\frac{ax}{a} = \frac{by}{a}.$$

The a's on the left cancel, leaving

$$x = \frac{by}{a}.$$

In effect, what you have done is to move the a from the numerator on the left to the denominator on the right. You have "cross-multiplied" it.

$$\text{ⓐ}x = \frac{by}{\text{ⓐ}}$$

Consider another equality,

$$\frac{x}{a} = by.$$

If you multiply both sides by a, then

$$\frac{ax}{\text{ⓐ}} = aby$$

The a's on the left cancel, leaving

$$x = aby.$$

In effect, what you have done is to move the a from the denominator on the left to the numerator on the right. You have "cross-multiplied."

$$\frac{x}{\text{ⓐ}} = \text{ⓐ}by$$

Consequently, if you ever want to solve an equation for just one of its parts, all that you have to do is to "cross-multiply" in such a way as to have the component you wish to solve for alone in the numerator on one side of the equation. To solve the preceding example, we cross-multiplied as follows.

$$\frac{(6.0 \text{ atm}) (2.0 \text{ liters})}{300°K} = \frac{(9.0 \text{ atm}) (? \text{ liters})}{250°K}$$

or

$$\frac{(6.0 \text{ atm}) (250°K) (2.0 \text{ liters})}{(9.0 \text{ atm}) (300°K)} = ? \text{ liters}$$

As you can see from the above, any units for pressure and volume are fine as long as they are the same for the final state of the gas as for the initial. The temperature, however, must be in °K. Question 4 on the pretest illustrates this point.

If 300 ml of oxygen gas at a temperature of 350°K and a pressure of 780 torr has its volume reduced to 0.150 liter and its pressure raised to 3.0 atm, what will its temperature be?

First you must set up your table with everything in the same units.

$P' = 780$ torr $P'' =$ _____ torr   2280

$V' = 300$ ml $V'' =$ ____ ml   150

$T' = 350°$K $T'' = ?$ °___   K

Then you have the relationship

$$\frac{P'V'}{T'} = \text{\_\_\_\_\_}.$$ $\dfrac{P''V''}{T''}$

Putting in the values from the table (and since the dimensions all match, you needn't bother to put them in),

$$\frac{(780)(300)}{350} = \frac{\text{_____}}{?\,°\text{K}}.$$ (2280)(150)

Cross-multiplying gives you

$$?\,°\text{K} = \frac{(350)\text{_____}}{(780)(300)}$$ (2280)(150)

$$= \text{\_\_\_}°\text{K}.$$ 511

To get this temperature to °C, you must _____ 273.   subtract

$$?\,°\text{C} = \text{\_\_\_}°\text{C}$$ 238

For convenience chemists have selected a temperature and a pressure that they call either standard conditions (SC) or standard temperature and pressure (STP). This is 1.00 atm and

273°K (__°C). You may recall that this is the temperature and   0
pressure we used to get a value for R. At STP, then, 1.0 mole of a

gas will occupy _____ liters.   22.4
Question 5 on the pretest refers to these conditions. What must the pressure be in millimeters Hg to have a gas at 18°C occupy the same volume as it would at STP? This problem is a little strange because there is no volume given. You simply know that it is the

same in the initial state as in the final state. You can assume that they are both 1 liter since they will cancel in any event.

(STP)

$P' = 1.0$ atm	$P'' = ?$ ___
$V' = 1$ liter	$V'' = 1$ liter
$T' =$ ___	$T'' = 291°K$

atm

273°K

Putting these values into

$\dfrac{P''V''}{T''}$

$$\frac{P'V'}{T'} = \underline{\quad\quad},$$

you have

atm

$$\frac{(1.0)\,(1)}{273} = \frac{(?\ \underline{\quad})\,(1)}{291}.$$

Cross-multiplying gives you

1.07 atm

$$\frac{(1.0)\,(1)\,(291)}{(1)\,(273)} = \underline{\quad\quad\quad}.$$

To get the pressure in millimeters Hg, you solve the following equation.

$\dfrac{760 \text{ mm Hg}}{\text{atm}}$

810

? mm Hg = 1.07 atm × _____

= ___ mm Hg

This is all there is to changing a sample of gas from one set of parameters to another.

SECTION C Determining an Unknown Parameter

The second type of gas law problem does not involve changing the conditions of a sample of gas. Instead, you are asked to determine one of the four parameters which define a gas (pressure, volume, temperature, or moles) having the other three given. This means that you will have three separate pieces of information given. It is sometimes hard to choose which one of the three to start with. One easy way to decide is to look at the ideal gas equation

$PV = nRT$

and see which two pieces of information given are both on the same side of the equation. You can start with both of these. You always will then use the conversion factor R, the ideal gas constant, which is

liter, atm

mole, °K

$$\frac{0.0820\,(\underline{\quad})\,(\underline{\quad})}{(\underline{\quad})\,(\underline{\quad})}.$$

Let's try question 6 on the pretest to see how this goes. How many moles of an ideal gas will occupy 2.0 liters at a temperature of 27°C and a pressure of 0.5 atm? The question asked is "How

many _____?" The information given is pressure equals _____,
moles,   0.5 atm

volume equals _____, and the temperature must be converted
2.0 liters

to °__. It is then _____. The two pieces of given information that
K,   300°K

are both on the same side of the equation of state are *P* and __.
V

You may therefore start with both of these.

? moles = (0.5 atm) (_____)
2.0 liters

Next you can use the conversion factor for *R* to cancel these two

unwanted dimensions. You multiply by _____, its
$$\frac{(mole)\,(°K)}{0.0820\,(liter)\,(atm)}$$

inverted form, leaving the unwanted dimension ___. You can
°K

cancel this by multiplying by _____, which is the reciprocal of the
$$\frac{1}{300°K}$$

information given. Now the only dimension left is the dimension of the answer, which must be

$$?\text{ moles} = (0.5\,\cancel{atm})\,(2.0\,\cancel{liters}) \times \frac{(mole)\,\cancel{(°K)}}{0.0820\,\cancel{(liter)}\,\cancel{(atm)}} \times \frac{1}{300\cancel{°K}}$$

= 0.04 mole.

The above example is particularly easy since all the parameters in the information given are in the same dimensions as in the gas law conversion factor, *R*. The next example is not so simple. If 3.0 moles of an ideal gas are at a temperature of −23°C and a pressure of 380 torr, what volume in milliliters will the gas occupy?"

The question asked is "How many _____?" The information
milliliters

given is the temperature (which must be in °K) _____; the
250°K

pressure, _____; and the number of moles, _____. The two
380 torr   3.0 moles

given parameters which are on the same side of the ideal gas

equation are __ and __, so you start with these two.
n   *T*

? ml = (3.0 moles) (_____)
250°K

Now you can cancel the two unwanted dimensions by

multiplying by the ideal gas constant, *R*, which is _____.
$$\frac{0.0820\,(liter)\,(atm)}{(mole)\,(°K)}$$

The unwanted dimensions now are _____ and _____.
liters,   atmospheres

However, the units of pressure in the information given are _____.
torrs

In order to cancel the unwanted dimension, you must convert atmospheres to torrs by multiplying by the conversion factor

_____. Now you can use your third piece of information given
$$\frac{760\ torr}{atm}$$

$$\frac{1}{380 \text{ torr}}$$

liters

milliliters, $\dfrac{ml}{10^{-3} \text{ liter}}$

by multiplying by its reciprocal, _____. The unwanted dimension now remaining is _____, and since you want your answer to be in _____, you multiply by _____. The answer is thus

$$? \text{ ml} = (3.0 \text{ moles})(250°K) \times \frac{0.0820 \text{ (liter) (atm)}}{\text{(mole) (°K)}} \times \frac{760 \text{ torr}}{\text{atm}}$$

$$\times \frac{1}{380 \text{ torr}} \times \frac{ml}{10^{-3} \text{ liter}}$$

$$= 1.2 \times 10^5 \text{ ml.}$$

Sometimes you will have n in the information given, but as weight of gas rather than moles of gas. However, the conversion is very simple, as question 7 in the pretest demonstrates.

When 16.0 g of He gas is contained in a 2.0 liter vessel at a temperature of 100°K, what will be its pressure in atmospheres?

atmospheres

100°K   2.0 liters

16.0 g

T, V n

T, n

16.0 g He

The question asked is "How many _____?" The information given is the temperature, _____; the volume, _____; and the amount of He, _____.Of the four variables, *P, V, T,* and *n*, the pieces of information given are __, __ and, indirectly, __. The two which are on the same side of the ideal gas equation are __ and __, so you start with these.

$$? \text{ atm} = (100°K) (\underline{\hspace{2cm}})$$

It is obvious that if you are going to cancel the amount of He by using the gas law conversion factor, *R*, the amount of He will have to have the dimensions _____. It is simple to get from grams to moles; all you do is multiply by _____, the inverted form of the molecular weight. Now you can multiply by the gas law conversion factor *R*, _____, to cancel the two unwanted dimensions.

moles

$$\frac{\text{mole He}}{4.0 \text{ g He}}$$

$$\frac{0.0820 \text{ (liter) (atm)}}{\text{(mole) (°K)}}$$

liters

$$\frac{1}{2.0 \text{ liters}}$$

The unwanted dimension remaining is _____, but since you have this in the third piece of the information given, you multiply by _____. Now the only dimension left is the dimension of the answer, which must be

$$? \text{ atm} = (100°\cancel{K}) (16.0 \cancel{\text{ g He}}) \times \frac{\text{mole He}}{4.0 \cancel{\text{ g He}}}$$

$$\times \frac{0.0820 \text{ (}\cancel{\text{liter}}\text{)(atm)}}{\text{(mole }\cancel{\text{He}}\text{) (}\cancel{°K}\text{)}} \times \frac{1}{2.0 \cancel{\text{ liters}}}$$

16

$$= \underline{\hspace{0.7cm}} \text{ atm.}$$

Question 8 on the pretest shows you another way that the information can be given to you. What is the density of O_2 gas in grams per liter at 27°C and 152 cm Hg pressure?

Probably the best way to do this problem is to reword the question to ask how many grams of O_2 there are in _____ at 27°C and a pressure of 152 cm Hg. Now the question asked is "How many _____?" The information given is the volume, _____; the temperature, _____ (it must be in °K); and the pressure, _____. The two parameters which are on the same side of the equation of state are __ and __, so you write

$$? \text{ g } O_2 = (152 \text{ cm Hg}) (\text{_____}).$$

Before you can cancel the pressure term, you must convert it to the dimension _____ by multiplying by _____. Now you can use your gas law conversion factor and you multiply by its inverted form, _____. You can cancel the unwanted dimension °K by using the third piece of information given; you multiply by the reciprocal of the temperature, _____. The unwanted dimension now is _____, and you can convert this to the dimensions of your answer by multiplying by _____, the molecular weight. The answer will therefore be

$$? \text{ g } O_2 = (152)\,(1.0)\left(\frac{1}{76}\right)\left(\frac{1}{0.0820}\right)\left(\frac{1}{300}\right)(32.0) \text{ g } O_2$$

$$= 2.6 \text{ g}/1.0 \text{ liter, so } d = 2.6 \text{ g/liter.}$$

Note that you could have solved the problem directly for density, which is a conversion factor, by starting with the gas law conversion factor and working from there.

$$? \, d_{O_2} = \frac{? \text{ g } O_2}{\text{liter } O_2} = \frac{(\text{mole})\,(°K)}{0.0820\,(\text{liter})\,(\text{atm})} \times \frac{32.0 \text{ g } O_2}{\text{mole } O_2}$$

$$\times \frac{152 \text{ cm Hg}}{300°K} \times \frac{\text{atm}}{76 \text{ cm Hg}}$$

You set the gas law conversion factor so that it gives liters in the denominator, and cancel with the information given.

SECTION D Determining Molecular Weight

By reversing the process of the last two examples, it is possible to determine the molecular weight of an ideal gas if you know its pressure, volume, temperature, and weight. Question 9 on the pretest is a straightforward example of this calculation.

Right margin answers:

1.0 liter

grams of O_2

1.0 liter,   300°K

152 cm Hg

P, V

1.0 liter

atmospheres, $\dfrac{\text{atm}}{76 \text{ cm Hg}}$

$\dfrac{(\text{mole})\,(°K)}{0.0820\,(\text{liter})\,(\text{atm})}$

$\dfrac{1}{300°K}$

mole O_2

$\dfrac{32.0 \text{ g } O_2}{\text{mole } O_2}$

What is the molecular weight of an unknown gas if 3.0 liters at a pressure of 1.0 atm and a temperature of 27°C weigh 2.4 g? You can do this problem by directly solving for the conversion factor for molecular weight, grams per mole. The question is therefore, "How many grams of unknown per mole?"

$$? \text{ molecular weight } X = \frac{? \text{ g } X}{\text{mole } X}$$

Hidden in the problem will always be one conversion factor, the weight of unknown per volume of unknown. This is, of course, its density. In the example above it is

2.4

3.0

$$\frac{\underline{\quad} \text{ g } X}{\underline{\quad} \text{ liters}}.$$

You also have two pieces of information given: the pressure,

1.0 atm,   300°K

$$\frac{0.0820 \, (\text{liter}) \, (\text{atm})}{(\text{mole}) \, (°\text{K})}$$

$\underline{\hspace{2cm}}$, and the temperature, $\underline{\hspace{1.5cm}}$. And, of course, you will use the gas law conversion factor R, $\underline{\hspace{3cm}}$.

It is best to start with the conversion factor given in the problem, the density.

$$\frac{? \text{ g } X}{\text{mole } X} = \frac{2.4 \text{ g } X}{3.0 \text{ liters}}$$

The unwanted dimension, liters, can be cancelled by using the gas

$$\frac{0.0820 \, (\text{liter}) \, (\text{atm})}{(\text{mole}) \, (°\text{K})}$$

atmospheres,   °K

$$\frac{1}{1.0 \text{ atm}}, \quad 300°\text{K}$$

law conversion factor R, $\underline{\hspace{3cm}}$. You now have both dimensions of the answer, but the two unwanted dimensions,

$\underline{\hspace{2.5cm}}$ and $\underline{\hspace{1cm}}$, must be cancelled. Using the values from

the information given, you multiply by $\underline{\hspace{1.5cm}} \times \underline{\hspace{1.5cm}}$, and the answer will be

$$\frac{? \text{ g } X}{\text{mole } X} = \frac{2.4 \text{ g } X}{3.0 \text{ liters}} \times \frac{0.0820 \, (\text{liter}) \, (\text{atm})}{(\text{mole}) \, (°\text{K})} \times \frac{1}{1.0 \text{ atm}} \times 300 °\text{K}$$

$$= \frac{20 \text{ g } X}{\text{mole } X}.$$

If you like to remember formulas, you can see that the above is really

$$\frac{? \text{ g } X}{\text{mole } X} = \frac{(\text{weight sample})}{V} \times 0.0820 \times \frac{T}{P}$$

$$= \frac{(\text{g}) \, RT}{PV}.$$

But make sure that you express everything in °K, liters, and atmospheres!

We can do question 10 on the pretest by using this formula. What is the molecular weight of a gas whose density is 4.50 g/liter at $-73°C$ and a pressure of 380 mm Hg?

g (the weight of gas) = 4.50 g

V (the volume of gas) = _____ liter   1.00

T (the temperature)   = ___°K   200

P (the pressure)   = _____ atm   0.500

$$\frac{?\,g\,X}{\text{mole }X} = \frac{(g)\,RT}{PV}$$

$$= \frac{4.50 \times \underline{\hspace{1cm}} \times 200}{\underline{\hspace{1cm}} \times 1.00}$$ 0.0820

                                                       0.500

$$= \frac{\underline{\hspace{0.5cm}}\,g\,X}{\text{mole }X}$$ 148

SECTION E   Mixtures of Gases— Dalton's Law

When you mix several gases in a container, each gas acts as if the other gases are not present. Each gas exerts its own pressure (called the partial pressure, p). Consequently, the total pressure in the container equals the sum of all the partial pressures. This is called Dalton's law. Mathematically, it is

$$P = p_1 + p_2 + p_3 + \text{etc.}$$

The p_1, p_2, p_3 refer to the partial pressures (individual pressures) of gas 1, gas 2, gas 3, etc.

Question 11 illustrates this phenomenon. What will be the pressure in a 4.0 liter container which is at 27°C and which contains 2.0 moles of H_2 gas and 5.0 moles of He gas? You must determine the pressure of each gas separately, then simply add all the pressures together to get the total pressure.

? atm H_2 = (2.0 moles H_2) (300°K) × 0.0820 _____ $\dfrac{\text{(liter) (atm)}}{\text{(mole } H_2) \text{ (°K)}}$

                    × _____ $\dfrac{1}{4.0 \text{ liters}}$

              = ____ atm H_2 12.3

? atm He = (You calculate this, then check your answer with

              = _____ atm ___.)   30.8, He

To get the total pressure in the container you must ____ the partial   add
pressures, and when you express this to the correct number of significant figures you get

P_{total} = _____ atm.   43 or 4.3×10^1

One place that this comes up experimentally is when you collect a gas over some liquid. Every liquid has a certain amount of its own gas phase associated with it. The amount (pressure) depends on the temperature. This vapor pressure of liquids can be looked up in a table, either in your text or in the *Handbook of Chemistry and Physics* [Ohio: The Chemical Rubber Co.]. Experimentally, you will measure the total pressure. However, this total pressure is due to the gas plus the vapor of the liquid. In order to get the pressure of

subtract

the gas, you will have to _____ the vapor pressure from the total pressure which is experimentally determined.

Question 12 in the pretest is a good example of this type of problem. What is the molecular weight of a gas if 70.0 ml collected over water at a temperature of 27°C has a total pressure of 756.5 mm Hg? The vapor pressure of water at this temperature is 26.5 mm Hg. The weight of dry gas is 0.080 g. First you calculate

subtracting

the partial pressure of the unknown gas by _____ the partial pressure of water.

−26.5 mm Hg

730.0

$$p_X = 756.5 \text{ mm Hg} \underline{\hspace{2cm}}$$

$$= \underline{\hspace{2cm}} \text{ mm Hg}$$

If you want to use the memorized formula for molecular weight,

$$\frac{? \text{ g } X}{\text{mole } X} = \frac{(g) \, RT}{PV},$$

atmospheres

$$\frac{\text{atm}}{760.0 \text{ mm Hg}}$$

0.9605 atm

you must convert the pressure to _____.

$$p_X \text{ atm} = 730.0 \text{ mm Hg} \times \underline{\hspace{2cm}}$$

$$= \underline{\hspace{1cm}}$$

To save one step, you can express the volume immediately in liters. To get the molecular weight, you have

300

0.0700

29

$$\frac{? \text{ g } X}{\text{mole } X} = \frac{(0.080)(0.0820)(\underline{\hspace{0.5cm}})}{(0.9605)(\underline{\hspace{0.7cm}})}$$

$$= \frac{\underline{\hspace{0.3cm}} \text{ g } X}{\text{mole } X}. \quad \text{(Notice the number of significant figures.)}$$

A somewhat more complicated problem would be the following. A 6.0 liter container at 127°C has a total pressure of 12.0 atm. It contains 0.50 g of He. How many moles of O_2 does it contain if that is the only other gas present?

The first thing you must determine is the partial pressure of He.

4.0 g He

$$\frac{(\text{liter})(\text{atm})}{(\text{mole He})(°\text{K})}$$

0.68

$$p_{He} \text{ atm} = (0.50 \text{ g He})(400°\text{K})\left(\frac{\text{mole He}}{\underline{\hspace{1cm}}}\right)$$

$$\times 0.0820 \left(\underline{\hspace{1.5cm}}\right)\left(\frac{1}{6.0 \text{ liters}}\right)$$

$$= \underline{\hspace{0.5cm}} \text{ atm}$$

Now you can determine the partial pressure of O_2 by _____

the partial pressure of He from the _____ pressure.

$$p_{O_2} = 12.0 \text{ atm} - \text{_____} = 11.3 \text{ atm}$$

Now it is very simple to calculate the number of moles of O_2.

? moles O_2 = (11.3 atm) (6.0 liters) (_____)

$$\times (\text{\_\_\_\_\_})$$

$$= \text{\_\_\_ moles } O_2$$

SECTION F Diffusion of Gases

The rate at which a gas diffuses is inversely proportional to its molecular weight (or as it was first noted, to its density). This is called Graham's law. The diffusion of two different gases (at the same pressure and temperature) can be mathematically stated as

$$\frac{r_1}{r_2} = \frac{\sqrt{M_2}}{\sqrt{M_1}} = \sqrt{\frac{M_2}{M_1}},$$

where r_1 and r_2 are the rates of diffusion of gases 1 and 2 and M_1 and M_2 are their respective molecular weights.

Question 13 on the pretest is a typical example of the type of problem you will come across. A sample of N_2 gas diffuses through a small hole at a rate of 2.6 ml per minute. At what rate would Cl_2 gas diffuse through the same hole under the same conditions?

The molecular weight of N_2 is _____ g/mole; the molecular

weight of Cl_2 is _____ g/mole; and the rate of diffusion of N_2 is

_____ ml/min. Using Graham's law, you have

$$\frac{r_{Cl_2}}{r_{N_2}} = \sqrt{\frac{M_{N_2}}{M_{Cl_2}}}$$

Putting in the specified values, you have

$$\frac{r_{Cl_2}}{2.6 \text{ ml/min}} = \sqrt{\frac{28.0}{71.0}}$$

$$= \sqrt{0.394} = 0.628$$

$$r_{Cl_2} = 2.6 \times 0.628 \text{ ml/min}$$

$$= 1.6 \text{ ml/min.}$$

(If you cannot determine square roots on your calculator, Section B in Chapter 1 shows how to do it with logarithms.)

---

subtracting

total

0.68 atm

$$\frac{(\text{mole } O_2) \, (°K)}{0.0820 \, (\text{liter}) \, (\text{atm})}$$

$$\frac{1}{400°K}$$

2.1

28.0

71.0

2.6

Question 14 on the pretest is another way of presenting the problem. Here the rates of diffusion are known and you want to determine the molecular weight

"If 20.0 g of an unknown gas diffuses through a pinhole in the same time that it takes 40.0 g of CH_4 to diffuse through the same pinhole under the same conditions, what is the molecular weight of the unknown gas?" The rate is amount/time, and since both gases diffuse in the same time, we can say that the rates are expressed simply as the amount.

40.0   Rate CH_4 = _____

20.0   Rate X = _____

16.0   Molecular weight CH_4 = _____ g/mole

Therefore, we can write Graham's law as

X

r_X, CH_4

$$\frac{r_{CH_4}}{\_\_} = \frac{\sqrt{M\_\_}}{\sqrt{M\_\_}},$$

and, putting in the specified values, we have

40.0

20.0, 4.00

8.00

64.0

$$\frac{\_\_}{\_\_} = \frac{\sqrt{M_X}}{\_\_}$$

$$\_\_ = \sqrt{M_X}$$

$$\_\_ = M_X \quad \text{(in grams per mole).}$$

PROBLEM SET

You may omit those problems marked with an asterisk since they involve complicated calculations.

1. If 333 ml of an ideal gas at a temperature of 25°C and a pressure of 750 torr has its temperature lowered to −11°C and its pressure lowered to 730 mm Hg, what will its new volume be in milliliters?

2. What will the temperature be in °C if 2.0 liter of an ideal gas at 0°C is expanded to 2.5 liters? Its initial pressure was 750 mm Hg and its final pressure is 740 mm Hg.

3. A sample of an ideal gas occupying 0.50 liter at 0.99 atm pressure and a temperature of 273°K is expanded until its volume is 755 ml and the temperature 0°C. What will be its pressure in torrs?

4. What will be the volume at STP for an ideal gas which occupies 333 ml at 298°K and 755 mm Hg pressure?

*5. How many cubic feet will a sample of gas which occupies 2.5 ft³ at 68°F occupy at 300°F if the pressure is held constant?

6. What pressure in atmospheres will be exerted by 0.312 mole of an ideal gas at 42°C in a volume of 58.4 ml?

7. How many moles of a gas will occupy 5.0 liters at STP?

8. At what temperature must you hold 12.8 g of O_2 gas to have a pressure of 0.20 atm and a volume of 0.50 liter?

9. What is the density in grams/liters of Cl_2 gas at STP?

10. How many grams will 6.0 liters of N_2 at −23°C and 2.0 atm pressure weigh?

11. What is the molecular weight of a gas if 268 ml at 69°C and 17.9 torr pressure weigh 0.0156 g?

12. What is the molecular weight of an ideal gas if its density at 289.3°K and a pressure of 0.980 atm is 2.66 g/liter?

13. It was found that at 25°C the density of N_2O gas was 1.73 g/liter. At what pressure was it being kept (in atmospheres)?

*14. What volume in milliliters will 1.00 g of C_8H_{18} occupy at 735 torr and 210.2°F?

15. How many grams of acetylene, C_2H_2, will occupy 1.00 liter at 24°C and a pressure of 742 mm Hg?

16. What will the total pressure be if a 3.0 liter container at −123°C contains 2.8 g N_2, 2.2 g CO_2, and 0.02 g H_2?

17. How many moles of a gas are present if, when the gas is collected over water at 26°C, it has a total pressure of 755 torr and a volume of 3.5 liters? The vapor pressure of water at this temperature is 25 mm Hg.

*18. A sample of H_2 gas collected over water at 25°C and 1.0 atm total pressure occupies 0.10 liter. What volume would the dry gas occupy at this same pressure and temperature? The partial pressure of water at 25°C is 23.8 mm Hg.

19. A sample of O_2 diffuses through a small opening at the rate of 3.00 ml per minute. At what rate would H_2 diffuse through the same hole under the same conditions?

20. An unknown gas was found to diffuse through a pinhole at the rate of 3.0 g per minute. Using the same pinhole, He was found to diffuse at the rate of 12.0 g in 2.0 minutes. What is the molecular weight of the unknown gas?

PROBLEM SET ANSWERS

You should be able to do Problems 1–5. If you cannot, redo Section B. If you do not get four correct answers from Problems 6–10, check Section C. Problems 11 and 12 are simple and you should get them both. They are covered in Section D. You should be able to do at least four problems in the group 13–18 and both Problems 19 and 20, which are covered in Section F.

1. $P' = 750$ torr $P'' = 730$ torr $\dfrac{(750)\,(333)}{298} = \dfrac{(730)\,(?\text{ ml})}{262}$

   $V' = 333$ ml $V'' = ?$ ml

   $T' = 298°$K $T'' = 262°$K $?$ ml $= 301$ ml

2. $P' = 750$ mm Hg $P'' = 740$ mm Hg $\dfrac{(750)\,(2.0)}{273} = \dfrac{(740)\,(2.5)}{?\,°\text{K}}$

   $V' = 2.0$ liters $V'' = 2.5$ liters

   $T' = 273°$K $T'' = ?\,°$K $?\,°$K $= 337°$K

   $?\,°$C $= 64°$C

3. $P' = 0.99$ atm $P'' = ?$ atm $\dfrac{(0.99)\,(0.50)}{273} = \dfrac{(?\text{ atm})\,(0.755)}{273}$

   $V' = 0.50$ liter $V'' = 0.755$ liter

   $T' = 273°$K $T'' = 273°$K $?$ atm $= 0.66$ atm

   $?$ torr $= 5.0 \times 10^2$ torr

4. $P' = 755$ mm Hg $P'' = 760$ mm Hg $\dfrac{(755)\,(333)}{298} = \dfrac{(760)\,(?\text{ ml})}{273}$

   $V' = 333$ ml $V'' = ?$ ml

   $T' = 298°$K $T'' = 273°$K $?$ ml $= 303$ ml

5. $P' = 1$ $P'' = 1$ $\dfrac{(1)\,(2.5)}{293} = \dfrac{(1)\,(?\text{ ft}^3)}{422}$

   $V' = 2.5$ ft^3 $V'' = ?$ ft^3

   $T' = 293°$K $T'' = 422°$K $?$ ft$^3 = 3.6$ ft^3

6. $?$ atm $= (0.312 \;\text{mole})\,(315°\text{K}) \times \dfrac{0.0820\;\text{(liter)}\,\text{(atm)}}{\text{(mole)}\,\text{(°K)}} \times \dfrac{1}{0.0584\;\text{liter}}$

   $= 138$ atm

7. $? \text{ moles} = (5.0 \text{ liters}) \, (1.0 \text{ atm}) \times \dfrac{(\text{mole}) \, (°K)}{0.0820 \, (\text{liter}) \, (\text{atm})} \times \dfrac{1}{273 °K}$

    $= 0.22 \text{ mole}$

8. $? °K = (0.50 \text{ liter}) \, (0.20 \text{ atm}) \times \dfrac{(\text{mole}) \, (°K)}{0.0820 \, (\text{liter}) \, (\text{atm})}$

    $\times \dfrac{32.0 \text{ g } O_2}{\text{mole } O_2} \times \dfrac{1}{12.8 \text{ g } O_2}$

    $= 3.0 °K$

9. $? \text{ g } Cl_2 = 1.00 \text{ liter} \, (1.00 \text{ atm}) \times \dfrac{(\text{mole}) \, (°K)}{0.0820 \, (\text{liter}) \, (\text{atm})} \times \dfrac{71.0 \text{ g } Cl_2}{\text{mole } Cl_2}$

    $\times \dfrac{1}{273 °K}$

    $= 3.17 \text{ g } Cl_2$

    $\left(\rule{0pt}{2.8em}\right.$ You might also remember that at STP there are 22.4 liters per mole.

    $\dfrac{? \text{ g}}{\text{liter}} = \dfrac{\text{mole}}{22.4 \text{ liters}} \times \dfrac{71.0 \text{ g}}{\text{mole}} = \dfrac{3.17 \text{ g}}{\text{liter}} \left.\rule{0pt}{2.8em}\right)$

10. $? \text{ g } N_2 = (6.0 \text{ liters}) \, (2.0 \text{ atm}) \times \dfrac{(\text{mole}) \, (°K)}{0.0820 \, (\text{liter}) \, (\text{atm})} \times \dfrac{1}{250 °K}$

    $\times \dfrac{28.0 \text{ g } N_2}{\text{mole } N_2}$

    $= 16 \text{ g } N_2$

11. $\dfrac{? \text{ g } X}{\text{mole } X} = \dfrac{0.0156 \text{ g}}{0.268 \text{ liter}} \times \dfrac{0.0820 \, (\text{liter}) \, (\text{atm})}{(\text{mole}) \, (°K)} \times \dfrac{1}{17.9 \text{ torr}}$

    $\times \dfrac{760 \text{ torr}}{\text{atm}} \times 342 °K$

    $= 69.3 \text{ g/mole}$

    If you prefer to remember the formula, molecular weight $= \dfrac{g \, RT}{PV}$, then

    $P \, (\text{atm}) = 17.9 \text{ torr} \times \dfrac{\text{atm}}{760 \text{ torr}} = 0.0236 \text{ atm.}$

    $\dfrac{? \text{ g } X}{\text{mole } X} = \dfrac{(0.0156) \, (0.0820) \, (342)}{(0.0236) \, (0.268)} = 69.2 \text{ g/mole}$

12. $\dfrac{? \text{ g } X}{\text{mole } X} = \dfrac{2.66 \text{ g } X}{\text{liter } X} \times \dfrac{0.0820 \, (\text{liter}) \, (\text{atm})}{(\text{mole}) \, (°K)} \times \dfrac{289.3 °K}{0.980 \text{ atm}}$

    $= \dfrac{64.4 \text{ g } X}{\text{mole } X}$

13. $? \text{ atm} = \dfrac{1.73 \text{ g N}_2\text{O}}{\text{liter}} \times \dfrac{0.0820 \text{ (liter) (atm)}}{\text{(mole) (}^\circ\text{K)}} \times \dfrac{\text{mole N}_2\text{O}}{44.0 \text{ g N}_2\text{O}} \times 298^\circ\text{K}$

$= 0.961 \text{ atm}$

14. First get the temperature in °C.

$^\circ\text{C} = \dfrac{5}{9}(^\circ\text{F}-32) = \dfrac{5}{9}(178.2) = 99.0^\circ\text{C}$

Then convert it to °K.

$^\circ\text{K} = 273 + {^\circ\text{C}} = 372^\circ\text{K}$

$? \text{ ml} = (1.00 \text{ g C}_8\text{H}_{18})(372^\circ\text{K}) \times \dfrac{0.0820 \text{ (liter) (atm)}}{\text{(mole) (}^\circ\text{K)}} \times \dfrac{\text{mole C}_8\text{H}_{18}}{114.0 \text{ g C}_8\text{H}_{18}}$

$\times \dfrac{760 \text{ torr}}{\text{atm}} \times \dfrac{1}{735 \text{ torr}} \times \dfrac{10^3 \text{ ml}}{\text{liter}}$

$= 277 \text{ ml}$

15. $? \text{ g C}_2\text{H}_2 = (1.00 \text{ liters})(742 \text{ mm Hg}) \times \dfrac{\text{atm}}{760 \text{ mm Hg}}$

$\times \dfrac{\text{(mole) (}^\circ\text{K)}}{0.0820 \text{ (liter) (atm)}} \times \dfrac{26.0 \text{ g C}_2\text{H}_2}{\text{mole C}_2\text{H}_2} \times \dfrac{1}{297^\circ\text{K}}$

$= 1.04 \text{ g C}_2\text{H}_2$

16. $? \text{ atm N}_2 = (150^\circ\text{K})(2.8 \text{ g N}_2) \times \dfrac{\text{mole N}_2}{28.0 \text{ g N}_2}$

$\times \dfrac{0.0820 \text{ (liter) (atm)}}{\text{(mole) (}^\circ\text{K)}} \times \dfrac{1}{3.0 \text{ liters}}$

$= 0.41 \text{ atm}$

$? \text{ atm CO}_2 = (150^\circ\text{K})(2.2 \text{ g CO}_2) \times \dfrac{\text{mole CO}_2}{44.0 \text{ g CO}_2}$

$\times \dfrac{0.0820 \text{ (liter) (atm)}}{\text{(mole) (}^\circ\text{K)}} \times \dfrac{1}{3.0 \text{ liters}}$

$= 0.205 \text{ atm}$

$? \text{ atm H}_2 = (150^\circ\text{K})(0.02 \text{ g H}_2) \times \dfrac{\text{mole H}_2}{2.0 \text{ g H}_2}$

$\times \dfrac{0.082 \text{ (liter) (atm)}}{\text{(mole) (}^\circ\text{K)}} \times \dfrac{1}{3.0 \text{ liter}}$

$= 0.041 \text{ atm}$

$P_{\text{total}} = p_{\text{N}_2} + p_{\text{CO}_2} + p_{\text{H}_2}$

$= 0.41 + 0.205 + 0.041 = 0.66 \text{ atm}$

17. p_{gas} = 755 torr − 25 torr = 730 torr

? moles = (3.5 ~~liters~~)(730 ~~torr~~) × $\dfrac{\text{atm}}{760 \text{ torr}}$

$\times \dfrac{(\text{mole}) (°K)}{0.0820 (\text{liter}) (\text{atm})} \times \dfrac{1}{299°K}$

= 0.14 mole

18. First determine the number of moles of gas as in Problem 17.

? moles = (0.10 ~~liter~~)(760 ~~mm Hg~~ − 23.8 ~~mm Hg~~)

$\times \dfrac{\text{atm}}{760 \text{ mm Hg}} \times \dfrac{(\text{mole}) (°K)}{0.0820 (\text{liter}) (\text{atm})} \times \dfrac{1}{298°K}$

= 0.0040 mole

Then determine the volume that this many moles will occupy.

? liters = (0.0040 ~~mole~~)(298°K)

$\times \dfrac{0.0820 (\text{liter}) (\text{atm})}{(\text{mole}) (°K)} \times \dfrac{1}{1.0 \text{ atm}}$

= 0.098 liter

19. $\dfrac{r_{H_2}}{r_{O_2}} = \sqrt{\dfrac{M_{O_2}}{M_{H_2}}}$

$\dfrac{r_{H_2}}{3.00 \text{ ml/min}} = \sqrt{\dfrac{32}{2}} = \sqrt{16} = 4.0$

r_{H_2} = 3.00 ml/min × 4.0 = 12 ml/min

20. $\dfrac{r_{He}}{r_X} = \dfrac{\sqrt{M_X}}{\sqrt{M_{He}}}$

$\dfrac{12.0 \text{ g/2.0 min}}{3.0 \text{ g/min}} = \dfrac{\sqrt{M_X}}{\sqrt{4.0}}$

$\dfrac{6.0}{3.0} = \dfrac{\sqrt{M_X}}{2.0}$

$4.0 = \sqrt{M_X}$

16 g/mole = M_X

Chapter 11
Stoichiometry Involving Gases

PRETEST

These questions should be fairly easy to answer if you remember that in all stoichiometry problems you head towards a "bridge" conversion factor in moles. You should be able to do these in 20 minutes.

1. How many liters of O_2 gas at 2.0 atm pressure and 27°C will be consumed by the complete reaction of 8.0 g H_2 according to the reaction

    $2H_2 + O_2 = 2H_2O$?

2. How many grams of water can be produced by the complete reaction of 5.0 liters of O_2 at 6.0 atm pressure and 27°C according to the same reaction as in question 1?

3. How many liters of H_2 gas at 123°C and 740 mm Hg pressure will be required to react with 5.0 liters of O_2 at 2.0 atm pressure and 27°C according to the same reaction as in question 1?

4. How many liters of CO will be formed by the reaction of 5.0 liters of O_2 if both gases are at 2.0 atm and 300°K? The reaction is

    $2C + O_2 = 2CO$.

5. How many liters of HCl gas at 2.0 atm pressure and 300°K will be needed to produce 5.0 liters of CO_2 gas at 1520 torr and 27°C according to the equation

    $CaCO_3 + 2HCl = CaCl_2 + CO_2 + H_2O$?

6. What is the percent purity of an iron ore sample if 1.25 g of sample, when reacted with an excess of HCl, produces 220 ml of H_2 gas collected over water at a total pressure of 1.0 atm and 26°C? The vapor pressure of water at this temperature is 25 mm Hg. The reaction is

    $2Fe + 6HCl = 2FeCl_3 + 3H_2$.

7. How many liters of Cl_2 gas at STP will be produced by the reaction of 30.0 ml of a solution of HCl that contains 12.0 moles of HCl per liter of solution? There is an excess of $KMnO_4$. The reaction is

    $2KMnO_4 + 16HCl = 2KCl + 2MnCl_2 + 5Cl_2 + 8H_2O$.

PRETEST ANSWERS

If you didn't answer all the questions correctly, locate them in the body of the chapter and see where you went wrong. If you are totally lost as to how to approach these problems, you had better go through the entire chapter. It is quite short. If you got them all correct, go straight to the problem set at the end of the chapter.

1. 25 liters O_2

2. 43 g H_2O

3. 27 liters H_2

4. 10 liters CO

5. 10 liters HCl

6. 25.8% pure

7. 2.51 liters Cl_2

When you have chemical reactions involving gases, it is possible to have amounts of reactants or products not expressed in weights or moles, but rather as the volume (or pressure or temperature) of a gas. Chapter 10 shows how to convert from the variables which define a gas to get moles or grams.

Actually, there are two main types of problems. The first is where all three parameters which define the gas (P, V, and T) are given. We will examine this type in Section A.

SECTION A   Problems Where P, V, and T Are Given

It is easy to recognize this type of problem since you will find pressure, volume, and temperature of the gas in the information given. You must solve the problem in two steps. First, from the

moles

P, V, and T you can calculate the _____ of the gas. Once you have this, you use it as your information given in the second step, which is to solve for the substance of interest. Question 2 on the pretest is an example of this type.

How many grams of H_2O can be produced by the complete reaction of 5.0 liters of O_2 gas at 6.0 atm pressure and 27°C according to the reaction

$$2H_2 + O_2 = 2H_2O?$$

5.0 liters

As you see, all three variables are given. The volume is _____,

temperature,   300

the pressure is 6.0 atm, and the _____ is _____°K. You

moles

therefore solve for the number of _____ of O_2.

(mole) (°K), $\dfrac{1}{300°K}$

$$\text{Moles } O_2 = (5.0 \text{ liters}) (6.0 \text{ atm}) \times \frac{\rule{2cm}{0.4pt}}{0.0820 \text{ (liter) (atm)}} \times \rule{1.5cm}{0.4pt}$$

1.2

$$= \rule{1cm}{0.4pt} \text{ moles } O_2$$

The 1.2 moles O_2 is now your information given so now you can do step two.

2 moles H_2O

$$? \text{ g } H_2O = 1.2 \text{ moles } O_2 \times \frac{\rule{2cm}{0.4pt}}{\text{mole } O_2} \times \frac{18.0 \text{ g } H_2O}{\text{mole } H_2O}$$

$$= 43 \text{ g } H_2O \quad \text{(to two significant figures)}$$

You can see how simple step two is. Your information given is in moles, which brings you immediately to your "bridge" conversion factor from the balanced equation.

Question 6 on the pretest is simply a more complex example of this type. What is the percent purity of an iron ore sample if 1.25 g of sample, when reacted with an excess of HCl, produces 220 ml of H_2 gas collected over water at a total pressure of 1.0 atm and 26°C? The vapor pressure of water at this temperature is 25 mm Hg. The reaction is

$2Fe + 6HCl = 2FeCl_3 + 3H_2$.

When you look at the problem, you will immediately note that all three variables, __, __ and __, are given for the H_2 gas. Your first step, then, will be to calculate the number of _____ of H_2. Write down these values for the variables.

$T = $ ____°K

$V = $ _____ liter

$p_{H_2} = $ _____ mm Hg   (You get this by subtracting the partial pressure of _____ from the _____ pressure expressed in mm Hg.)

? moles $H_2 = ($_____$) (735$ mm Hg$)$

$$\times \text{\underline{\hspace{1cm}}} \times \frac{(\text{moles}) (°K)}{0.0820 (\text{liter}) (\text{atm})}$$

$$\times \text{\underline{\hspace{1cm}}}$$

$$= \text{\underline{\hspace{1cm}}} \text{ mole } H_2$$

You now know that 1.25 g of impure sample produces

_____ mole of H_2. This is the conversion factor given in the problem. Rewording the problem to ask how many grams of Fe there are in 100 g impure sample, you start with

? g Fe = _____.

To remove the unwanted dimension grams impure, you can multiply by _____, the conversion factor given in the problem. Now you can use your "_____," so you multiply by the conversion factor _____. Finally, to get the desired dimension grams Fe, you multiply by _____. The answer, then, is

$$? \text{ g Fe} = 100 \times \frac{8.68 \times 10^{-3}}{1.25} \times \frac{2}{3} \times 55.8 \text{ g Fe}$$

$$= 25.8 \text{ g Fe} = 25.8\%.$$

This problem is more difficult because it involves a percent purity, but in the first step you still must calculate the number of moles of the gas. It then becomes part of the conversion factor given in the problem.

P V T

moles

299

0.220

735

water,   total

0.220 liter

$\dfrac{\text{atm}}{760 \text{ mm Hg}}$

$\dfrac{1}{299°K}$

8.68×10^{-3}

8.68×10^{-3}

100 g imp.

$\dfrac{8.68 \times 10^{-3} \text{ mole } H_2}{1.25 \text{ g imp.}}$

bridge

$\dfrac{2 \text{ moles Fe}}{3 \text{ moles } H_2}$

$\dfrac{55.8 \text{ g Fe}}{\text{mole Fe}}$

SECTION B Problems Where One of the Variables, *P*, *V*, or *T*, Is Missing

The second type of problem that you will see when you are dealing with reactions of gases is one in which only two of the three variables of a gas are given. You are asked to find the third. Once again, you will have to do the problem in two steps. However, in this case the first step will be to use the balanced equation to determine the number of moles of the gas. Then in the second step, you use the gas law calculation to determine the missing variable.

Question 1 on the pretest is a very simple example of this type of problem. How many liters of O_2 gas at 2.0 atm pressure and 27°C will be consumed by the complete reaction of 8.0 g H_2 according to the reaction

$$2H_2 + O_2 = 2H_2O?$$

As you can see, the problem only gives two of the variables for the O_2 and asks you to determine the third. Consequently, your first step will be to determine the number of moles of O_2 from the equation.

$\dfrac{\text{mole } H_2}{2.0\,\text{g }H_2}$

$\qquad\qquad 2.0$

$$? \text{ moles } O_2 = 8.0 \text{ g } H_2 \times \underline{\qquad} \times \frac{\text{mole } O_2}{2 \text{ moles } H_2}$$

$$= \underline{\quad} \text{ moles } O_2$$

Now you can determine the volume of O_2 since you know the number of moles, the pressure, and the temperature.

$300°K$

$\dfrac{1}{2.0\,\text{atm}}$

$$? \text{ liters } O_2 = (2.0 \text{ moles}) \, (\underline{\qquad}) \times \frac{0.0820 \,(\text{liter})\,(\text{atm})}{(\text{mole})\,(°K)}$$

$$\times \underline{\qquad}$$

$$= 25 \text{ liters } O_2$$

Question 7 on the pretest is the same type, but it has some complications. How many liters of Cl_2 gas at STP will be produced by the reaction of 30.0 ml of a solution of HCl that contains 12.0 moles of HCl per liter of solution? There is an excess of $KMnO_4$. The reaction is

$$2KMnO_4 + 16HCl = 2KCl + 2MnCl_2 + 5Cl_2 + 8H_2O.$$

T

P

Only two of the three variables for the Cl_2 gas are given, __ and __. You see this by noting that the gas is at STP. Consequently, you will have to solve for the number of moles of Cl_2 from the equation.

$$? \text{ moles } Cl_2 = 30.0 \text{ ml soln.}$$

The conversion factor given in the problem is the number of moles

of HCl per liter of solution, _____.

Therefore, your first step will be to convert milliliters solution to

_____ solution so that you can use this conversion factor. You

multiply by _____. Now you can use the conversion

factor given in the problem. You multiply by _____.

Finally, you can use the "bridge" to get to the dimensions of the

answer. You multiply by _____. The answer, then, is

$$? \text{ moles } Cl_2 = 30.0 \times 10^{-3} \times 12.0 \times \frac{5 \text{ moles } Cl_2}{16}$$

$$= 0.112 \text{ mole } Cl_2.$$

Now you can determine the volume of Cl_2 gas.

$$? \text{ liters } Cl_2 = (0.112 \text{ mole}) \,(\underline{\quad}) \times \frac{0.0820 \,(\text{liter})\,(\text{atm})}{(\text{mole})\,(°K)}$$

$$\times \underline{\qquad}$$

$$= \underline{\quad} \text{ liters } Cl_2$$

Remember the conversion factor for STP, 22.4 liters/mole at
STP. If you use it, you can do the last step as follows.

$$? \text{ liters } Cl_2 = 0.112 \,\cancel{\text{mole}} \times \frac{22.4 \text{ liters}}{\cancel{\text{mole}}} = 2.51 \text{ liters.}$$

This is a very handy conversion factor, but *you must be at STP*!

SECTION C Problems Concerning Two Gases, One with *P*, *V*, and *T* Given, and the Other with Just Two Variables Given

It is possible to have a problem which involves two gases, one
with all three of its variables given and the other with only two;
you will be asked to calculate the third. This type of problem must
be done in three steps. First you calculate the number of moles of
the gas whose three variables are known. Then you use this
number of moles to determine the number of moles of the other
gas. Finally, you use the number of moles of the other gas and the
two variables given to calculate the missing variable.

$$\frac{12.0 \text{ moles HCl}}{\text{liter soln.}}$$

$$\frac{\text{liters}}{10^{-3} \text{ liter soln.}}$$

$$\frac{\text{ml soln.}}{\text{}}$$

$$\frac{12.0 \text{ moles HCl}}{\text{liter soln.}}$$

$$\frac{5 \text{ moles } Cl_2}{16 \text{ moles HCl}}$$

273°K

$$\frac{1}{1.00 \text{ atm}}$$

2.51

This seems very complicated, but an example should clear it up. Question 3 on the pretest asks how many liters of H_2 gas at 123°C and 740 mm Hg pressure will be required to react with 5.0 liters of O_2 at 2.0 atm pressure and 27°C according to the equation

$$2H_2 + O_2 = 2H_2O.$$

Looking at the problem, you see that you have all three variables given for ___ gas. Consequently, you start by calculating the number of _____ of O_2.

O_2

moles

$$? \text{ moles } O_2 = (5.0 \text{ liters}) (\text{_____}) \times \frac{(\text{mole}) (°K)}{0.0820 (\text{liter}) (\text{atm})}$$

2.0 atm

$\dfrac{1}{300°K}$

$$\times \text{_____}$$

$$= 0.41 \text{ mole } O_2$$

In the second step you calculate the number of moles of the other gas, H_2, from the balanced equation.

$\dfrac{2 \text{ moles } H_2}{\text{mole } O_2}$

$$? \text{ moles } H_2 = 0.41 \text{ mole } O_2 \times \text{_____}$$

0.82

$$= \text{\_\_\_\_} \text{ mole } H_2$$

Finally, the third step is to calculate the number of liters of H_2 at the temperature and pressure given in the problem.

396°K

$$? \text{ liters } H_2 = (0.82 \text{ mole}) (\text{_____}) \times \frac{0.0820 (\text{liter}) (\text{atm})}{(\text{mole}) (°K)}$$

$\dfrac{1}{740 \text{ mm Hg}}$

$$\times \text{_____} \times \frac{760 \text{ mm Hg}}{\text{atm}}$$

27

$$= \text{\_\_\_} \text{ liters } H_2$$

SECTION D Problems Where Both Gases Are at the Same Temperature and Pressure

In special cases you may have two gases, one completely described as to *P*, *V*, and *T*, the other missing its volume, and both at the same temperature and pressure. There is a big simplification you can make. We will look at question 4 in the pretest since both gases are at STP.

How many liters of CO will be formed by the reaction of 5.0 liters of O_2 if both gases are at 2.0 atm and 300°K? The reaction is

$$2C + O_2 = 2CO.$$

We will solve the problem in the normal manner, but not get the answers to all the multiplications. You will see as we go along.

First you determine the moles of O_2, since all three variables, P, V, and T, are given.

$$? \text{ moles } O_2 = (5.0 \cancel{\text{ liters}}) (2.0 \cancel{\text{ atm}}) \times \frac{(\text{mole}) \cancel{(°K)}}{0.0820 \cancel{(\text{liter})} \cancel{(\text{atm})}}$$

$$\times \frac{1}{300 \cancel{°K}}$$

$$= 5.0 \left(\frac{2.0}{(0.0820)(300)} \right) \text{mole } O_2$$

The second step is to determine the moles of CO using the balanced equation,

$$? \text{ moles CO} = 5.0 \left(\frac{2.0}{(0.0820)(300)} \right) \cancel{\text{mole } O_2} \times \frac{2 \text{ moles CO}}{\cancel{\text{mole } O_2}}$$

$$= 5.0 \times 2 \left(\frac{2.0}{(0.0820)(300)} \right) \text{ moles CO.}$$

The third step is to calculate the volume of CO from moles of CO.

$$? \text{ liters CO} = 5.0 \times 2 \left(\frac{2.0}{(0.0820)(300)} \right) \cancel{\text{moles CO}} \times 300 \cancel{°K}$$

$$\times \frac{0.0820 (\text{liter}) \cancel{(\text{atm})}}{\cancel{(\text{mole})} \cancel{(°K)}} \times \frac{1}{2.0 \cancel{\text{ atm}}}$$

$$= 5.0 \times 2 \left(\frac{2.0}{(0.0820)(300)} \right) \left(\frac{(300)(0.0820)}{2.0} \right) \text{ liter}$$

$$= 5.0 \times 2 \text{ liters CO} = 10 \text{ liters CO}$$

As you can see, both the temperature, 300°K, and the pressure, 2.0 atm, cancelled. The end result is that when the gases are at the same temperature and pressure, the volumes that react with one another are in the same relationship as the moles from the balanced equation. Because both gases are at the same temperature and pressure, you may write a new conversion factor from the balanced equation. That conversion factor is

$$\frac{2 \text{ liters CO}}{\text{liter } O_2} \cdot$$

You have simply substituted volume for moles in your "bridge," and now the solution of the problem is extremely simple.

$$? \text{ liters CO} = \underline{\hspace{3cm}}$$

The unwanted dimension in the information given, liter O_2, is in your new "bridge," and so you multiply to cancel the liter O_2.

$$? \text{ liters CO} = 5.0 \text{ liters } O_2 \times \underline{\hspace{3cm}}$$

5.0 liters O_2

$$\frac{2 \text{ liters CO}}{\text{liter } O_2}$$

The only dimension left is the dimension of the answer, which must be

$$? \text{ liters CO} = 5.0 \times 2 \text{ liters CO}$$

$$= 10 \text{ liters CO}.$$

This relationship between volumes of gases reacting at the same temperature and pressure leads to Gay-Lussac's law, which says that when gases combine they do so in simple, whole number multiples of the volumes.

Question 5 on the pretest is another example in which the two gases are at the same temperature and pressure. You have to look carefully to see that this is so.

How many liters of HCl gas at 2.0 atm pressure and 300°K will be needed to produce 5.0 liters of CO_2 at 1520 torr and 27°C according to the equation

$$CaCO_3 + 2HCl = CaCl_2 + CO_2 + H_2O?$$

If you convert 1520 torr to atmospheres,

$$\dfrac{\text{atm}}{760 \text{ torr}}$$

2.00

$$? \text{ atm} = 1520 \text{ torr} \times \underline{\hspace{1.5cm}}$$

$$= \underline{\hspace{1cm}} \text{ atm,}$$

and you convert 27°C to °K,

+273

300

$$? \text{ °K} = 27°C \underline{\hspace{1.5cm}}$$

$$= \underline{\hspace{1cm}}°K,$$

pressure

you see that both the HCl and the CO_2 are at the same _____

temperature

and _____. You can use the shortcut method.

5.0 liter CO_2

$$? \text{ liters HCl} = \underline{\hspace{2cm}}$$

Next you use your "bridge" conversion factor expressed in liters,

$$\dfrac{2 \text{ liters HCl}}{\text{liter } CO_2}$$

liter HCl

$$? \text{ liters HCl} = 5.0 \text{ liters } CO_2 \times \underline{\hspace{2cm}},$$

and the only dimension left is _____. Your answer must therefore be

$$? \text{ liters HCl} = 5.0 \times 2 \text{ liters HCl}$$

10 liters HCl

$$= \underline{\hspace{2cm}}.$$

PROBLEM SET

1. How many liters of Cl_2 gas at 40°C and 755 mm Hg pressure will be produced by the complete reaction of 20.0 g of MnO_2 according to the reaction

   $$MnO_2 + 4HCl = MnCl_2 + Cl_2 + 2H_2O?$$

2. How many liters of CO_2 at STP will be produced by the reaction of 2270 g of $CaCO_3$ following the reaction

   $$CaCO_3 = CaO + CO_2?$$

3. How many liters of O_2 gas at 1.0 atm pressure and 27°C will be consumed by the reaction of 100 g NH_3 according to the reaction

   $$4NH_3 + 5O_2 = 4NO + 6H_2O?$$

4. How many liters of NO gas collected over water at 29°C and 1.0 atm total pressure will be produced by the reaction of 100 g NH_3 according to the equation in problem 3? The vapor pressure of water at this temperature is 30 mm Hg.

5. How many milliliters of O_2 gas at 25°C and 243.1 torr pressure will be produced by the decomposition of 1.25 g of 97.5% pure $KClO_3$ according to the reaction

   $$2KClO_3 = 2KCl + 3O_2?$$

6. How many grams of C will react with 500 ml of O_2 gas at 30°C and a pressure of 740 mm Hg according to the reaction

   $$2C + O_2 = 2CO?$$

7. The Solvay process for the production of $NaHCO_3$ uses the following reaction,

   $$NaCl + NH_3 + H_2O + CO_2 = NaHCO_3 + NH_4Cl.$$

   How many pounds of $NaHCO_3$ will be produced by the complete reaction of 20.0 liters of NH_3 gas at 750 torr pressure and 27°C?

8. How many grams of $KMnO_4$ are required for the preparation of 3.0 liters Cl_2 gas at 0°C and a pressure of 0.50 atm according to the reaction

   $$2KMnO_4 + 16HCl = 2MnCl_2 + 5Cl_2 + 2KCl + 8H_2O?$$

9. How many milliliters of an HCl solution containing 10.0 moles of HCl per liter of solution are needed for the production of 15.0 liters of CO_2 at 2.0 atm pressure and -23°C following the reaction

   $$CaCO_3 + 2HCl = CaCl_2 + CO_2 + H_2O?$$

10. How many moles HCl are in a liter of solution if 50 ml of the solution are required to produce 50 ml of Cl_2 gas at 27°C and 1520 torr according to the reaction

    $MnO_2 + 4HCl = MnCl_2 + Cl_2 + 2H_2O$?

11. How many liters of NO gas will be produced when 5.0 liters of O_2 react according to the reaction

    $4NH_3 + 5O_2 = 4NO + 6H_2O$?

    Both gases are at 27°C and 2.0 atm pressure.

*12. How many gallons of water (density $= 1.00$ g/ml) are needed to yield 1.0×10^6 cu ft of H_2 gas at 20.0 atm pressure and 77°F according to the reaction

    $2H_2O + C = 2H_2 + CO_2$?

13. What will be the temperature in °C of the NO_2 gas formed from the reaction of 2.0 liters of O_2 gas at 0°C and 700 mm Hg according to the reaction

    $2NO + O_2 = 2NO_2$?

    The NO_2 has a volume of 10.0 liters and its pressure is 0.50 atm.

14. What will be the volume of NH_3 collected over water at a total pressure of 743 torr and a temperature of 25°C from the reaction of 0.15 mole of H_2 according to the equation

    $N_2 + 3H_2 = 2NH_3$?

    The vapor pressure of water at 25°C is 24 mm Hg.

15. What volume of air (4.0 volume % CO_2) at 1.05 atm pressure and 27°C will be required to yield 4.0 lb of $C_{12}H_{22}O_{11}$ according to the reaction

    $11H_2O + 12CO_2 = C_{12}H_{22}O_{11} + 12O_2$?

PROBLEM SET ANSWERS

Since you were able to handle Chapter 10 on gas laws and Chapter 8 on stoichiometry, this chapter should be easy for you. You should be able to do at least thirteen of the fifteen problems correctly.

1. ? moles Cl_2 = 20.0 g $\overline{MnO_2}$ × $\dfrac{\overline{\text{mole } MnO_2}}{86.9 \text{ g } \overline{MnO_2}}$ × $\dfrac{\text{mole } Cl_2}{\overline{\text{mole } MnO_2}}$

    = 0.230 mole Cl_2

    ? liters Cl_2 = (0.230 $\overline{\text{mole}}$) (313°K) × $\dfrac{0.0820 \text{ (liter) } \overline{\text{(atm)}}}{\overline{\text{(mole) (°K)}}}$

    $\times \dfrac{1}{755 \overline{\text{ mm Hg}}} \times \dfrac{760 \overline{\text{ mm Hg}}}{\overline{\text{atm}}}$

    = 5.94 liters Cl_2

2. $? \text{ moles } CO_2 = 2270 \text{ g } CaCO_3 \times \dfrac{\text{mole } CaCO_3}{100.1 \text{ g } CaCO_3} \times \dfrac{\text{mole } CO_2}{\text{mole } CaCO_3}$

   $= 22.70 \text{ moles } CO_2$

   $? \text{ liters } CO_2 = 22.70 \text{ moles } CO_2 \times \dfrac{22.4 \text{ liters } CO_2}{\text{mole } CO_2} \quad (\text{at STP})$

   $= 508 \text{ liters } CO_2$

3. $? \text{ moles } O_2 = 100 \text{ g } NH_3 \times \dfrac{\text{mole } NH_3}{17.0 \text{ g } NH_3} \times \dfrac{5 \text{ moles } O_2}{4 \text{ moles } NH_3}$

   $= 7.35 \text{ moles } O_2$

   $? \text{ liters } O_2 = (7.35 \text{ moles})(300°K) \times \dfrac{0.0820 \text{ (liter) (atm)}}{\text{(mole) (°K)}}$

   $\times \dfrac{1}{1.0 \text{ atm}}$

   $= 181 \text{ liters } O_2$

4. $? \text{ moles } NO = 100 \text{ g } NH_3 \times \dfrac{\text{mole } NH_3}{17.0 \text{ g } NH_3} \times \dfrac{4 \text{ moles } NO}{4 \text{ moles } NH_3}$

   $= 5.88 \text{ moles } NO$

   $p_{NO} = 760 \text{ mm Hg} - 30 \text{ mm Hg} = 730 \text{ mm Hg}$

   $? \text{ liters } NO = (5.88 \text{ moles})(302°K) \times \dfrac{0.0820 \text{ (liter) (atm)}}{\text{(mole) (°K)}}$

   $\times \dfrac{1}{730 \text{ mm Hg}} \times \dfrac{760 \text{ mm Hg}}{\text{atm}}$

   $= 152 \text{ liters } NO = 1.5 \times 10^2 \text{ liters } NO$

5. $? \text{ moles } O_2 = 1.25 \text{ g imp.} \times \dfrac{97.5 \text{ g } KClO_3}{100 \text{ g imp.}} \times \dfrac{\text{mole } KClO_3}{122.6 \text{ g } KClO_3}$

   $\times \dfrac{3 \text{ moles } O_2}{2 \text{ moles } KClO_3}$

   $= 1.49 \times 10^{-2} \text{ mole } O_2$

   $? \text{ ml } O_2 = (1.49 \times 10^{-2} \text{ mole})(298°K) \times \dfrac{0.0820 \text{ (liter) (atm)}}{\text{(mole) (°K)}}$

   $\times \dfrac{1}{243.1 \text{ torr}} \times \dfrac{760 \text{ torr}}{\text{atm}} \times \dfrac{10^3 \text{ ml}}{\text{liter}}$

   $= 1.14 \times 10^3 \text{ ml } O_2$

6. $? \text{ moles } O_2 = (0.500 \text{ liter}) (740 \text{ mm Hg}) \times \dfrac{\text{atm}}{760 \text{ mm Hg}}$

$$\times \dfrac{(\text{mole}) (°K)}{0.0820 (\text{liter}) (\text{atm})} \times \dfrac{1}{303°K}$$

$$= 0.0196 \text{ mole } O_2$$

$? \text{ g C} = 0.0196 \text{ mole } O_2 \times \dfrac{2 \text{ moles C}}{\text{mole } O_2} \times \dfrac{12.0 \text{ g C}}{\text{mole C}}$

$$= 4.70 \times 10^{-1} \text{ g C}$$

7. $? \text{ moles } NH_3 = (20.0 \text{ liters}) (750 \text{ torr}) \times \dfrac{\text{atm}}{760 \text{ torr}}$

$$\times \dfrac{(\text{mole}) (°K)}{0.0820 (\text{liter}) (\text{atm})} \times \dfrac{1}{300°K}$$

$$= 0.802 \text{ mole } NH_3$$

$? \text{ lb } NaHCO_3 = 0.802 \text{ mole } NH_3 \times \dfrac{\text{mole } NaHCO_3}{\text{mole } NH_3}$

$$\times \dfrac{84.0 \text{ g } NaHCO_3}{\text{mole } NaHCO_3} \times \dfrac{\text{lb } NaHCO_3}{454 \text{ g } NaHCO_3}$$

$$= 0.148 \text{ lb } NaHCO_3$$

8. $? \text{ moles } Cl_2 = (3.0 \text{ liters}) (0.50 \text{ atm}) \times \dfrac{(\text{mole}) (°K)}{0.0820 (\text{liter}) (\text{atm})}$

$$\times \dfrac{1}{273°K}$$

$$= 0.067 \text{ mole } Cl_2$$

$? \text{ g } KMnO_4 = 0.067 \text{ mole } Cl_2 \times \dfrac{2 \text{ moles } KMnO_4}{5 \text{ moles } Cl_2} \times \dfrac{158.0 \text{ g } KMnO_4}{\text{mole } KMnO_4}$

$$= 4.2 \text{ g } KMnO_4$$

9. $? \text{ moles } CO_2 = (15.0 \text{ liters}) (2.0 \text{ atm}) \times \dfrac{(\text{mole}) (°K)}{0.0820 (\text{liter}) (\text{atm})}$

$$\times \dfrac{1}{250°K}$$

$$= 1.46 \text{ moles } CO_2$$

$? \text{ ml soln.} = 1.46 \text{ moles } CO_2 \times \dfrac{2 \text{ moles HCl}}{\text{mole } CO_2}$

$$\times \dfrac{\text{liter soln.}}{10.0 \text{ moles HCl}} \times \dfrac{\text{ml soln.}}{10^{-3} \text{ liter soln.}}$$

$$= 292 \text{ ml soln.} = 2.9 \times 10^2 \text{ ml soln.}$$

10. $? \text{ moles } Cl_2 = (0.050 \text{ liter}) (1520 \text{ torr}) \times \dfrac{\text{atm}}{760 \text{ torr}}$

$$\times \dfrac{(\text{mole}) (°K)}{0.0820 (\text{liter}) (\text{atm})} \times \dfrac{1}{300°K}$$

$$= 4.1 \times 10^{-3} \text{ mole } Cl_2$$

$? \text{ moles HCl} = 1 \text{ liter soln.} \times \dfrac{4.1 \times 10^{-3} \text{ mole } Cl_2}{0.050 \text{ liter soln.}} \times \dfrac{4 \text{ moles HCl}}{\text{mole } Cl_2}$

$$= 0.33 \text{ mole HCl}$$

11.   Since both gases are at the same P and T, you may substitute liters for moles.

$? \text{ liters NO} = 5.0 \text{ liters } O_2 \times \dfrac{4 \text{ liters NO}}{5 \text{ liters } O_2}$

$$= 4.0 \text{ liters NO}$$

12.   First determine the volume in liters.

$? \text{ liters} = 1.0 \times 10^6 \text{ ft}^3 \times \dfrac{(12)^3 \text{ in.}^3}{\text{ft}^3} \times \dfrac{(2.54)^3 \text{ cm}^3}{\text{in.}^3}$

$$\times \dfrac{\text{ml}}{\text{cm}^3} \times \dfrac{10^{-3} \text{ liter}}{\text{ml}}$$

$$= 2.8 \times 10^7 \text{ liters}$$

Then find the temperature in °K.

$? °C = \dfrac{5}{9} (°F - 32) = \dfrac{5}{9} (77 - 32) = 25°C$

$? °K = (25 + 273) = 298°K$

$? \text{ moles } H_2 = (2.8 \times 10^7 \text{ liters}) (20.0 \text{ atm})$

$$\times \dfrac{(\text{mole}) (°K)}{0.0820 (\text{liter}) (\text{atm})} \times \dfrac{1}{298°K}$$

$$= 2.3 \times 10^7 \text{ moles } H_2$$

$? \text{ gal } H_2O = 2.3 \times 10^7 \text{ moles } H_2 \times \dfrac{2 \text{ moles } H_2O}{2 \text{ moles } H_2} \times \dfrac{18.0 \text{ g } H_2O}{\text{mole } H_2O}$

$$\times \dfrac{\text{ml } H_2O}{1.00 \text{ g } H_2O} \times \dfrac{\text{qt } H_2O}{946 \text{ ml } H_2O} \times \dfrac{\text{gal } H_2O}{4 \text{ qt } H_2O}$$

$$= 1.1 \times 10^5 \text{ gal } H_2O$$

13.　? moles O_2 = (2.0 liters) (700 mm Hg) $\times \dfrac{atm}{760\ mm\ Hg}$

$$\times \frac{(mole)\ (°K)}{0.0820\ (liter)\ (atm)} \times \frac{1}{273°K}$$

= 0.082 mole O_2

? moles NO_2 = 0.082 mole $O_2 \times \dfrac{2\ moles\ NO_2}{mole\ O_2}$

= 0.164 mole NO_2

? °K = (10.0 liters) (0.50 atm)

$$\times \frac{(mole)\ (°K)}{0.0820\ (liter)\ (atm)} \times \frac{1}{0.164\ mole}$$

= 372°K

? °C = 372 − 273 = 99°C

14.　? moles NH_3 = 0.15 mole $H_2 \times \dfrac{2\ moles\ NH_3}{3\ moles\ H_2}$

= 0.10 mole NH_3

To get the partial pressure NH_3

$$p_{NH_3} = 743\ torr - \left(25\ mm\ Hg \times \frac{1\ torr}{mm\ Hg} \right) = 718\ torr$$

? liters NH_3 = (0.10 mole) (298°K) $\times \dfrac{0.0820\ (liter)\ (atm)}{(mole)\ (°K)}$

$$\times \frac{760\ torr}{atm} \times \frac{1}{718\ torr}$$

= 2.6 liters NH_3

15.　? moles CO_2 = 4.0 lb $C_{12}H_{22}O_{11} \times \dfrac{454\ g\ C_{12}H_{22}O_{11}}{lb\ C_{12}H_{22}O_{11}}$

$$\times \frac{mole\ C_{12}H_{22}O_{11}}{342.0\ g\ C_{12}H_{22}O_{11}} \times \frac{12\ moles\ CO_2}{mole\ C_{12}H_{22}O_{11}}$$

= 64 moles CO_2

? liters air = (64 moles CO_2) (300°K) $\times \dfrac{0.0820\ (liter\ CO_2)\ (atm)}{(mole)\ (°K)}$

$$\times \frac{1}{1.05\ atm} \times \frac{100\ liters\ air}{4.0\ liters\ CO_2}$$

= 3.7×10^4 liters air

Chapter 12
Solution Concentration I :
Molarity, Formality, Molality

PRETEST

Before you try these questions, you will need three new conversion factors.

$$M = \text{molarity} = \frac{\text{moles solute}}{\text{liter solution}}$$

$$F = \text{formality} = \frac{\text{formula weights solute}}{\text{liter solution}}$$

$$m = \text{molality} = \frac{\text{moles solute}}{\text{kilogram solvent}}$$

The following problems are quite simple if you know the conversions. You should be able to finish them in 20 minutes.

1. How many grams of HCl are in 3.00 liters of 1.5 M HCl?

2. To what volume in liters must you dilute 11.7 g NaCl to prepare a 0.10 F solution of NaCl?

3. What is the molarity of an NaCl solution which was prepared by adding 23.4 g NaCl to 100 g water? The density of the resulting solution is 1.20 g/ml.

4. How many liters of a 0.70 M solution of HCl would you have to take in order to have 85.0 g HCl?

5. How many milliliters of a 6.0 M HCl solution must you dilute to 3.0 liters to have a 0.10 M HCl solution?

6. What is the formality of a solution prepared by adding enough water to 250 ml of a 6.0 F solution of KCl to bring the volume up to 5.0 liters?

7. What will the molarity be of a solution prepared by mixing 3.0 liters of 0.10 M HCl with 1.0 liter of 6.0 M HCl?

8. What is the molality of an alcohol solution which was prepared by dissolving 23.0 g of alcohol (C_2H_5OH) in 200 g water?

9. What is the molality of a 6.0 M solution of HCl whose density is 1.1 g/ml?

PRETEST ANSWERS

If you answered all nine questions correctly, skip Chapter 12 and go directly to the problem set at the end of the chapter. If you missed questions 1, 3, or 4, you had better go through Section A. If you missed question 2, check Section B for the definition of formality; if you missed questions 5 and 6, work through Section D in the chapter; and if you missed question 7, do Section E. Questions 8 and 9 come from Section C. Incidentally, you will find all of these questions worked out as examples in the chapter.

1. 164 g HCl (or 1.6×10^2 to the correct number of significant figures)

2. 2.0 liters

3. 3.89 *M* (If you got 4.80 *M*, you did the problem incorrectly; you did not add the weight of solvent to solute.)

4. 3.3 liters

5. 50 ml

6. 0.30 *F*

7. 1.6 *M*

8. 2.50 *m*

9. 6.8 *m*

Up to this point we have always described any solution we had as so much weight percent of one substance dissolved in another, or sometimes as so many milliliters of one substance in another. This is perfectly correct, but it is not always the most convenient way to do it. If you have a compound which is involved in a chemical reaction and you are going to use a "bridge" conversion factor from the balanced equation, you must always convert whatever

units the compound is given in to _____. It would really be much easier if, instead of expressing a concentration in weight percent, you had expressed the concentration in moles to start with.

	moles

What we are going to do in this chapter is to examine three different ways of expressing concentrations which are commonly used because they simplify your calculations. We will start with what is called molarity.

SECTION A   Molarity

When you work in a laboratory, it is much faster and easier to measure a volume of a solution than to weigh it on a balance. Consequently, a concentration expression was set up which would

have the amount of the substance of interest expressed in _____, because these are convenient, and the amount of solution expressed in the metric unit of volume, which is liters.

	moles

When concentrations are expressed this way, they are called molarity and abbreviated as M. Thus a 6.0 M solution of sugar will

have ____ moles of sugar in __ liter of solution or writing this as a fractional conversion factor,

	6.0, 1

$$\frac{6.0 \text{ moles sugar}}{\text{Liter solution}}.$$

If you have 3 moles of acetic acid dissolved in a liter of solution,

you call this a ____ or _____ solution.

	3 *M*, 3 molar

There are several new terms that you will be using when you work with solutions. First there is the *solute*, which is the substance that you dissolve. You dissolve the solute in a *solvent*, and the resulting mixture is called the *solution*. Generally, the solute is present in smaller amounts than solvent.

Now we can give molarity its exact definition,

$$\boxed{\text{Molarity} = M = \frac{\text{moles solute}}{\text{liter solution}}.}$$

As you can see, it is a conversion factor.

How about trying a few problems? They are really quite easy.

How many moles of HNO_3 are in 300 ml of a 6.0 *M* solution of HNO_3? Starting in the usual way, you write

$$? \, \underline{\hspace{2cm}} = \underline{\hspace{2cm}}.$$

	moles HNO_3, 300 ml soln.

$$\frac{6.0 \text{ moles } HNO_3}{\text{liter soln.}}$$

$$\frac{\text{liter solution}}{\text{ml solution}}$$

$$10^{-3} \text{ liter solution}$$

mole HNO_3

If you write 6.0 *M* HNO_3 as a conversion factor given in the problem, you have _____. In order to be able to use this conversion factor, you must convert the unwanted dimension, milliliters solution, to _____. You can do this by multiplying by _____. Now you can multiply by the molarity, the conversion factor given in the problem, and the only dimension left which does not cancel is _____. Therefore, the answer must be

$$? \text{ moles } HNO_3 = 300 \cancel{\text{ ml soln.}} \times \frac{10^{-3} \cancel{\text{ liter soln.}}}{\cancel{\text{milliliter soln.}}}$$

$$\times \frac{6.0 \text{ moles } HNO_3}{\cancel{\text{liter soln.}}}$$

$$= 1.8 \text{ moles } HNO_3.$$

Try problem 1 on the pretest, which asks how many grams of HCl are in 3.00 liters of 1.5 *M* HCl? Starting in the usual way, you have

grams HCl,   3.00 liters soln.

$$? \underline{\hspace{2cm}} = \underline{\hspace{3cm}}.$$

$$\frac{1.5 \text{ moles } HCl}{\text{liter soln.}}$$

mole HCl

grams HCl

$$\frac{36.5 \text{ g } HCl}{\text{mole } HCl}$$

The conversion factor given in the problem is the molarity of the HCl solution, _____. Multiplying by this conversion factor, the molarity will cancel the unwanted dimension, liter solution, and leave a new unwanted dimension, _____. However, the dimension of the answer is _____, so you must multiply by the conversion factor _____, which is the molecular weight of HCl. The answer will be

$$? \text{ g } HCl = 3.00 \cancel{\text{ liters soln.}} \times \frac{1.5 \cancel{\text{ moles } HCl}}{\cancel{\text{liter soln.}}} \times \frac{36.5 \text{ g } HCl}{\cancel{\text{mole } HCl}}$$

$$= 164 \text{ g } HCl.$$

Question 4 on the pretest is the same sort of problem turned the other way around. How many liters of a 0.70 *M* solution of HCl would you have to take in order to have 85.0 g HCl? (Remember to write your own setup as you fill in the answers to the blanks.)
Starting in the usual way, you have

liters soln.,   85.0 g HCl

$$? \underline{\hspace{2cm}} = \underline{\hspace{2cm}}.$$

grams HCl

$$\frac{0.70 \text{ mole } HCl}{\text{liter soln.}}$$

$$\frac{\text{mole } HCl}{36.5 \text{ g } HCl}$$

The unwanted dimension is _____. The conversion factor given in the problem is the molarity of the solution, _____. You can get to this from the unwanted dimension of the information given by multiplying by _____, the inverted form of the molecular weight. The unwanted dimension

left is _____, which appears in the conversion factor, the

molarity, given in the problem. Next you multiply by _____,

and the only remaining dimension is _____, which is the
dimension that the answer must have. Therefore, the answer is

$$? \text{ liters soln.} = 85.0 \text{ g HCl} \times \frac{\text{mole HCl}}{36.5 \text{ g HCl}} \times \frac{\text{liter soln.}}{0.70 \text{ mole HCl}}$$

$$= 3.3 \text{ liters soln.}$$

How can you determine the molarity of a solution? You can do
this most easily by using the same type of shortcut as in Chapter 4,
that we used for density, which is also a conversion factor. You
say that your answer must have the dimensions of molarity, which

are _____. Then you set up the problem so that all
dimensions except these two cancel. A simple example will
make this clear.

What is the molarity of an acetic acid solution if there are 3.0
moles of acetic acid in 4.0 liters of solution? Since the answer will

be molarity, its dimensions must be _____; so you
write

$$? M = \frac{? \text{ moles acetic acid}}{\text{liter soln.}} =$$

The conversion factor given in the problem is _____.
If you write this in, you see that you are left with exactly the
dimensions that the answer must have.

$$? M = \frac{? \text{ moles acetic acid}}{\text{liter soln.}} = \frac{3.0 \text{ moles acetic acid}}{4.0 \text{ liters soln.}}$$

$$= 0.75 \ M$$

Of course, you won't always get all the information presented to
you in such an easy form. However, what you must do is simply
to keep multiplying by the appropriate conversions which will get
you to the two dimensions of the answer.

Here is another problem. What is the molarity of a solution which
contains 8.0 g NaOH in 300 ml of solution? Since the question
asked is the molarity, then the answer will have the

dimensions _____. The conversion factor given in the

problem is _____.
You start by writing

$$? M = \frac{? \text{ moles NaOH}}{\text{liter soln.}} = \frac{8.0 \text{ g NaOH}}{300 \text{ ml soln.}}.$$

$$\frac{\text{mole HCl}}{\text{liter soln.}}$$
$$\frac{0.70 \text{ mole HCl}}{\text{liter soln.}}$$

$$\frac{\text{moles solute}}{\text{liter soln.}}$$

$$\frac{\text{moles acetic acid}}{\text{liter soln.}}$$

$$\frac{3.0 \text{ moles acetic acid}}{4.0 \text{ liters soln.}}$$

$$\frac{\text{mole NaOH}}{\text{liter soln.}}$$
$$\frac{8.0 \text{ g NaOH}}{300 \text{ ml soln.}}$$

moles

liters

$$\frac{\text{moles NaOH}}{40.0 \text{ g NaOH}}, \quad \frac{\text{ml soln.}}{10^{-3} \text{ liter soln.}}$$

Now you must convert grams NaOH to _____ NaOH and milliliters solution to _____ solution. You can do this by multiplying by the two conversion factors _____ and _____. The answer, then, is

$$? M = \frac{? \text{ moles NaOH}}{\text{liter soln.}} = \frac{8.0 \text{ g NaOH}}{300 \text{ ml soln.}} \times \frac{\text{mole NaOH}}{40.0 \text{ g NaOH}}$$

$$\times \frac{\text{ml soln.}}{10^{-3} \text{ liter soln.}}$$

$$= \frac{0.67 \text{ mole NaOH}}{\text{liter soln.}}$$

$$= 0.67 \text{ } M.$$

Sometimes unexpected difficulties are put in your way. Instead of having information about the volume of solution, you will have information about the amounts of solvent. In these cases you must add the amount of solute to the amount of solvent to get the amount of solution. Question 3 on the pretest is a typical example of this, asking you to find the molarity of an NaCl solution which was prepared by adding 23.4 g NaCl to 100 g water. The density of the resulting solution was 1.20 g/ml.

First you must determine the weight of the solution. It is equal

NaCl

123.4

to the weight of _____ plus the weight of water, so you have

_____ grams solution. Now you can get the conversion factor given in the problem as

$$\frac{23.4 \text{ g NaCl}}{\text{_____ g soln.}}$$

123.4

You also have another conversion factor given in the problem,

$$\frac{1.20 \text{ g soln.}}{\text{ml soln.}}$$

_____, the density of the solution. Now the problem is straightforward.

moles NaCl

$$? M = \frac{?\text{_____}}{\text{liter soln.}} = \frac{23.4 \text{ g NaCl}}{123.4 \text{ g soln.}}$$

moles

$$\frac{\text{mole NaCl}}{58.5 \text{ g NaCl}}$$

liter

$$\frac{1.20 \text{ g soln.}}{\text{ml soln.}}$$

milliliters solution

$$\frac{\text{ml soln.}}{10^{-3} \text{ liter soln.}}$$

Next you must convert grams NaCl to _____ NaCl, which can be done by multiplying by the conversion factor _____.

Then you must convert grams solution to _____ solution. You can do this in two steps. First you can multiply by _____ to remove the unwanted dimension grams solution. The new unwanted dimension in the denominator is _____, and you can get to the desired liter solution by multiplying by

_____.

The answer, then, is

$$? M = \frac{? \text{ mole NaCl}}{\text{liter soln.}}$$

$$= \frac{23.4 \text{ g NaCl}}{123.4 \text{ g soln.}} \times \frac{\text{mole NaCl}}{58.5 \text{ g NaCl}}$$

$$\times \frac{1.20 \text{ g soln.}}{\text{ml soln.}} \times \frac{\text{ml soln.}}{10^{-3} \text{ liter soln.}}$$

$$= \frac{3.89 \text{ moles NaCl}}{\text{liter soln.}}$$

$$= 3.89 \ M.$$

If you didn't add the weight of the NaCl to the weight of water to get the weight of solution, you would have found the incorrect answer, 4.80 *M*, mentioned in the pretest. This is another example of why you must completely label your dimensions.

This problem is about as complicated as they can become. Incidentally, the density of the solution is not the same as the density of solvent. However, for very dilute solutions in which the amount of solute is very small, you can assume that the addition of a little solute won't change the density of the solvent significantly.

You must be very careful if you are given a problem of this type with the amounts of solute and solvent specified in volumes. In many cases the total volume of a solution is not simply the sum of the volumes of the components. If this sort of problem arises, you must have the densities of the solute and solvent available so that you can determine their weights, which are additive. This type of problem is so seldom encountered that we won't even do an example.

SECTION B   Formality

By this point in the chapter some chemists and chemistry teachers might question our choice of words, saying that we have been using "molarity" when we should have been using "formality." As you may know, some substances, when they dissolve in water, break up into charged ions. Thus HCl, dissolved in water, exists as H^+ ions and Cl^- ions. No molecules of HCl are present. Perhaps you will recall that we mentioned this distinction in Chapter 5. At that point we said that you could express molecular weights for

these ionized compounds as gram-formula _____. The   weights
conversion factors would be similar to molecular weight, which is g/mole. Formula weight is expressed as g/gram-formula weight. Thus, while we were talking about the *molecular weight* of HCl when it was dissolved in water, some chemists would say that you can only talk about the *formula weight* since no molecules are present. The formula weight is, of course, identical with the

molecular weight. The end result is that when you talk about a 1.0 M solution of HCl, meaning that there is 1.0 mole of HCl per liter, you could call this a 1.0 F solution of HCl, which means that there is 1.0 formula weights of HCl per liter.

$$F = \text{formality} = \frac{\text{formula weights solute}}{\text{liter solution}}$$

Question 2 on the pretest asks to what volume in liters you must dilute 11.7 g NaCl to prepare a 0.10 F solution of NaCl. Starting in the usual way, you have

liters soln.

$$? \underline{\hspace{2cm}} = 11.7 \text{ g NaCl.}$$

The conversion factor given in the problem is that you have a 0.10 F solution. This can be written as

formula wt

$$\frac{0.10 \underline{\hspace{2cm}} \text{NaCl}}{\text{Liter soln.}} .$$

The formula weight of NaCl (which is exactly equal to what we have been calling molecular weight) is

Formula wt

$$\frac{58.5 \text{ g NaCl}}{\underline{\hspace{2cm}} \text{NaCl}} .$$

Now you can solve the problem. In order to cancel the

$$\frac{\text{formula wt NaCl}}{58.5 \text{ g NaCl}}$$

formula wt NaCl

unwanted dimension you multiply by \underline{\hspace{2cm}}. This leaves the unwanted dimension \underline{\hspace{2cm}}. You can cancel this dimension and get to the dimensions of your answer by

$$\frac{\text{liter soln.}}{0.10 \text{ formula wt NaCl}}$$

multiplying by \underline{\hspace{2cm}}. The answer is therefore

$$? \text{ liters soln.} = 11.7 \text{ g NaCl} \times \frac{\text{formula wt NaCl}}{58.5 \text{ g NaCl}}$$

$$\times \frac{\text{liter soln.}}{0.10 \text{ formula wt NaCl}}$$

$$= 2.0 \text{ liters soln.}$$

Since molarity and formality problems are calculated in exactly the same way, you might simply interchange M (molarity) and F (formality) when you are faced with problems containing F (formality).

SECTION C Molality

There is one more concentration unit which deals with moles of solute. This is *molality*, which is defined as

$$\text{Molality} = m = \frac{\text{moles solute}}{1000 \text{ g solvent}} = \frac{\text{moles solute}}{\text{kg solvent}} .$$

Notice that the moles of solute are not related to volume of solution, but rather to weight of solvent. This type of relationship finds use in only a special class of problems. These problems are related to what are called the *colligative properties* of solutions, which will be covered in Chapter 16. For the moment it is only necessary that you know how to handle simple problems expressing the concentration in molality.

We can start right out with question 8 on the pretest, which asks for the molality of an alcohol solution prepared by dissolving 23.0 g alcohol (C_2H_5OH) in 200 g water. The specific question

asked is "_____" Therefore the dimensions that

the answer will have are moles alcohol/___ water, and you can write

$$? \, m = \frac{? \text{ moles alcohol}}{\text{kg water}} =$$

The conversion factor given in the problem is _____.

You can convert grams alcohol to _____ alcohol and

convert grams water to _____ water by multiplying by two

conversion factors, _____ and _____. And so the answer is

$$? \, m = \frac{? \text{ moles alcohol}}{\text{kg water}}$$

$$= \frac{23.0 \, \cancel{\text{g alcohol}}}{200 \, \cancel{\text{g water}}} \times \frac{\text{mole alcohol}}{46.0 \, \cancel{\text{g alcohol}}} \times \frac{10^3 \, \cancel{\text{g water}}}{\text{kg water}}$$

$$= 2.50 \, m.$$

There are many ways to present the conversion factor given in the problem. Take, for example, the problem of finding the molality of a 30.0 wt % solution of ethylene glycol in water. (The formula for ethylene glycol is $C_2H_6O_2$.) Here you must use the fact that the

sum of the percentages always adds up to _____, so that if the

solution is 30.0 wt % ethylene glycol, it must be ____ wt % water. As a matter of fact, you can make a new conversion factor from this:

there are _____ g ethylene glycol per _____ g water, or, writing this

as a fraction, _____.

Now the solution to the problem is the same as in the first example.

$$? \, m = \frac{? \, _____ \text{ ethylene glycol}}{_____} =$$

What is the molality?

kg

23.0 g alcohol
200 g water

moles

kilograms

$\dfrac{\text{mole alcohol}}{46.0 \text{ g alcohol}}, \quad \dfrac{10^3 \text{ g water}}{\text{kg water}}$

100

70.0

30.0, 70.0

$\dfrac{30.0 \text{ g ethylene glycol}}{70.0 \text{ g water}}$

moles

kg water

If you now put in the conversion factor that you developed from the information given in the problem, you have

$$? m = \frac{? \text{ moles ethylene glycol}}{\text{kg water}} = \frac{30.0 \text{ g ethylene glycol}}{70.0 \text{ g water}}.$$

moles

kilograms

$$\frac{\text{mole ethylene glycol}}{62.0 \text{ g ethylene glycol}}, \quad \frac{10^3 \text{ g water}}{\text{kg water}}$$

You must then convert grams ethylene glycol to _____ ethylene glycol and grams water to _____ water. This is done using the two conversion factors, _____ and _____. And so the answer is

$$? m = \frac{30.0 \text{ g ethylene glycol}}{70.0 \text{ g water}}$$

$$\times \frac{\text{mole ethylene glycol}}{62.0 \text{ g ethylene glycol}} \times \frac{10^3 \text{ g water}}{\text{kg water}}$$

$$= 6.91 \ m.$$

It is possible to go from a concentration given in molarity to a concentration in molality, but you must know the density of the solution. To find the molality you must know the weights of solute and solvent separately. If you have the density of the solution, you can calculate how much a liter of it weighs; if you know the number of moles of solute in a liter (the molarity), you can calculate the weight of solute in a liter; then, if you subtract the weight of solute in a liter from the total weight of a liter of solution, you will have the weight of solvent in a liter of solution. This seems very complicated but only because it must be done in several steps. Working question 9 on the pretest will clarify this type of problem.

What is the molality of a 6.0 M solution of HCl whose density is 1.1 g/ml? First find the weight of a liter of the solution.

$$? \text{ g soln.} = 1.0 \text{ liter soln.}$$

$$\frac{1.1 \text{ g soln.}}{\text{ml soln.}}$$

milliliters

$$\frac{\text{ml soln.}}{10^{-3} \text{ liter soln.}}, \quad \frac{1.1 \text{ g soln.}}{\text{ml soln.}}$$

1100

You can do this using the density of the solution, _____,

after first converting liter solution to _____ solution. The answer will be

$$? \text{ g soln.} = 1.0 \text{ liter soln.} \times \text{_____} \times \text{_____}$$

$$= \text{_____ g soln.}$$

Next you must calculate the weight of HCl in 1.0 liter solution.

$$? \text{ g HCl} = 1.0 \text{ liter soln.}$$

Since you know that the solution is 6.0 M HCl, you have a

$$\frac{6.0 \text{ moles HCl}}{\text{liter soln.}}$$

mole HCl

conversion factor given; it is _____. If you multiply by this, you cancel liter solution but have the new unwanted dimension, _____. However, it is very simple to get from this unwanted dimension to the dimension of the answer by multiplying by _____. And so the answer is

$$? \text{ g HCl} = 1.0 \text{ } \cancel{\text{liter soln.}} \times \frac{6.0 \text{ moles } \cancel{\text{HCl}}}{\cancel{\text{liter soln.}}} \times \frac{36.5 \text{ g HCl}}{\cancel{\text{mole HCl}}}$$

$$= 219 \text{ g HCl.}$$

You now know the weight of one liter of solution and the weight of HCl which it contains. You can find the weight of water in the solution by subtraction.

$$? \text{ g water} = 1100 \text{ g soln.} - 219 \text{ g HCl} = 881 \text{ g water}$$

You now know that 1.0 liter of the solution containing ___

moles of HCl also contains _____ grams water. This is the conversion factor that you need to solve the problem. Now you start the usual way.

$$? \text{ } m = \frac{? \text{ _____}}{\text{_____}} =$$

You put in the conversion factor, which you have calculated.

$$? \text{ } m = \frac{? \text{ moles HCl}}{\text{kg water}} = \text{_____.}$$

All you must do now is convert _____ water to _____ water.

Therefore you multiply by _____. And so the answer is

$$? \text{ } m = \frac{6.0 \text{ moles HCl}}{881 \text{ } \cancel{\text{g water}}} \times \frac{10^3 \text{ } \cancel{\text{g water}}}{\text{kg water}}$$

$$= 6.8 \text{ } m.$$

The same type of problem could be asked the other way around. You are given a solution of known molality and are asked to calculate the molarity. In this case you must determine the weight of solute per kg solvent and add this weight to the 1000 g solvent. Then, using the density of solution, calculate what volume of solution this composite weight of solution equals. You then know the number of moles of solute per this calculated volume of solution and can easily calculate the molarity.

These complicated problems occur infrequently. We won't spend any more time on them, but will go on to more usual calculations.

SECTION D   Preparing Dilute Solutions from Concentrated Ones

Very frequently in a laboratory you will have available a solution of known molarity or molality and will want to use it to prepare a certain volume of a more dilute solution. Calculating how much to use is very simple, except that you will be working with two

---

Right margin answers:

$$\frac{36.5 \text{ g HCl}}{\text{mole HCl}}$$

6.0

881

$$\frac{\text{moles HCl}}{\text{kg water}}$$

$$\frac{6.0 \text{ moles HCl}}{881 \text{ g water}}$$

grams,   kilograms

$$\frac{10^3 \text{ g water}}{\text{kg water}}$$

different solutions: one the initial solution that you have and the other the final solution that you wish to prepare. You must be very careful to label your dimensions so that you do not confuse the solutions. We will work through question 5 on the pretest, which will make this very clear.

How many milliliters of a 6.0 *M* solution would you dilute to 3.0 liters in order to have a 0.10 *M* HCl solution? You will probably find it easiest to call the 6.0 *M* HCl solution the "initial solution" and the 0.10 *M* HCl solution the "final solution." There are two molarities given in the problem and therefore two conversion factors,

$$\frac{6.0 \text{ moles HCl}}{\text{Liter init. soln.}} \quad \text{and} \quad \frac{0.10 \text{ mole HCl}}{\text{Liter} \underline{\hspace{1cm}}}$$

final soln.

The information given in the problem is that the volume of the

3.0 liters final soln.

final solution will be _____. Thus, starting in the usual way, you write

init. soln.,   final soln.

? ml _____ = 3.0 liters _____.

Notice how careful you must be to distinguish between the two solutions.

liter final soln.

$\dfrac{0.10 \text{ mole HCl}}{\text{liter final soln.}}$

mole HCl

$\dfrac{\text{mole HCl}}{\text{liter init. soln.}}$

$\dfrac{6.0 \text{ moles HCl}}{\text{liter init. solution}}$

milliliters init. solution

$\dfrac{\text{ml init. soln.}}{10^{-3} \text{ liter init. soln.}}$

To cancel the unwanted dimension _____, you

multiply by the conversion factor _____. The unwanted

dimension now is _____, but this appears in the other

conversion factor, and so you can multiply by _____.

The unwanted dimension now is _____, but you want

your answer to have the dimensions _____. You

can make this conversion by multiplying by _____. Your setup should therefore appear as

$$? \text{ ml init. soln.} = 3.0 \text{ liters final soln.} \times \frac{0.10 \text{ mole HCl}}{\text{liter final soln.}}$$

$$\times \frac{\text{liter init. soln.}}{6.0 \text{ moles HCl}} \times \frac{\text{ml init. soln.}}{10^{-3} \text{ liter init. soln.}}$$

$$= 50 \text{ ml init. soln.}$$

These problems are exceptionally easy if you scrupulously write down the complete dimension of each solution to which you are referring.

Now let's do question 6 in the pretest. What is the formality of a solution prepared by adding enough water to 250 ml of a 6.0 *F* solution of KCl to bring the volume up to 5.0 liters? Although this problem concerns formality, we are going to solve it as if it were molarity since the two are identical.

The problem contains two conversion factors: one is

_____, the molarity (formality) of the initial solution; the

other conversion factor given is that there are 250 ml of _____

solution in 5.0 liters _____ solution, which can be written

fractionally as _____.

Since you are interested in the molarity of the final solution, you write

$$? M \text{ final solution} = \frac{? \text{ moles KCl}}{\text{liter final soln.}}.$$

Now you can pick either of the conversion factors given in the problem to start. It must have one of the dimensions of the answer in the correct place. Say you start with the first,

$$\frac{6.0 \text{ moles KCl}}{\text{Liter init. soln.}}$$

This gives you moles KCl in the numerator but has an unwanted

dimension in the denominator, which is _____. However, your second conversion factor allows you to cancel this. You

therefore multiply by _____. (Notice how 250 ml was changed to 0.250 liter. You will start doing this by yourself soon. It saves writing one more conversion factor.)

The dimensions which remain are the dimensions of the answer.

$$? F = ? M = \frac{? \text{ moles KCl}}{\text{liter final soln.}} = \frac{6.0 \text{ moles KCl}}{\cancel{\text{liter init. soln.}}}$$

$$\times \frac{0.250 \ \cancel{\text{liter init. soln.}}}{5.0 \text{ liters final soln.}}$$

$$= 0.30 \ F$$

Sometimes the concentration of the initial solution isn't given to you in molarity, molality, or formality, but this doesn't really make the problem much more difficult. For example, let's find how many grams of a 35 wt % solution of NaOH must be diluted to 25.0 ml to have a 0.10 *M* solution of NaOH.

In this case the initial NaOH solution is shown as a weight percent. The conversion factor that you have from this is

_____. The concentration of the final solution, however, is given in molarity. The conversion factor you have from

this is _____. The information given in the problem is

that you want to prepare 25.0 ml _____.

Starting in the usual way, then,

$$? \text{ g init. soln.} = 25.0 \text{ ml final soln.}$$

$$\frac{6.0 \text{ moles KCl}}{\text{liter init. soln.}}$$
initial

final
$$\frac{250 \text{ ml init. soln.}}{5.0 \text{ liters final soln.}}$$

liters init. soln.

$$\frac{0.250 \text{ liter init. soln.}}{5.0 \text{ liters final soln.}}$$

$$\frac{35 \text{ g NaOH}}{100 \text{ g init. soln.}}$$

$$\frac{0.10 \text{ mole NaOH}}{\text{liter final soln.}}$$
final soln.

One of the conversion factors has a volume of final solution in it,

liters

but it is expressed in _____ final solution. However, you can save

10^3

yourself one conversion factor by expressing 1 liter as _____ ml.
Then you can multiply by

$$\frac{0.10 \text{ mole NaOH}}{10^3 \text{ ml final soln.}}$$

moles NaOH

The unwanted dimension now is _____, but the other

grams

conversion factor given in the problem uses _____ NaOH.

grams

Consequently, you must convert moles NaOH to _____ NaOH by

$$\frac{40.0 \text{ g NaOH}}{\text{mole NaOH}}$$

multiplying by the conversion factor _____. Now you can multiply

$$\frac{100 \text{ g init. soln.}}{35 \text{ g NaOH}}$$

by the conversion factor given in the problem, _____.

grams initial solution

The only dimension remaining is _____, and so the
answer must be

$$? \text{ g init. soln.} = 25.0 \text{ ml final soln.} \times \frac{0.10 \text{ mole NaOH}}{10^3 \text{ ml final soln.}}$$

$$\times \frac{40.0 \text{ g NaOH}}{\text{mole NaOH}} \times \frac{100 \text{ g init. soln.}}{35 \text{ g NaOH}}$$

$$= 0.29 \text{ g init. soln.}$$

For practice we will try another problem which is just a little
more complicated. Instead of wanting to know the weight of initial
solution, you will want the volume.

How many milliliters of a 10 wt % solution of NaOH whose
density is 1.1 g/ml will you need to prepare 300 ml of a 0.20 *M*
solution of NaOH? There are three conversion factors given in the
problem, and you might start by writing these down.

$$\frac{10 \text{ g NaOH}}{100 \text{ g \_\_\_\_\_}}$$

init. soln.

$$\frac{1.1 \text{ g \_\_\_\_\_}}{\text{ml \_\_\_\_\_}}$$

init. soln.

init. soln.

$$\frac{0.20 \text{ mole NaOH}}{\text{liter _____}}$$

final soln.

Since the solute is expressed in grams in one of the conversion
factors, you will also undoubtedly need a conversion factor which

mole

converts grams NaOH to _____ NaOH. This conversion factor is

molecular

the _____ weight and can be written as

$$\frac{40.0 \text{ g NaOH}}{\text{\_\_\_\_\_ NaOH}}$$

mole

Now you are ready to start. The question asked is "How many

milliliters _____ solution?" The information given is 0.300 liter

_____ solution. (Notice how 300 ml was immediately written as
0.300 liters; this will save one conversion factor.) Starting in the
usual way, you have

>   ? ml init. soln. = 0.300 liter final soln.

From the conversion factors that you have written down, you

pick one with _____ in the denominator. So you

multiply by _____. The unwanted dimension now is

_____. Since the other conversion factor given in the

problem is in _____ NaOH, you must multiply by the molecular

weight conversion factor, _____. Now you can multiply by

_____. The unwanted dimension now is

_____, but the last conversion factor, the density,
will get you to the dimensions of the answer. You multiply by

_____, and the only dimension remaining is the
dimension of the answer, which must be

> ? ml init. soln. = $0.300 \, \overline{\text{liter final soln.}} \times \dfrac{0.20 \, \overline{\text{mole NaOH}}}{\overline{\text{liter final soln.}}}$

> $\times \, \dfrac{40.0 \, \text{g NaOH}}{\overline{\text{mole NaOH}}} \times \dfrac{100 \, \overline{\text{g init. soln.}}}{10 \, \overline{\text{g NaOH}}}$

> $\times \, \dfrac{\text{ml init. soln.}}{1.1 \, \overline{\text{g init. soln.}}}$

> = 22.0 ml init. soln.

initial

final

$\dfrac{\text{liter final solution}}{0.20 \, \text{mole NaOH}}$
$\dfrac{\text{liter final soln.}}{\text{mole NaOH}}$

$\dfrac{\text{grams}}{40.0 \, \text{g NaOH}}$
$\dfrac{\text{mole NaOH}}{100 \, \text{g init. soln.}}$
$\dfrac{10 \, \text{g NaOH}}{\text{grams initial solution}}$

$\dfrac{\text{ml init. soln.}}{1.1 \, \text{g init. soln.}}$

SECTION E   Mixtures of Solutions
of Different Molarity

In a laboratory you will occasionally mix two or more solutions of
the same substance which are of different concentration. In order
to calculate the concentration of the resulting mixture you must
know the total amount of solute present and the total volume of
the mixture.

   You may prefer to do this calculation in a series of steps. First
you calculate the amount of solute in each sample taken. Next you
add these together. Then you add together the volumes of the
samples. Using these two composite amounts, you can calculate
the concentration of the mixture. We will do question 7 on the
pretest as an example of this type.

What will be the molarity of a solution prepared by mixing 3.0 liters of 0.10 M HCl with 1.0 liter of 6.0 M HCl?

First you calculate the number of moles of HCl present in each solution.

$$? \text{ moles HCl} = 3.0 \underline{\text{ liters}}_{\,0.1\,M} \times \frac{0.1 \text{ mole HCl}}{\underline{\text{liter}}_{\,0.1\,M}}$$

$$= 0.30 \text{ mole HCl}$$

$$? \text{ moles HCl} = 1.0 \underline{\text{ liter}}_{\,6.0\,M} \times \frac{6.0 \text{ mole HCl}}{\underline{\text{liter}}_{\,6.0\,M}}$$

$$= 6.0 \text{ moles HCl}$$

(Notice how the solutions are labeled to distinguish between them.)

Total moles = 0.30 + 6.0 = 6.3 moles HCl

Then you calculate the total volume of the final solution, which is equal to the sum of the volume of the two solutions.

Total volume = 3.0 liters + 1.0 liter = 4.0 liters

You now have the number of moles of HCl and the volume of solution and can easily calculate the molarity.

$$? \, M = \frac{? \text{ mole HCl}}{\text{liter soln.}} = \frac{6.3 \text{ moles HCl}}{4.0 \text{ liters soln.}} = 1.6 \, M$$

PROBLEM SET

These problems are relatively simple. Some have nonmetric units to refresh your memory on these conversions. Check the answer to each problem as you finish it to see that you are on the right track.

1. How many moles of ethyl alcohol, C_2H_5OH, are present in 65 ml of a 1.5 M solution?

2. How many liters of a 6.0 M solution of acetic acid, CH_3COOH, will contain 0.0030 mole of acetic acid?

3. How many milliliters of a 0.10 F solution of HCl will contain 14.6 g HCl?

4. How many grams of a 0.50 F solution of HCl would contain 7.3 g HCl? The density of the solution is 1.2 g/ml.

5. What is the molarity of the solution which was prepared by dissolving 1.96 g H_2SO_4 in enough water to bring the volume to 250 ml?

6. What is the weight percent formic acid, HCOOH, in a 0.40 M solution of formic acid whose density is 1.15 g/ml?

7. What is the formality of a 25 wt % solution of NaOH whose density is 1.3 g/ml?

8. How many grams of acetic acid, CH_3COOH, must you take to prepare 600 ml of a 1.50 M solution?

9. How many grams of sodium formate, HCOONa, are in 5.0 ml of a 1.4 F solution of sodium formate?

10. How many ounces of NaCl are required to prepare 4.0 liters of 0.10 F NaCl?

11. How many pounds of KOH are in 5.0 gal of 0.10 F KOH solution?

12. How many grams of ethyl alcohol, C_2H_5OH, must be added to 30.0 g water to prepare a 0.10 m solution?

13. What is the molality of a 40 wt % solution of NaOH?

14. How many grams of a 3.0 m solution of KOH will provide 85 g of KOH?

15. What is the wt % $AlCl_3$ in a 0.20 m solution of $AlCl_3$?

16. To what volume must you bring 25.0 ml of a 6.00 M solution of sugar in order to prepare a 0.150 M solution of sugar?

17. How many milliliters of 5.0 M formic acid, HCOOH, must you use to prepare 450 ml of 0.35 M formic acid?

18. What is the molarity of a solution prepared by adding 3.0 liters of water (solvent) to 1.7 liters of 0.50 M sugar?

19. What is the molality of a 0.10 M solution of ethylene glycol, $C_2H_6O_2$, whose density is 0.90 g/ml?

20. What will be the molarity of the solution resulting from mixing 50 ml of 10.0 M HCl with 2.0 liters of 0.10 M HCl?

PROBLEM SET ANSWERS

You should be able to do at least nine out of the first eleven problems correctly. If you cannot, redo Sections A and B in the chapter. Problems 12, 13, 14, 15, and 19 are covered in Section C in the chapter, and if you miss more than two of these, redo the section. You should be able to do problems 16 and 17. If you cannot, check back to Section D. Problems 18 and 20 are covered in Section E, and you should be able to do at least one of them.

1. ? moles alcohol = 65 ml soln. $\times \dfrac{10^{-3} \text{ liter soln.}}{\text{ml soln.}} \times \dfrac{1.5 \text{ moles alcohol}}{\text{liter soln.}}$

   = 0.098 mole alcohol

2. ? liters soln. = 0.0030 mole acetic acid $\times \dfrac{\text{liter soln.}}{6.0 \text{ moles acetic acid}}$

   = 5.0×10^{-4} liter soln.

3. ? ml soln. = 14.6 g HCl $\times \dfrac{\text{formula wt HCl}}{36.5 \text{ g HCl}} \times \dfrac{10^{3} \text{ ml soln.}}{0.10 \text{ formula wt HCl}}$

   = 4.0×10^{3} ml soln.

4. ? g soln. = 7.3 g HCl $\times \dfrac{\text{formula wt HCl}}{36.5 \text{ g HCl}}$

   $\times \dfrac{10^{3} \text{ ml soln.}}{0.50 \text{ formula wt HCl}} \times \dfrac{1.2 \text{ g soln.}}{\text{ml soln.}}$

   = 4.8×10^{2} g soln.

5. ? $M = \dfrac{? \text{ moles } H_2SO_4}{\text{liter soln.}}$

   $= \dfrac{1.96 \text{ g } H_2SO_4}{250 \text{ ml soln.}} \times \dfrac{\text{ml soln.}}{10^{-3} \text{ liter soln.}} \times \dfrac{\text{mole } H_2SO_4}{98.1 \text{ g } H_2SO_4}$

   = 0.0799 *M*

6. ? g HCOOH = 100 g soln. $\times \dfrac{\text{ml soln.}}{1.15 \text{ g soln.}} \times \dfrac{10^{-3} \text{ liter soln.}}{\text{ml soln.}}$

   $\times \dfrac{0.40 \text{ mole HCOOH}}{\text{liter soln.}} \times \dfrac{46.0 \text{ g HCOOH}}{\text{mole HCOOH}}$

   = 1.6 g or 1.6%

7. ? $F = \dfrac{? \text{ formula wt NaOH}}{\text{liter soln.}}$

   $= \dfrac{25 \text{ g NaOH}}{100 \text{ g soln.}} \times \dfrac{\text{formula wt NaOH}}{40.0 \text{ g NaOH}} \times \dfrac{1.3 \text{ g soln.}}{\text{ml soln.}} \times \dfrac{\text{ml soln.}}{10^{-3} \text{ liter soln.}}$

   = 8.1 *F*

8. ? g acetic acid = 0.600 liter soln. × $\dfrac{1.50 \text{ moles acetic acid}}{\text{liter soln.}}$

    × $\dfrac{60.0 \text{ g acetic acid}}{\text{mole acetic acid}}$

    = 54.0 g acetic acid

9. ? g HCOONa = 5.0 ml soln. × $\dfrac{10^{-3} \text{ liter soln.}}{\text{ml soln.}}$

    × $\dfrac{1.4 \text{ formula wt HCOONa}}{\text{liter solution}}$ × $\dfrac{68.0 \text{ g HCOONa}}{\text{formula wt HCOONa}}$

    = 0.48 g HCOONa

10. ? oz NaCl = 4.0 liters soln. × $\dfrac{0.10 \text{ formula wt NaCl}}{\text{liter soln.}}$

    × $\dfrac{58.5 \text{ g NaCl}}{\text{formula wt NaCl}}$ × $\dfrac{\text{lb NaCl}}{454 \text{ g NaCl}}$ × $\dfrac{16 \text{ oz NaCl}}{\text{lb NaCl}}$

    = 0.82 oz NaCl

11. ? lb KOH = 5.0 gal soln. × $\dfrac{4 \text{ qt soln.}}{\text{gal soln.}}$ × $\dfrac{0.946 \text{ liter soln.}}{\text{qt soln.}}$

    × $\dfrac{0.10 \text{ formula wt KOH}}{\text{liter soln.}}$ × $\dfrac{56.1 \text{ g KOH}}{\text{formula wt KOH}}$ × $\dfrac{\text{lb KOH}}{454 \text{ g KOH}}$

    = 0.23 lb KOH

12. ? g ethyl alcohol = 30.0 g water × $\dfrac{\text{kg water}}{10^3 \text{ g water}}$

    × $\dfrac{0.10 \text{ mole ethyl alcohol}}{\text{kg water}}$ × $\dfrac{46.0 \text{ g ethyl alcohol}}{\text{mole ethyl alcohol}}$

    = 0.14 g ethyl alcohol

13. ? m = $\dfrac{? \text{ moles NaOH}}{\text{kg solvent}}$

    = $\dfrac{40 \text{ g NaOH}}{60 \text{ g solvent}}$ × $\dfrac{\text{mole NaOH}}{40.0 \text{ g NaOH}}$ × $\dfrac{10^3 \text{ g solvent}}{\text{kg solvent}}$

    = 17 m

14. This is a hard one! First you must determine the weight of KOH associated with 1 kg of solvent (1000 g).

    ? g KOH = 1 kg solvent × $\dfrac{3.0 \text{ moles KOH}}{\text{kg solvent}}$ × $\dfrac{56.1 \text{ g KOH}}{\text{mole KOH}}$

    = 168 g KOH

Then you can set up a new conversion factor which says that there are 168 g KOH per (1000+168) g solution. Now you can solve the problem.

$$? \text{ g soln.} = 85 \text{ g KOH} \times \frac{1168 \text{ g soln.}}{168 \text{ g KOH}}$$

$$= 5.9 \times 10^2 \text{ g soln.}$$

15. This problem is similar to Problem 14. First you must determine the weight of $AlCl_3$ associated with a kilogram of solvent.

$$? \text{ g } AlCl_3 = \text{kg solvent} \times \frac{0.20 \text{ mole } AlCl_3}{\text{kg solvent}} \times \frac{133.5 \text{ g } AlCl_3}{\text{mole } AlCl_3}$$

$$= 26.7 \text{ g } AlCl_3$$

Then you set up a new conversion factor which says there are 26.7 g $AlCl_3$ per (1000+26.7) g solution. Now you can solve the problem.

$$? \text{ g } AlCl_3 = 100 \text{ g soln.} \times \frac{26.7 \text{ g } AlCl_3}{1026.7 \text{ g soln.}}$$

$$= 2.6 \text{ g } AlCl_3 \text{ or } 2.6\%$$

16. $? \text{ liters final soln.} = 0.0250 \text{ liter init. soln.} \times \dfrac{6.00 \text{ moles sugar}}{\text{liter init. soln.}}$

$$\times \frac{\text{liter final soln.}}{0.150 \text{ mole sugar}}$$

$$= 1.00 \text{ liter final soln.}$$

17. $? \text{ ml init. soln.} = 450 \text{ ml final soln.} \times \dfrac{0.35 \text{ mole HCOOH}}{10^3 \text{ ml final soln.}}$

$$\times \frac{\text{liter init. soln.}}{5.0 \text{ moles HCOOH}} \times \frac{\text{ml init. soln.}}{10^{-3} \text{ liter init. soln.}}$$

$$= 32 \text{ ml init. soln.}$$

18. $? M = \dfrac{? \text{ moles sugar}}{\text{liter final soln.}}$

$$= \frac{0.50 \text{ mole sugar}}{\text{liter init. soln.}} \times \frac{1.7 \text{ liters init. soln.}}{4.7 \text{ liters final soln.}}$$

$$= 0.18 \frac{\text{mole sugar}}{\text{liter final soln.}}$$

$$= 0.18 M$$

(Notice that the volume of the final solution is the volume of the initial solution + volume solvent added.)

19. This is one of those problems in which you must calculate the weight of solute in a liter of solution, then the weight of a liter of solution, then the weight of solute by subtraction.

$$? \text{ g ethylene glycol} = 1.0 \text{ liter soln.} \times \frac{0.10 \text{ mole ethylene glycol}}{\text{liter soln.}}$$

$$\times \frac{62.0 \text{ g ethylene glycol}}{\text{mole ethylene glycol}}$$

$$= 6.2 \text{ g ethylene glycol}$$

$$? \text{ g soln.} = 1.0 \text{ liter soln.} \times \frac{\text{ml soln.}}{10^{-3} \text{ liter soln.}} \times \frac{0.90 \text{ g soln.}}{\text{ml soln.}}$$

$$= 900 \text{ g soln.}$$

? g solvent = 900 g soln. − 6.2 g ethylene glycol = 893.8 g solvent

The new conversion factor that you have thus calculated is

$$\frac{0.10 \text{ mole ethylene glycol}}{893.8 \text{ g solvent}}.$$

$$? \, m = \frac{? \text{ moles ethylene glycol}}{\text{kg solvent}}$$

$$= \frac{0.10 \text{ mole ethylene glycol}}{893.8 \text{ g solvent}} \times \frac{10^3 \text{ g solvent}}{\text{kg solvent}}$$

$$= 0.11 \, m$$

20. Here you calculate the number of moles of HCl in each of the samples separately.

$$? \text{ moles HCl} = 0.050 \text{ liter}_{10\,M} \times \frac{10 \text{ moles HCl}}{\text{liter}_{10\,M}} = 0.50 \text{ mole HCl}$$

$$? \text{ moles HCl} = 2.0 \text{ liters}_{0.10\,M} \times \frac{0.10 \text{ mole HCl}}{\text{liter}_{0.10\,M}} = 0.20 \text{ mole HCl}$$

Total mole HCl = 0.70 mole HCl

Total volume = 2.05 liters

$$? \, M = \frac{? \text{ moles HCl}}{\text{liter total}} = \frac{0.70 \text{ mole HCl}}{2.05 \text{ liters total}}$$

$$= 0.34 \, M$$

Chapter 13
What is Equivalency?
A Bookkeeping Trick

There will be no pretest on this chapter. If you have had some exposure to the concepts of equivalent weight, oxidation number, and redox reactions, you might see how successful you are with the problems at the end of the chapter. However, it's more than likely that you will have to work through the chapter nevertheless.

Actually, the whole idea of equivalency is unnecessary. The new International System of Units (SI) doesn't consider equivalents (see Appendix VI). And even a great many freshman chemistry textbooks scarcely mention it any more. If your instructor feels that it is not necessary to learn these concepts (at least not now), he or she will tell you to skip this chapter and the following chapter entirely.

The whole purpose of equivalency is to give you a shortcut for solving stoichiometry problems, and most especially stoichiometry problems where the reactants are in solution. Whether the effort to learn the shortcut will be too great for the time saved using it remains to be seen.

SECTION A   Equivalency in Acid-base Reactions

Rather than starting with a definition, we will start out right away by showing how equivalency works. Consider the reaction

$$HCl + NaOH = NaCl + H_2O.$$

This balanced equation tells you that 1 mole of HCl reacts with

1

___ mole of NaOH. Since the molecular weight of HCl is 36.5 g/mole

40.0

and the molecular weight of NaOH is _____ g/mole, then you know

40.0

that 36.5 g of HCl will react with _____ g NaOH. In this reaction,

40.0

then, there are 36.5 g HCl per _____ g NaOH; or we say that

40.0

36.5 g HCl is equivalent to _____ g NaOH.
   Now consider another reaction with NaOH,

$$H_2SO_4 + 2NaOH = Na_2SO_4 + 2H_2O.$$

This balanced equation tells you that 1 mole of H_2SO_4 reacts with

2

___ moles of NaOH. The molecular weight of H_2SO_4 is 98.1 g/mole.

You could therefore say that 98.1 g of H_2SO_4 will react with

_____ g NaOH. But how many grams of H_2SO_4 will react with   80.0

40.0 g NaOH? Only _____ g H_2SO_4. (Notice that 49.05 to three   49.0
significant figures is 49.0. If you have forgotten how to round off
numbers ending with a 5, refer back to Section C in Chapter 1.)

In this reaction, then, there are 49.0 g H_2SO_4 per _____ g NaOH,   40.0

or we will say that 49.0 g H_2SO_4 is equivalent to _____ g NaOH.   40.0
The equivalent weight is the weight that will react with 40.0 g
NaOH. In this reaction, then, we can say that the equivalent weight

of H_2SO_4 is _____ g H_2SO_4 per equivalent H_2SO_4; or, written as a   49.0
fraction,

$$\frac{1 \text{ eq } H_2SO_4}{49.0 \text{ g } H_2SO_4}.$$

From the molecular weight of H_2SO_4 you know the conversion
factor

$$\frac{98.1 \text{ g } H_2SO_4}{\text{_____ } H_2SO_4},$$ mole

and combining these two conversion factors you see that there are

$$\frac{\text{__ eq } H_2SO_4}{\text{mole } H_2SO_4}.$$ 2

  Finally, consider one more reaction with NaOH.

$$H_3PO_4 + 3NaOH = Na_3PO_4 + 3H_2O$$

The balanced equation states that __ mole H_3PO_4 reacts with   1

__ moles NaOH; or, in terms of weight, 98.0 g H_3PO_4 react with   3

_____ g NaOH. To get the equivalent weight you must know the   120.0
amount of H_3PO_4 which reacts with only 40.0 g NaOH. It is

obvious that only _____ g H_3PO_4 are required to react with 40.0 g   32.7
NaOH, so you can write the conversion factor

$$\frac{1 \text{ eq } H_3PO_4}{\text{_____ g } H_3PO_4};$$ 32.7

or, as was done in the case of H_2SO_4, you can write that

$$\frac{\text{__ eq } H_3PO_4}{\text{mole } H_3PO_4}$$ 3

Let us examine these three reactions again.

$$HCl + \underline{1}\ NaOH = NaCl + H_2O \qquad \frac{1\ eq\ HCl}{mole\ HCl}$$

$$H_2SO_4 + \underline{2}\ NaOH = Na_2SO_4 + 2H_2O \qquad \frac{2\ eq\ H_2SO_4}{mole\ H_2SO_4}$$

$$H_3PO_4 + \underline{3}\ NaOH = Na_3PO_4 + 3H_2O \qquad \frac{3\ eq\ H_3PO_4}{mole\ H_3PO_4}$$

You will notice that the number of equivalents of acid per mole of acid in each case is the same as the coefficient which precedes the

NaOH | _____ in the balanced equation.

Equivalency *balances the equation* for you. The result of this, as you will see, is that you can always say that there is *one equivalent* of one substance *per one equivalent* of another substance. Before, when you were working stoichiometry problems, you set up a "bridge" conversion factor from the balanced equation. For example, it was

$$\frac{2\ moles\ NaOH}{mole\ H_2SO_4} \quad or \quad \frac{3\ moles\ NaOH}{mole\ H_3PO_4}.$$

You now have a new "bridge"; it is

$$\frac{1\ eq\ NaOH}{1\ eq\ H_2SO_4} \quad or \quad \frac{1\ eq\ NaOH}{1\ eq\ H_3PO_4}.$$

There is always "one equivalent of anything at all per one equivalent of anything else."

So you have invented a labor-saving device which eliminates the need to balance the equation, if you can only figure out the number of equivalents per mole without a balanced equation—and you can! Let's see how that is done.

As we said before,

HCl has 1 equivalent per mole.

H_2SO_4 has 2 equivalents per mole.

H_3PO_4 has 3 equivalents per mole.

The number of equivalents per moles is the same as the number of

H's | ____ in the acid. This is very simple, but, unfortunately, it isn't always the case. Consider the reaction

$$H_2SO_4 + NaOH = NaHSO_4 + H_2O.$$

In this particular reaction between H_2SO_4 and NaOH, 1 mole of

1, 98.1 | H_2SO_4 reacts with __ mole of NaOH. Therefore, ____ g of H_2SO_4 would react with 40.0 g NaOH. Consequently, in this particular

1 | reaction there is only __ equivalent of H_2SO_4 per mole H_2SO_4.

Look at the reaction. The H_2SO_4 loses only __ H when it becomes $NaHSO_4$. If we say that only one H reacts, now we can say exactly how to determine the number of equivalents per mole of an acid.

1

It is the number of ___ which react (are lost). You don't have to balance the equation, but you *must know the product*.

H's

As a matter of fact, you can never say how many equivalents there are per mole of anything unless you know the products of the reaction. You cannot even talk about equivalency unless there is a chemical reaction.

For practice, determine the number of equivalents per mole for the following acids.

HNO_3 producing $NaNO_3$ has $\dfrac{\text{__ eq } HNO_3}{\text{mole } HNO_3}$

1

because only __ H reacts.

1

H_3BO_3 producing Na_2HBO_3 has $\dfrac{\text{__ eq } H_3BO_3}{\text{mole } H_3BO_3}$

2

because __ H's react.

2

H_2CO_3 producing $NaHCO_3$ has $\dfrac{\text{__ eq } H_2CO_3}{\text{mole } H_2CO_3}$

1

because only __ H reacts.

1

H_2CO_3 producing Na_2CO_3 has $\dfrac{\text{__ eq } H_2CO_3}{\text{mole } H_2CO_3}$

2

because __ H's react.

2

Chemists frequently get a little careless in specifying what the product is when a strong acid reacts with a base since, given enough base, the acid will lose all its H's. Of course, an acid with only one H can have only one H to react. At this time you may not know which acids are strong and which are weak; so in all examples in this text you will either be told the product or be told that the reaction "goes to completion" (that all the H's react).

Now you know how to determine the number of equivalents per mole for acids which lose their H's. However, there are many types of compounds other than acids. Next we will consider bases and start with NaOH as an example.

$$NaOH + HCl = NaCl + H_2O$$

From this balanced equation you can see that you have the conversion factor

$\dfrac{\text{__ mole HCl}}{\text{mole NaOH}}$.

1

1 | You also now know that since the HCl has only __ H to react, you can write

1 | $$\frac{\text{__ eq HCl}}{\text{mole HCl}}.$$

1 | Since there is always 1 equivalent of one substance per __ equivalent of another, you can write the conversion factor

1 | $$\frac{\text{__ eq NaOH}}{\text{eq HCl}}.$$

Multiplying all these conversion factors together gives

1 | $$\frac{\text{__ eq NaOH}}{\text{mole NaOH}}$$

Let us look at the balanced equation again.

$$NaOH + \underline{1}\ HCl = NaCl + H_2O$$

1 | The coefficient in front of the HCl is __, and the number of

1 | equivalents of NaOH per mole NaOH is __. Once again you see that the equivalency is simply balancing the equation for you.
Here is another example.

$$Ca(OH)_2 + 2HCl = CaCl_2 + 2H_2O$$

Before we even examine it, how about taking a guess; there are

2 | $$\frac{\text{__ eq Ca(OH)}_2}{\text{mole Ca(OH)}_2}.$$

If you got the correct answer, it was because you looked at the

2, HCl | balanced equation and saw the coefficient __ in front of the ____.
Now let's examine it and see how this is so. The balanced equation tells you that

2 | $$\frac{\text{__ mole HCl}}{\text{mole Ca(OH)}_2}.$$

But you know that since HCl has only 1 H, there is

1 | $$\frac{\text{__ eq HCl}}{\text{mole HCl}}.$$

By the equivalency method we have set up, there is

1 | $$\frac{\text{__ eq Ca(OH}_2)}{\text{eq HCl}}.$$

Multiplying all of these conversion factors together you get

2 | $$\frac{\text{__ eq Ca(OH)}_2}{\text{mole Ca(OH)}_2},$$

just as you suspected from the balanced equation.

We will look at one more base, $Al(OH)_3$.

$$Al(OH)_3 + 3HCl = AlCl_3 + 3H_2O$$

Without examining the reaction any further, you should know from the balanced equation that there are

$$\frac{\underline{\hspace{1cm}} Al(OH)_3}{mole\ Al(OH)_3}.$$

3 eq

You know this because there is a coefficient of __ in front of the HCl.

3

It is all well and good to get the number of equivalents per mole by inspecting the balanced equation. However, the entire purpose of using equivalency is not to have to balance an equation. How, then, can you get the number of equivalents of base per mole of base? In the case of acids, you counted the number of H's which react (are lost) going from acid to product. In the case of bases,

you must count the number of _____ which are lost.

OH's

Looking at this last example, then, $Al(OH)_3$ produces $AlCl_3$. It has lost __ OH's. Therefore you know that there are

3

$$\frac{\underline{\hspace{0.6cm}} eq\ Al(OH)_3}{mole\ Al(OH)_3}.$$

3

You have one less complication with bases than you have with acids. In all the problems that you will face, the base will have all its OH's reacting.

Incidentally, thus far you have not been given the exact story as to what is reacting. Acids do not lose an H atom; it is an H^+ ion which reacts, and this is an H atom minus 1 electron so that it has a +1 charge. Bases do not lose an OH group; they lose an OH^- ion, a combination of an O atom and an H atom which has picked up 1 electron and so has a −1 charge.

Let us see if we can make some sense of this whole picture. We are saying that one equivalent of a substance is that amount of substance which can either produce or consume 1 H^+ ion, or the amount of a substance which can produce or consume some other substance which can produce or consume 1 H^+ ion.

Now you can see why you can count OH^-'s to get the number of equivalents per mole. All we have to do is look at the reaction

$$H^+ + OH^- = H_2O.$$

You can see that 1 OH^- is equivalent to 1 H^+.

You can also determine the number of equivalents per mole for salts, which have neither H^+ nor OH^- present. To do this we take an ion of the salt and see how many H^+ ions it will react with. For example, NaCl exists in solution as Na^+ ions and Cl^- ions. We can say that the Cl^- in the NaCl is capable of combining with the H^+ ion.

$$Cl^- + H^+ = HCl$$

Since 1 Cl^- is equivalent to 1 H^+, there is 1 equivalent NaCl per mole of NaCl.

Let's also look at the salt Na_2SO_4. Here we can say that the $SO_4^=$ ion can combine with H^+ ions.

$$SO_4^= + 2H^+ = H_2SO_4$$

Since one $SO_4^=$ ion is equivalent to 2 H^+'s, we can say that there are 2 equivalents Na_2SO_4 per mole Na_2SO_4.

3 | What about a salt like $Al_2(SO_4)_3$? Here you have __ $SO_4^=$ ions present in one molecule, so you must write

6 | $$3SO_4^= + \_H^+ = 3H_2SO_4,$$

and so there are

6 | $$\frac{\_ \text{ eq } Al_2(SO_4)_3}{\text{mole } Al_2(SO_4)_3}.$$

The easy way to do this mentally is to say that there are 3 $SO_4^=$

2 | ions and each is equivalent to __ H^+ ions so the whole molecule

6 | must be equivalent to 3×2 or 6 H^+. Therefore, there are __ eq $Al_2(SO_4)_3$ per mole $Al_2(SO_4)_3$.

You will find some examples like these in the problem set at the end of the chapter and can get some practice then.

Unhappily, these acid-base reactions which produce a salt are not the only type of chemical reaction. There is another very important type in which the number of reacting H^+ ions does not control how the equation balances. These reactions involve an electron transfer, and we are going to have to develop some method of finding the number of equivalents per mole based on the electrons transferred.

SECTION B Oxidation Number

There is one type of chemical reaction in which there is a transfer of electrons from one substance to another. The substance which appears to give up electrons is said to be oxidized; the substance which seems to gain electrons is said to be reduced; and a reaction must have both an electron acceptor and an electron donor to proceed. Consequently, this class of reactions is called "reduction-oxidation," or "redox."

Since no H^+ ions are involved in this electron transfer, we will have to determine the equivalency on the basis of electrons. However, as the equation below shows, 1 electron is equivalent to 1 H^+ ion.

$$H^+ + 1 \text{ electron}^- = H$$

(An electron has, of course, a -1 charge.)

We must develop a method of counting the number of electrons in a molecule or group first before and then after the reaction. Knowing this number, we can then tell how many electrons have been transferred in the course of the reaction. We do this by using what is called the "oxidation number" of the atoms in the group. This oxidation number is rather arbitrary and is really only a book-keeping method based on a set of rules which have been set up. The rules are given below.

(1)   The oxidation number of H is $+1$.

(2)   The oxidation number of any group of atoms is equal to the charge of the group. A group may be just a single atom.

(3)   The sum of the oxidation numbers of all the atoms in a group equals the oxidation number of the group.

What, then, is the oxidation number of Cl in HCl? Since HCl has no charge, the oxidation number of the group HCl must

therefore equal __. The oxidation number of H is ___. Therefore,   0, $+1$

the oxidation number of Cl must be ___, so that the sum of the -1

oxidation numbers equals __.   0

Here is another example. What is the oxidation number of O in

OH$^-$? The charge on the group is ___. [You will notice that the -1
charge on a group is written as a superscript (above the line)
following the group. It will show plus $(+)$ or minus $(-)$ and the
number of charges. The number 1 need not be written. Sometimes
the number of charges will be indicated by repeating the $+$ or $-$.
For example, Fe^{++} is Fe^{+2} or even $(SO_4)^=$ is $(SO_4)^{-2}$.] The sum of

the oxidation numbers must equal ___. The oxidation number of -1

H is ___. Consequently, the oxidation number of the O must be $+1$

___. You have $+1-2=-1$. -2

What is the oxidation number of iron metal, Fe? Here the group

consists of only one atom, Fe, and the charge on it is __; that is,   0

it has no charge. Consequently, the oxidation number of Fe is __.   0

What is the oxidation number of Fe^{+3}? Here the charge on the

group (one atom) is ___, so the oxidation number of Fe^{+3} must $+3$

be ___. $+3$

You can see that when you have a group which contains just a
single atom, the oxidation number of the atom will equal the
charge. We often talk about the "valence" of an ion; this is really
"electrovalence" and means the charge on the ion. The oxidation
number of an ion containing only one atom equals the
electrovalence, and any substance in its elemental state will have
an oxidation number of zero. Thus, iron metal, Fe, will have an

oxidation number of __; O_2, because it is a pure element will have   0

an oxidation number of __; and even H_2 will have an oxidation   0

0

-2

0

-1

0, 0

+1

-2

+1

-2

-2, -2

-8

+6

-2

-6

+4

number of __. The fact that it is pure element takes precedence over the rule saying H is +1.

What is the oxidation number of O in H_2O? It must be ____

since each H is +1 and the charge on the group is __. You therefore have $2(+1) - 2 = 0$.

The oxidation number of O is almost always −2, the few exceptions being the peroxides, like H_2O_2, where its oxidation

number is ____, and the superoxides, as in the compound KO_2, where the oxidation number is −1/2. The peroxides and superoxides occur only with metals of Group I and II and H. Thus, in almost all cases you can assign an oxidation number of −2 to oxygen.

What is the oxidation number of Cl in HClO? The sum of the

oxidation numbers must be __ since the charge on the group is __.

The oxidation number of H is ____, and the oxidation number of O

is the usual ____. Consequently, the oxidation number of the Cl

must be ____. You have $+1 + 1 - 2 = 0$.

What is the oxidation number of S in the $SO_4^=$ ion? The charge

on the group is ____, so the sum of the oxidation numbers must be

____. The oxidation number of one O is ____, so the total

oxidation number for four O's will be ____. Consequently, the

oxidation number for the S must be ____ in order to make the sum equal −2.

$$-8 + 6 = -2$$

What is the oxidation number of S in the $SO_3^=$ ion? The sum of

the oxidation numbers of the group must add up to ____, and since the group contains three O's, which will have a combined

oxidation number of ____, then the S must have an oxidation

number of ____.

You will notice that S has an oxidation number of +6 in $SO_4^=$ and +4 in $SO_3^=$. Many elements will show different oxidation numbers in different groups. You will find a Periodic Table in Appendix I, which shows most of the common oxidation numbers for the more usual elements. You will note that the maximum positive oxidation number that an element can show is, with only a few exceptions, the same as its group number in the Periodic Table. You may also note that the elements in Groups IA, IIA and IIIA can only show an oxidation number of +1, +2, +3 respectively (except B). Elements in all other groups can show several different oxidation numbers. Even H, if it is combined with a very metallic element such as Na or Ca can show a second oxidation number, −1.

Since the use of oxidation numbers is really a bookkeeping trick to keep track of the transfer of electrons, you shouldn't have any feeling about some strange values. For example, find the oxidation number of C in CH_2O (formaldehyde). Since there is no charge on the group, the sum of the oxidation numbers must be __. Since   0

each H has an oxidation number of ____, you will have a total of   +1

____ for the two H's present. The oxidation number for O is the   +2

usual ____, so the oxidation number for the C must be __.   −2, 0

Zero isn't so strange, but there are even a few compounds where the oxidation number is a fraction. Consider Fe_3O_4. The

four oxygens have a total oxidation number of ____, and the entire   −8
group has no charge, so the sum of the oxidation numbers must be

__. Consequently, three Fe's must have a total oxidation number of   0

____. That means that each Fe must have an oxidation number of   +8

____. $+\dfrac{8}{3}$

Now figure out the oxidation number for Mo in the compound Mo_2B. The B will show its −5 oxidation number. The oxidation

number of Mo is ____. $+\dfrac{5}{2}$

Incidentally, the usual method for noting the oxidation number is to write it underneath the atom. So there is no confusion, we also will put it in a circle in this text. Thus we could write

$$\text{H} \quad \text{Cl}$$
$$\text{(+1)} \quad \text{(−1)}$$

or

$$\text{Fe}^{+3}$$
$$\text{(+3)}$$

Recall that the electrovalence is written above the line and to the right.

SECTION C   Equivalency in Redox Reactions

At this stage you are probably thinking, "This is all very interesting, but so what?". All of this is important because in Redox reactions two atoms will change their oxidation numbers: one will increase its oxidation number (that is, become oxidized), and the other will have its oxidation number lowered (that is, become reduced). The oxidized substance gives up electrons

$$\text{Mg} = \text{Mg}^{+2} + 2 \text{ electrons}^-,$$
$$\text{(0)} \quad \text{(+2)}$$

and the reduced substance picks up electrons

$$2H^+ + 2 \text{ electrons}^- = H_2.$$

$$2 \times \textcircled{+1} \qquad\qquad \textcircled{0}$$

The entire equation is

$$Mg + 2H^+ = Mg^{+2} + H_2.$$

The number of electrons that Mg gave up, 2 electrons, is equal to the change in its oxidation number from 0 to $+2$. Therefore, if you can count the change in oxidation number from product to reactant, you have a count of the number of electrons transferred. Since 1 electron is equivalent to 1 H^+, you now have the number of equivalents per mole. In the case of Mg going to Mg^{+2}, there

2 was a change of ___ in the oxidation numbers of the Mg atom;

2 then there must be ___ equivalents of Mg per mole of Mg.

We will try a few of these for practice. How many equivalents per mole are there for MnO_4^- in the reaction

$$MnO_4^- \longrightarrow Mn^{+2}?$$

First you determine the oxidation number of Mn in MnO_4^-. Since

−8 the four oxygens have a total oxidation number of ___, and the

−1 charge of the group is ___, the oxidation number of the Mn must

+7 be ___. Then you determine the oxidation number of the Mn in

+2 Mn^{+2}; it must be ___.

To get the number of equivalents per mole, you must determine the change in oxidation number. It goes from $+7$ to $+2$, so the

5, gained change is ___. This means that the Mn atom must have _____ 5

reduced electrons. Since it gains electrons, we say it is _____. You now know the number of equivalents per mole and can write

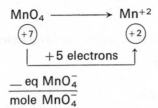

5 $$\frac{\_ \text{ eq } MnO_4^-}{\text{mole } MnO_4^-}$$

Here is another example. How many equivalents per mole are there for Fe in the reaction

$$Fe \longrightarrow Fe^{+3}?$$

You will normally do this by simply writing the oxidation numbers beneath the atom.

$$Fe \longrightarrow Fe^{+3}$$

$$\bigcirc \qquad\qquad \bigcirc$$

0, +3 ___   ___

Then put in the number of electrons gained or lost.

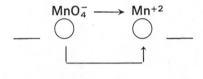

| | -3 electrons |

You now know that in this reaction there are

$$\frac{\_\_ \text{ eq Fe}}{\text{mole Fe}}$$

3

You try to set up the first example in this form.

$$MnO_4^- \longrightarrow Mn^{+2}$$

$+7, +2$

$+5$ electrons

Here is another example for you to try.

$$S \longrightarrow H_2SO_4$$

$0, +6$

-6 electrons

You can write that there are

$$\frac{\_\_ \text{ eq S}}{\text{mole S}}$$

6

Another would be

$$S \longrightarrow SO_2$$

$0, +4$

-4 electrons

You can write that in this reaction there are

$$\frac{\_\_ \text{ eq S}}{\text{mole S}}$$

4

As is true for determining the equivalency in acid-base reactions, you *must know the product* to determine the number of equivalents per mole. You also must count the total number of electrons involved. Here is an example which will bring this point out. How many equivalents are there per mole for $Cr_2O_7^{-2}$ in the reaction where it produces Cr^{+3}?

The reaction is

$$Cr_2O_7^{-2} \longrightarrow Cr^{+3}.$$

However, each time a $Cr_2O_7^{-2}$ ion breaks up, it is bound to yield two Cr^{+3} ions since there are two Cr atoms in the $Cr_2O_7^{-2}$ ion. Therefore, you must write

$$Cr_2O_7^{-2} \longrightarrow 2Cr^{+3},$$

and you must get the total number of electrons involved. This means that you will have to take the oxidation number for the two Cr's in the $Cr_2O_7^{-2}$ and also the two Cr^{+3} ions formed. Thus

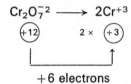

$$+6 \text{ electrons}$$

Notice how the two Cr's were counted together in $Cr_2O_7^{-2}$. The seven oxygens together will have a total oxidation number of

−14, −2

_____, and since the charge on the group is _____, the two Cr's

+12

together must have an oxidation number of _____. In the case of the Cr^{+3} ion, you know that one Cr^{+3} will have an oxidation

+3

number of _____. But since there are two of them, the total will be

+6

_____. The number of equivalents of $Cr_2O_7^{-2}$ per mole of $Cr_2O_7^{-2}$ is

6

$$[+12 - (+6)] = \_\_.$$

Try another problem like this one. How many equivalents per mole are there for $C_2O_4^{-2}$ when it produces CO_2? The reaction will be

$$C_2O_4^{-2} \longrightarrow CO_2,$$

but it is obvious that the breakdown of one $C_2O_4^{-2}$ will produce

two, 2

_____ C's. Consequently, you will have to write a __ in front of the CO_2. Now you can count the number of electrons totally involved.

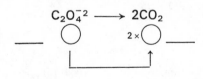

+6, +4

−2 electrons

Therefore, you can now write that there are

2

$$\frac{\_\_ \text{ eq } C_2O_4^{-2}}{\text{mole } C_2O_4^{-2}}.$$

Perhaps you are curious to see how all this works to balance an equation for you. Consider the reaction between MnO_4^- and $C_2O_4^{-2}$. It will require H^+ in order to combine with the extra oxygens and form H_2O. However, we are interested in the number of moles of the MnO_4^- and the number of moles of the $C_2O_4^{-2}$ in the balanced equation.

The MnO_4^- produces Mn^{+2} and the $C_2O_4^{-2}$ produces CO_2. As you did previously, write the following.

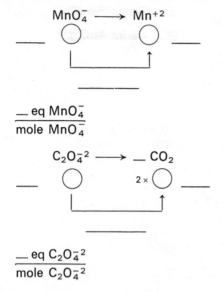

$MnO_4^- \longrightarrow Mn^{+2}$	$+7, \ +2$
	$+5$ electrons
	5
$\dfrac{\_ \text{eq } MnO_4^-}{\text{mole } MnO_4^-}$	
$C_2O_4^{-2} \longrightarrow \_ CO_2$	2
	$+6, \ +4$
	-2 electrons
$\dfrac{\_ \text{eq } C_2O_4^{-2}}{\text{mole } C_2O_4^{-2}}$	2

Now the unbalanced equation is

$$MnO_4^- + C_2O_4^{-2} + H^+ \longrightarrow Mn^{+2} + CO_2 + H_2O.$$

Since there are __ equivalents per mole for the MnO_4^-, you would   5

expect to find a 5 in front of the _____ in the balanced equation. $C_2O_4^{-2}$

Since there are __ equivalents per mole of $C_2O_4^{-2}$, you would   2

expect to find the coefficient in front of the _____ to be __. $MnO_4^-, \ 2$
    Here is the balanced equation,

$$2MnO_4^- + 5C_2O_4^{-2} + 16H^+ = 2Mn^{+2} + 10CO_2 + 8H_2O,$$

which is exactly what you expected. One equivalent of MnO_4^- will react with one equivalent of $C_2O_4^{-2}$ as was true in the acid-base reactions.

After all this, you may be confused about when to count H^+ ions or their equivalents and when to count by change in oxidation number. It is very simple. If any atom changes its oxidation number going from reactant to product, you count the change in oxidation number. If no atom changes its oxidation number, you count the number of H^+ ions which react or some equivalent to the number of H^+ ions.

PROBLEM SET

Part I Oxidation Number

Give the oxidation number of the atom named in the compound shown.

1. Cl in Cl_2
2. S in H_2SO_4
3. S in SO_3^{-2}
4. S in $Na_2S_2O_3$
5. C in CH_4

6. C in CO_2
7. C in CH_3COO^-
8. Fe in Fe_3O_4
9. As in AsO_4^{-3}

10. H in AlH_3
11. V in VCl_5
12. N in N_2O_3
13. N in NH_3

Part II Equivalency

Give the number of equivalents of the compound shown per mole of compound when it undergoes the reaction shown.

14. H_3PO_4 in the reaction $H_3PO_4 \longrightarrow PO_4^{-3}$

15. H_3PO_4 in the reaction $H_3PO_4 \longrightarrow KH_2PO_4$

16. H_3PO_4 in the reaction $H_3PO_4 \longrightarrow H_3PO_3$

17. $Mg(OH)_2$ in the reaction $Mg(OH)_2 \longrightarrow MgCl_2$

18. Fe in the reaction $Fe \longrightarrow Fe^{++}$

19. H^+ in the reaction $H^+ \longrightarrow H_2$

20. $S_2O_8^{-2}$ in the reaction $S_2O_8^{-2} \longrightarrow SO_4^{-2}$

21. NH_4OH in the reaction $NH_4OH \longrightarrow NH_4Cl$

22. NH_4OH in the reaction $NH_4OH \longrightarrow NO_3^-$

23. CH_4 in the reaction $CH_4 \longrightarrow CO_2$

24. $C_6H_{12}O_6$ in the reaction $C_6H_{12}O_6 \longrightarrow CO_2$

25. $HClO_4$ in the reaction $HClO_4 \longrightarrow NaCl$

PROBLEM SET ANSWERS

You should be able to do all the first thirteen problems right. If you can't find the reason for any error, you had better redo Section B. Problems 14–25 are mixed between acid-base and redox equivalency, and you should be able to tell which is operative. If you missed more than two problems in this group, you had better redo both Section A and Section C.

1. 0 (any pure element is zero)
2. +6
3. +4

4. +2 (each S)
5. −4
6. +4

7. 0

8. $+\dfrac{8}{3}$

9. $+5$

10. -1 (Al must be $+3$, it is in Group IIIA.)

11. $+5$

12. $+3$

13. -3

14. $\dfrac{3 \text{ eq } H_3PO_4}{\text{mole } H_3PO_4}$ (acid-base reaction; 3 H^+'s react)

15. $\dfrac{1 \text{ eq } H_3PO_4}{\text{mole } H_3PO_4}$ (acid-base reaction; 1 H^+ reacts)

16. $\dfrac{2 \text{ eq } H_3PO_4}{\text{mole } H_3PO_4}$ (redox reaction; $H_3PO_4 \longrightarrow H_3PO_3$)
$\qquad\qquad\qquad\qquad\qquad\qquad$ $(+5)$ $\qquad\quad$ $(+3)$

17. $\dfrac{2 \text{ eq } Mg(OH)_2}{\text{mole } Mg(OH)_2}$ (acid-base reaction; 2 OH^-'s react)

18. $\dfrac{2 \text{ eq } Fe}{\text{mole } Fe}$ (redox reaction; $Fe \longrightarrow Fe^{+2}$)
$\qquad\qquad\qquad\qquad\qquad\qquad$ (0) $\qquad$ $(+2)$

19. $\dfrac{1 \text{ eq } H^+}{\text{mole } H^+}$ (redox reaction; $H^+ \longrightarrow \frac{1}{2}H_2$)
$\qquad\qquad\qquad\qquad\qquad\qquad$ $(+1)$ $\qquad$ (0)

20. $\dfrac{2 \text{ eq } S_2O_8^{-2}}{\text{mole } S_2O_8^{-2}}$ (redox reaction; $S_2O_8^{-2} \longrightarrow 2SO_4^{-2}$)
$\qquad\qquad\qquad\qquad\qquad\qquad$ $(+14)$ $\qquad$ $2\times(+6)$

21. $\dfrac{1 \text{ eq } NH_4OH}{\text{mole } NH_4OH}$ (acid-base reaction; 1 OH^- reacts)

22. $\dfrac{8 \text{ eq } NH_4OH}{\text{mole } NH_4OH}$ (redox reaction; $NH_4OH \longrightarrow NO_3^-$)
$\qquad\qquad\qquad\qquad\qquad\qquad$ (-3) $\qquad\qquad$ $(+5)$

23. $\dfrac{8 \text{ eq } CH_4}{\text{mole } CH_4}$ (redox reaction; $CH_4 \longrightarrow CO_2$)
$\qquad\qquad\qquad\qquad\qquad\qquad$ (-4) $\qquad$ $(+4)$

24. $\dfrac{24 \text{ eq } C_6H_{12}O_6}{\text{mole } C_6H_{12}O_6}$ (redox reaction; $C_6H_{12}O_6 \longrightarrow 6CO_2$)
$\qquad\qquad\qquad\qquad\qquad\qquad$ (0) $\qquad$ $6\times(+4)$

25. $\dfrac{8 \text{ eq } HClO_4}{\text{mole } HClO_4}$ (redox reaction; $HClO_4 \longrightarrow NaCl$)
$\qquad\qquad\qquad\qquad\qquad\qquad$ $(+7)$ $\qquad$ (-1)

Chapter 14
Solution Concentration II : Normality

PRETEST

Before you try these questions, you will need one new conversion factor.

$$N = \text{normality} = \frac{\text{equivalents solute}}{\text{liter solution}}$$

Knowing this conversion factor, these questions should be quite simple, and you should be able to finish in 20 minutes.

1. How many grams of H_2SO_4 would you need to take in order to prepare 3.0 liters of a 6.0 N solution of H_2SO_4? The reaction considered produces Na_2SO_4.

2. How many liters of a 0.50 N solution of H_3PO_4 would you have to take in order to have 32.7 g H_3PO_4? The reaction considered is $H_3PO_4 \longrightarrow HPO_4^{-2}$.

3. How many grams of $KMnO_4$ must you dilute to 500 ml in order to prepare a 0.10 N solution of $KMnO_4$? The reaction considered is $MnO_4^- \longrightarrow Mn^{+2}$.

4. What is the normality of an $FeCl_3$ solution if 81 g $FeCl_3$ is diluted to 1.5 liters? The reaction considered produces Fe metal.

5. What is the normality of a 4.0 M $Na_2Cr_2O_7$ solution? The reaction is $Cr_2O_7^{-2} \longrightarrow Cr^{+3}$.

6. To what volume in liters must you dilute 500 ml of 18 M H_2SO_4 in order to have a 0.10 N H_2SO_4 solution? The reaction considered is the complete reaction of the acid to form SO_4^{-2}.

7. What is the normality of a 3.0 wt % solution of C_2H_5OH whose density is 0.90 g/ml? The reaction considered produces CO_2.

PRETEST ANSWERS

If you correctly answered questions 1–5, you can go immediately to the problem set at the end of the chapter. If you got 6 and 7 right also you're really sharp, but if you missed them, they are worked out in the main section of the chapter and you can check to see where you went wrong. If you missed any of questions 1–5, you had better go through the entire chapter. You will find all of these problems used as examples in the chapter.

1. 8.8×10^2 g H_2SO_4 (882.9 has too many significant figures.)

2. 1.3 liters solution

3. 1.6 g $KMnO_4$

4. 1.0 *N*

5. 24 *N*

6. 1.8×10^2 liters

7. 7.0 *N*

SECTION A Determining Normality from Molarity

There is still another way that chemists express the concentration of a substance in a solution. It is called *normality* and can be defined as

$$\text{Normality} = N = \frac{\text{equivalents solute}}{\text{liter solution}}.$$

As you will see in the following chapter on solution stoichiometry, this is a particularly convenient expression for concentration, since it allows you to solve problems involving the reaction of solutions very rapidly.

When you want to calculate the concentrations in normality, you must almost always know the number of equivalents of solute per mole of solute. (This subject was covered in Chapter 13.)

The simplest sort of problem is to determine the normality of a solution of known molarity. What, for example, is the normality of a 4.0 M solution of $Na_2Cr_2O_7$? The reaction is $Cr_2O_7^{-2} \longrightarrow Cr^{+3}$.

The first thing to do is to determine the number of equivalents of $Na_2Cr_2O_7$ per mole $Na_2Cr_2O_7$ from the reaction shown in the

2

problem. There are two Cr's in $Cr_2O_7^{-2}$, so you must put a __ in front of the Cr^{+3}. The oxidation number of the two Cr's in $Cr_2O_7^{-2}$

+12, +6

is _____. Two times the oxidation number of Cr^{+3} is ____, and so

6

there are __ electrons transferred. Consequently, you can write

6

$$\frac{\_\text{ eq } Na_2Cr_2O_7}{\text{mole } Na_2Cr_2O_7}.$$

The molarity of the solution is given in the problem and can be written as the conversion factor

4.0 moles

$$\frac{_____ \, Na_2Cr_2O_7}{\text{liter soln.}}$$

You are now ready to solve the problem.

$$? \, N = \frac{? \text{ eq } Na_2Cr_2O_7}{\text{liter soln.}} =$$

If you multiply both of the conversion factors you have only the dimensions of the answer left.

$$? \, N = \frac{? \text{ eq } Na_2Cr_2O_7}{\text{liter soln.}} = \frac{6 \text{ eq } Na_2Cr_2O_7}{\text{mole } Na_2Cr_2O_7}$$
$$\times \frac{4.0 \text{ moles } Na_2Cr_2O_7}{\text{liter soln.}}$$

$$= 24 \, N$$

That was question 5 on the pretest; now you can see how simple this type of problem is.

SECTION B   Problems Concerning Weight of Solute

These problems are exactly the same as problems dealing with weights of solute in molarity concentrations, except that you will need one more conversion factor, the number of equivalents of solute per mole of solute. Here is question 1 on the pretest. How many grams of H_2SO_4 would you need to prepare 3.0 liters of a 6.0 *N* solution of H_2SO_4? The reaction considered produces Na_2SO_4.

First you must determine the number of equivalents of H_2SO_4

per mole H_2SO_4. Since _____ H^+ ions react, you can write   two

$$\frac{\text{— eq } H_2SO_4}{\text{mole } H_2SO_4}.$$ 2

The conversion factor given in the problem is the normality of the solution,

$$\frac{6.0 \text{ ——} H_2SO_4}{\text{————}}$$ eq

                                                                                     liter soln.

Since the problem deals with weight, you also will need the molecular weight conversion factor

$$\frac{98.1 \text{ ——} H_2SO_4}{\text{———— } H_2SO_4}$$ g

                                                                                     mole

Now you can set up the problem.

? _____ = 3.0 liters soln.   g H_2SO_4

The unwanted dimension _____ can be cancelled by   liter soln.
                                                                                     $\frac{6 \text{ eq } H_2SO_4}{\text{liter soln.}}$
multiplying by _____. The new unwanted dimension is   eq H_2SO_4

_____. This can be cancelled by multiplying by the conversion $\frac{\text{mole } H_2SO_4}{2 \text{ eq } H_2SO_4}$

factor _____. This then leaves the unwanted dimension, $\frac{98.1 \text{ g } H_2SO_4}{\text{mole } H_2SO_4}$
mole H_2SO_4, which can be directly converted to the dimensions of

the answer by multiplying by _____. The answer will then be

$$? \text{ g } H_2SO_4 = 3.0 \text{ liters soln.} \times \frac{6 \text{ eq } H_2SO_4}{\text{liter soln.}}$$

$$\times \frac{\text{mole } H_2SO_4}{2 \text{ eq } H_2SO_4} \times \frac{98.1 \text{ g } H_2SO_4}{\text{mole } H_2SO_4}$$

$$= 882.9 \text{ g } H_2SO_4$$

$$= 8.8 \times 10^2 \text{ g } H_2SO_4 \quad \text{(This is taken to the correct}$$
number of significant figures. If you have forgotten this, you will find it all in Section C, Chapter 1.)

Question 2 in the pretest is similar, but turned around to ask for volume. How many liters of 0.50 N H_3PO_4 would you have to take in order to have 32.7 g of H_3PO_4? The reaction considered is $H_3PO_4 \longrightarrow HPO_4^{-2}$.

First you must write down the conversion factor giving the

$\dfrac{\text{2 eq } H_3PO_4}{\text{mole } H_3PO_4}$

$\dfrac{\text{0.50 eq } H_3PO_4}{\text{liter soln.}}$

$\dfrac{\text{98.0 g } H_3PO_4}{\text{mole } H_3PO_4}$

liters solution

32.7 g H_3PO_4

number of equivalents per mole. It is _____ because only two H^+ ions react. The conversion factor given in the problem is the normality of the solution. This is _____. And, finally, since the problem concerns weight, you will need the molecular weight conversion factor _____.

The question asked is "How many _____?" The information given is _____. The solution is then very simple. Set it up and check your results with the answer shown below.

$$? \text{ liters } = 32.7 \text{ g } H_3PO_4 \times \frac{\text{mole } H_3PO_4}{98.0 \text{ g } H_3PO_4} \times \frac{\text{2 eq } H_3PO_4}{\text{mole } H_3PO_4}$$

$$\times \frac{\text{liter soln.}}{0.50 \text{ eq } H_3PO_4}$$

$$= 1.334 \text{ liters}$$

$$= 1.3 \text{ liters} \quad \text{(to the correct number of significant figures)}$$

Question 3 on the pretest is the same type as question 1. You can make the shortcut of immediately expressing 500 ml as

0.500

____ liter. Set up the problem and check your answer against what is shown below.

$$? \text{ g } KMnO_4 = 0.500 \text{ liter soln.} \times \frac{0.10 \text{ eq } KMnO_4}{\text{liter soln.}}$$

$$\times \frac{\text{mole } KMnO_4}{5 \text{ eq } KMnO_4} \times \frac{158.0 \text{ g } KMnO_4}{\text{mole } KMnO_4}$$

$$= 0.500 \times 0.10 \times \frac{1}{5} \times 158.0 \text{ g } KMnO_4$$

$$= 1.6 \text{ g } KMnO_4 \quad \text{(If you didn't find 5 eq per mole, check Chapter 13)}$$

SECTION C Determining the Normality from Weight

You will frequently be asked to determine the normality of a solution prepared by dissolving a certain weight of solute in sufficient solvent to produce a given volume of solution. You will have, as a conversion factor given in the problem, the weight solute per volume solution. You will also need a conversion factor from the molecular weight, the grams of solute per mole of solute. Finally, you will need to know the number of equivalents of solute per mole of solute. The answer will always be the normality, which is expressed dimensionally as _____.

$$\frac{eq\ solute}{liter\ soln.}$$

Question 4 on the pretest is this type of problem. What is the normality of an $FeCl_3$ solution made by diluting 81 g $FeCl_3$ to 1.5 liters? The reaction considered produces Fe metal.

The information given in the problem is

$$\frac{\underline{\quad\quad}\ FeCl_3}{1.5\ \underline{\quad\quad\quad}}.$$

81 g

liters soln.

The molecular weight of solute is

$$\frac{162.3\ \underline{\quad\quad}}{\underline{\quad\quad\quad}}.$$

g $FeCl_3$

mole $FeCl_3$

The number of equivalents per mole is

$$\frac{\underline{\ }\ eq\ FeCl_3}{\underline{\quad\quad}\ FeCl_3}.$$

3

mole

There are three equivalents per mole because the Fe goes from a

____ oxidation number in $FeCl_3$ to a __ oxidation number in Fe metal.

+3,   0

Now the solution of the problem is very simple. You put down the dimensions that the answer will have

$$?\ N = \frac{?\ eq\ FeCl_3}{liter\ soln.} =$$

Then you start with any conversion factor which has one of these dimensions in it. For example,

$$\frac{81\ g\ FeCl_3}{1.5\ liters\ soln.}$$

You now have the desired dimension, liter solution, in the denominator. The unwanted dimension is _____, so next

grams $FeCl_3$

you multiply by _____. This leaves the unwanted

$$\frac{mole\ FeCl_3}{162.3\ g\ FeCl_3}$$

dimension _____, which can be converted to the dimensions

$$\frac{mole\ FeCl_3}{3\ eq\ FeCl_3}$$

of the answer by multiplying by _____. This leaves the two dimensions of the answer, which must then be

$$\frac{3\ eq\ FeCl_3}{mole\ FeCl_3}$$

$$?\ N = \frac{?\ eq\ FeCl_3}{liter\ soln.}$$

$$= \frac{81\ \cancel{g\ FeCl_3}}{1.5\ liters\ soln.} \times \frac{\cancel{mole\ FeCl_3}}{162.3\ \cancel{g\ FeCl_3}} \times \frac{3\ eq\ FeCl_3}{\cancel{mole\ FeCl_3}}$$

$$= 1.0\ N.$$

This type of problem can become complicated if the weight of solute isn't given in grams per metric volume. Question 7 on the pretest is an example of the complications you may face. What is the normality of a 3.0 wt % solution of C_2H_5OH whose density is 0.90 g/ml? The reaction produces CO_2.

The first thing to do is to determine the number of equivalents of solute per mole. The reaction is

$$C_2H_5OH \longrightarrow CO_2$$

-4	As you did in Chapter 13, you determine the total oxidation number for the two C's in C_2H_5OH. It is ___. Then you take two
$+4$	times the oxidation number of each C in CO_2, that is $2\times$ ___ or a
$+8$, 12	total of ___. In order to go from -4 to $+8$, you will lose ___ electrons. Therefore, the conversion factor is
12	$$\dfrac{\text{___ eq } C_2H_5OH}{\text{mole } C_2H_5OH}$$

Next you write down the other conversion factors given in the problem. They are

C_2H_5OH	$\dfrac{3.0 \text{ g} \underline{\hspace{2cm}}}{100 \text{ g} \underline{\hspace{1cm}}}$
soln.	

and

soln.	$\dfrac{0.90 \text{ g} \underline{\hspace{1cm}}}{\text{ml} \underline{\hspace{1cm}}}.$
soln.	

You will also probably need the conversion factor from the molecular weight of C_2H_5OH,

C_2H_5OH	$\dfrac{46.0 \text{ g} \underline{\hspace{2cm}}}{\underline{\hspace{2cm}}}.$
mole C_2H_5OH	

Now you are ready to set up the problem.

eq C_2H_5OH	$? N = \dfrac{? \underline{\hspace{2cm}}}{\underline{\hspace{2cm}}} =$
liter soln.	

The only conversion factor with one of these two dimensions is

$\dfrac{12 \text{ eq } C_2H_5OH}{\text{mole } C_2H_5OH}$	_____, so you put this in. The unwanted dimension is
$\dfrac{\text{mole } C_2H_5OH}{\text{mole } C_2H_5OH}$	_____, which you can cancel by multiplying by the
$\dfrac{\text{mole } C_2H_5OH}{46.0 \text{ g } C_2H_5OH}$	conversion factor from the molecular weight, _____.
	To cancel the new unwanted dimension, you can multiply by
$\dfrac{3.0 \text{ g } C_2H_5OH}{100 \text{ g soln.}}$	_____, which leaves the unwanted dimension
grams soln., $\dfrac{0.90 \text{ g soln.}}{\text{ml soln.}}$	_____. Next, then, you multiply by _____. However, the dimension called for in the answer isn't milliliters solution, but

_____. You therefore multiply by the simple conversion

_____. The only dimensions left now are those of the answer, which must be

$$? N = \frac{? \text{ eq } C_2H_5OH}{\text{liter soln.}}$$

$$= \frac{12 \text{ eq } C_2H_5OH}{\text{mole } C_2H_5OH} \times \frac{\text{mole } C_2H_5OH}{46.0 \text{ g } C_2H_5OH} \times \frac{3.0 \text{ g } C_2H_5OH}{100 \text{ g soln.}}$$

$$\times \frac{0.90 \text{ g soln.}}{\text{ml soln.}} \times \frac{\text{ml soln.}}{10^{-3} \text{ liter soln.}}$$

$$= 7.0 \ N.$$

This problem is a good example of the advantages of writing down all the conversion factors given in the problem before you start on your setup. If you do this, everything falls into place.

SECTION D Preparing Dilute Solutions from Concentrated Ones

If you have a concentrated solution of a known normality and wish to dilute this to a certain amount of solution at a lower normality, the procedure is identical with that in Chapter 12, Section D, where molarity was considered. If, however, you have a solution of known molarity and wish to dilute this to a solution of a certain normality, then you have a slightly different calculation.

Question 6 in the pretest is a typical example. To what volume in liters must you dilute 500 ml of 18 M H_2SO_4 in order to have a 0.10 N H_2SO_4 solution? The reaction considered is the complete reaction of the acid to form SO_4^{-2}.

As before, the first thing you must do is to determine the equivalents per mole. In this question, you are told that the acid will completely react to SO_4^{-2}. This means that it loses both its H's. Consequently, the conversion factor is

$$\frac{\_\_ \text{ eq } H_2SO_4}{\_\_\_\_\_ H_2SO_4}.$$

Now you must get conversion factors for the two solutions, the initial 18 M solution and the 0.10 N final solution. These will be

$$\frac{18 \_\_\_\_\_ H_2SO_4}{\text{liter} \_\_\_\_ \text{ soln.}}$$

and

$$\frac{0.10 \_\_\_ H_2SO_4}{\text{liter} \_\_\_\_\_ \text{ soln.}}$$

Right margin:

liter
$$\frac{\text{ml soln.}}{10^{-3} \text{ liter soln.}}$$

2

mole

moles

init.

eq

final

final

0.500 liter init. soln.

$$\frac{18 \text{ moles } H_2SO_4}{\text{liter init. soln.}}$$

$$\frac{2 \text{ eq } H_2SO_4}{\text{mole } H_2SO_4}$$

$$\frac{\text{liter final soln.}}{0.10 \text{ eq } H_2SO_4}$$

The question asked is "How many liters _____ solution?" The information given is _____. Notice how we made an immediate change from 500 ml to 0.500 liter. Now the problem is very simple.

? liter final soln. = 0.500 liter init. soln. × _____

Now you can convert to equivalents H_2SO_4 by multiplying by

_____, and, finally, you can get to the dimensions of the

answer by multiplying by _____. The only dimension which has not cancelled is the dimension of the answer.

$$? \text{ liters final soln.} = 0.500 \cancel{\text{ liter init. soln.}} \times \frac{18 \text{ moles } \cancel{H_2SO_4}}{\cancel{\text{liter init. soln.}}}$$

$$\times \frac{2 \cancel{\text{ eq } H_2SO_4}}{\cancel{\text{mole } H_2SO_4}} \times \frac{\text{liter final soln.}}{0.10 \cancel{\text{ eq } H_2SO_4}}$$

$$= 180 \text{ liters final soln.}$$

$$= 1.8 \times 10^2 \text{ liters final soln. (expressed to show the correct number of significant figures)}$$

The above is essentially the same method you used to dilute a solution to a desired molarity (Chapter 12, Section D), except for the one added conversion factor, which gets you from moles solute

equivalents

to _____ solute. If you are ever asked to mix two solutions of different normalities and determine the normality of the resulting solution, you can use exactly the same procedure as in Section E, Chapter 12.

You are now ready to do the problem set.

PROBLEM SET

These problems are divided into groups according to the section of the chapter which covers their solution. The problems are in order of increasing complexity in each group. Those that are marked with an asterisk (*) are overly complicated, and if you cannot do them, just check the setup in the answers to the problems which appear on the following pages. Incidentally, some of the solutions are imaginary since the substances are not sufficiently soluble to prepare solutions of the concentrations used in the problems.

Section A

1. What is the normality of a 3.7 M solution of H_3PO_4 if the reaction considered is the production of NaH_2PO_4?

2. How many moles of $KMnO_4$ are in 250 ml of a 0.40 N solution of $KMnO_4$? The reaction produces Mn^{+2}.

Section B

3. How many grams of H_3AsO_4 are there in 15 liters of 0.12 N solution of H_3AsO_4? The reaction considered produces Na_2HAsO_4. The molecular weight of H_3AsO_4 is 141.9 g/mole.

4. How many grams of $Mg(OH)_2$ are required to prepare 150 ml of a 3.00 N solution of $Mg(OH)_2$ if the reaction considered produces Mg^{+2}? The molecular weight of $Mg(OH)_2$ is 58.3 g/mole.

*5. What is the weight percent of $K_2S_2O_8$ (molecular weight = 270.4 g/mole) in a 0.12 N solution which has a density of 1.05 g/ml? The product of the reaction is $SO_4^=$.

*6. How many milliliters of acetic acid, CH_3COOH, were added to enough water to prepare 3.0 liters of solution if it was found that the solution was 0.20 N? The reaction considered produced CO_2. The density of acetic acid is 1.05 g/ml and its molecular weight is 60.0 g/mole.

Section C

7. What is the normality of a solution which contains 15.8 g $Na_2S_2O_3$ (molecular weight = 158.2 g/mole) in 125 ml of solution? The reaction considered produces $S_2O_5^{-2}$ ions.

8. What is the normality of a $CaCO_3$ solution if the reaction considered produces $CaCl_2$? The solution was prepared by dissolving 0.10 mole of $CaCO_3$ in 125 ml of water.

9. What is the normality of a 19.6 wt % solution of H_2SO_4 (molecular weight = 98.1 g/mole), if the product of the reaction is $NaHSO_4$? The density of the solution is 1.1 g/ml.

*10. What is the normality of a solution of C_2H_5OH (molecular weight $=$ 46.0 g/mole) which was prepared by dissolving 6.0 ml of C_2H_5OH in enough water to make 50 ml of solution? The reaction product is $C_2H_4O_2$, and the density of C_2H_5OH is 0.80 g/ml.

Section D

11. To what volume must you dilute 300 ml of a 6.00 *N* solution of NaOH to have a 0.200 *N* solution? The reaction considered is the same for both solutions.

12. To what volume must you dilute 1.30 liters of a 2.00 *M* solution of $Na_2C_2O_4$ in order to have a 0.300 *N* solution? The reaction product is CO_2.

13. How many milliliters of 6.00 *M* H_2SO_4 must you dilute to prepare 500 ml of a 0.200 *N* solution? The reaction considered is the complete neutralization (reaction) of H_2SO_4.

Section D (see Chapter 12, Section E)

14. What will be the normality of a solution prepared by mixing 250 ml of a 0.15 *N* solution of H_2SO_4 with 0.500 liter of a 2.0 *N* solution of H_2SO_4? The reaction considered in all cases is the complete neutralization of the H_2SO_4.

15. What will be the normality of a solution of H_3PO_4 if it was prepared by mixing 4.0 liters of 6.0 *M* H_3PO_4 with 2.0 liters of 1.5 *M* H_3PO_4? The reaction considered yields Na_2HPO_4.

Problems for the Show-offs

The following three problems are extremely complex. You will probably never have to face anything like this in your entire career in chemistry. Only those of you who really enjoy flexing your muscles with dimensional analysis should even bother to try them.

16. What was the percent purity of a sample of HCOOH if a 0.20 *N* solution resulted when 2.5 g of the impure material was dissolved in enough water to produce 500 ml of solution? The reaction considered produces CO_2. The molecular weight of HCOOH is 46.0 g/mole.

17. What is the normality of a 5.0 *m* (molal) solution of sucrose, $C_{12}H_{22}O_{11}$, whose density is 1.4 g/ml? The reaction considered is
$$C_{12}H_{22}O_{11} \longrightarrow C_2H_5OH.$$
The molecular weight of sucrose is 342.0 g/mole.

18. What is the molality of a 0.10 *N* solution of $FeCl_3$ whose density is 1.05 g/ml? The reaction produces metallic Fe. The molecular weight of $FeCl_3$ is 162.3 g/mole.

PROBLEM SET ANSWERS

If you were able to get ten correct solutions, you are sufficiently adept at solving normality problems. If you have particular trouble with a section, go back and work through that section in the chapter. In all cases the way the number of equivalents per mole was determined is noted with the solution.

1. $? N = \dfrac{? \text{ eq } H_3PO_4}{\text{liter soln.}} = \dfrac{3.7 \text{ moles } H_3PO_4}{\text{liter soln.}} \times \dfrac{1 \text{ eq } H_3PO_4}{\text{mole } H_3PO_4}$

    $= 3.7 \ N$

    [One H^+ ion is lost (reacts).]

2. $? \text{ moles } KMnO_4 = 0.250 \text{ liter soln.} \times \dfrac{0.40 \text{ eq } KMnO_4}{\text{liter soln.}}$

    $\times \dfrac{\text{mole } KMnO_4}{5 \text{ eq } KMnO_4}$

    $= 0.020 \text{ mole } KMnO_4$

    (Mn goes from an oxidation number of $+7$ to $+2$.)

3. $? \text{ g } H_3AsO_4 = 15 \text{ liter soln.} \times \dfrac{0.12 \text{ eq } H_3AsO_4}{\text{liter soln.}} \times \dfrac{\text{mole } H_3AsO_4}{2 \text{ eq } H_3AsO_4}$

    $\times \dfrac{141.9 \text{ g } H_3AsO_4}{\text{mole } H_3AsO_4}$

    $= 1.3 \times 10^2 \text{ g } H_3AsO_4$

    (Two H^+ ions have reacted.)

4. $? \text{ g } Mg(OH)_2 = 0.150 \text{ liter soln.} \times \dfrac{3.00 \text{ eq } Mg(OH)_2}{\text{liter soln.}} \times \dfrac{\text{mole } Mg(OH)_2}{2 \text{ eq } Mg(OH)_2}$

    $\times \dfrac{58.3 \text{ g } Mg(OH)_2}{\text{mole } Mg(OH)_2}$

    $= 13.1 \text{ g } Mg(OH)_2$

    (Two OH^- ions react.)

5. $? \text{ g } K_2S_2O_8 = 100 \text{ g soln.} \times \dfrac{\text{ml soln.}}{1.05 \text{ g soln.}} \times \dfrac{0.12 \text{ eq } K_2S_2O_8}{10^3 \text{ ml soln.}}$

    $\times \dfrac{\text{mole } K_2S_2O_8}{2 \text{ eq } K_2S_2O_8} \times \dfrac{270.4 \text{ g } K_2S_2O_8}{\text{mole } K_2S_2O_8}$

    $= 1.5 \text{ g } K_2S_2O_8 = 1.5\%$

    (Two S's go from $+14$ oxidation number to $+12$.)

6. $? \text{ ml CH}_3\text{COOH} = 3.0 \text{ liters soln.} \times \dfrac{0.20 \text{ eq CH}_3\text{COOH}}{\text{liter soln.}}$

$$\times \dfrac{\text{mole CH}_3\text{COOH}}{8 \text{ eq CH}_3\text{COOH}} \times \dfrac{60.0 \text{ g CH}_3\text{COOH}}{\text{mole CH}_3\text{COOH}}$$

$$\times \dfrac{\text{ml CH}_3\text{COOH}}{1.05 \text{ g CH}_3\text{COOH}}$$

$$= 4.3 \text{ ml CH}_3\text{COOH}$$

$$(C_2H_4O_2 \longrightarrow 2CO_2)$$

⓪ $2\times\text{(+4)}$

$$| -8 \quad \text{electrons} \uparrow$$

7. $? N = \dfrac{? \text{ eq Na}_2\text{S}_2\text{O}_3}{\text{liter soln.}}$

$$= \dfrac{15.8 \text{ g Na}_2\text{S}_2\text{O}_3}{0.125 \text{ liter soln.}} \times \dfrac{\text{mole Na}_2\text{S}_2\text{O}_3}{158.2 \text{ g Na}_2\text{S}_2\text{O}_3}$$

$$\times \dfrac{4 \text{ eq Na}_2\text{S}_2\text{O}_3}{\text{mole Na}_2\text{S}_2\text{O}_3}$$

$$= 3.20 \, N$$

(The oxidation number for two S's has gone from $+4$ to $+8$.)

8. $? N = \dfrac{? \text{ eq CaCO}_3}{\text{liter soln.}}$

$$= \dfrac{0.10 \text{ mole CaCO}_3}{0.125 \text{ liter soln.}} \times \dfrac{2 \text{ eq CaCO}_3}{\text{mole CaCO}_3}$$

$$= 1.6 \, N$$

($\text{CO}_3^{=}$ is equivalent to 2 H^+.)

9. $? N = \dfrac{? \text{ eq H}_2\text{SO}_4}{\text{liter soln.}}$

$$= \dfrac{1 \text{ eq H}_2\text{SO}_4}{\text{mole H}_2\text{SO}_4} \times \dfrac{\text{mole H}_2\text{SO}_4}{98.1 \text{ g H}_2\text{SO}_4}$$

$$\times \dfrac{19.6 \text{ g H}_2\text{SO}_4}{100 \text{ g soln.}} \times \dfrac{1.1 \text{ g soln.}}{\text{ml soln.}} \times \dfrac{\text{ml soln.}}{10^{-3} \text{ liter soln.}}$$

$$= 2.2 \, N$$

(One H^+ ion reacts.)

10. $? N = \dfrac{? \text{ eq C}_2\text{H}_5\text{OH}}{\text{liter soln.}}$

$$= \dfrac{6.0 \text{ ml C}_2\text{H}_5\text{OH}}{0.050 \text{ liter soln.}} \times \dfrac{0.80 \text{ g C}_2\text{H}_5\text{OH}}{\text{ml C}_2\text{H}_5\text{OH}}$$

$$\times \dfrac{\text{mole C}_2\text{H}_5\text{OH}}{46.0 \text{ g C}_2\text{H}_5\text{OH}} \times \dfrac{4 \text{ eq C}_2\text{H}_5\text{OH}}{\text{mole C}_2\text{H}_5\text{OH}}$$

$$= 8.3 \, N$$

$$(C_2H_5OH \longrightarrow C_2H_4O_2)$$

$$\overset{\text{(-4)}}{} \qquad \overset{\text{(0)}}{}$$

$$\boxed{}\,{-4 \text{ electrons}}\uparrow$$

11. ? liters final soln. $= 0.300 \,\cancel{\text{liter init. soln.}} \times \dfrac{6.00 \,\cancel{\text{eq NaOH}}}{\cancel{\text{liter init. soln.}}}$

$$\times \dfrac{\text{liter final soln.}}{0.200 \,\cancel{\text{eq NaOH}}}$$

$$= 9.00 \text{ liter final soln.}$$

12. ? liters final soln. $= 1.30 \,\cancel{\text{liters init. soln.}} \times \dfrac{2.00 \,\cancel{\text{moles Na}_2\text{C}_2\text{O}_4}}{\cancel{\text{liter init. soln.}}}$

$$\times \dfrac{2 \,\cancel{\text{eq Na}_2\text{C}_2\text{O}_4}}{\cancel{\text{mole Na}_2\text{C}_2\text{O}_4}} \times \dfrac{\text{liter final soln.}}{0.300 \,\cancel{\text{eq Na}_2\text{C}_2\text{O}_4}}$$

$$= 17.3 \text{ liters final soln.}$$

(Two C's go from an oxidation number of $+6$ to $+8$.)

13. ? ml init. soln. $= 0.500 \,\cancel{\text{liter final soln.}} \times \dfrac{0.200 \,\cancel{\text{eq H}_2\text{SO}_4}}{\cancel{\text{liter final soln.}}} \times \dfrac{\cancel{\text{mole H}_2\text{SO}_4}}{2 \,\cancel{\text{eq H}_2\text{SO}_4}}$

$$\times \dfrac{\cancel{\text{liter init. soln.}}}{6.00 \,\cancel{\text{mole H}_2\text{SO}_4}} \times \dfrac{\text{ml init. soln.}}{10^{-3} \,\cancel{\text{liter init. soln.}}}$$

$$= 8.33 \text{ ml init. soln.}$$

(H_2SO_4 loses 2 H^+ ions.)

14. As was the case with mixtures of solutions of different molarities, you must find the number of moles (or equivalents, in this case) present in each solution separately. Then you add to get the total number of moles (equivalents in this case) and divide this by the sum of the volumes.

? eq H_2SO_4 $= 0.250 \,\cancel{\text{liter}_{\,0.15\ N}} \times \dfrac{0.15 \text{ eq } H_2SO_4}{\cancel{\text{liter}_{\,0.15\ N}}} = 0.0375 \text{ eq } H_2SO_4$

? eq H_2SO_4 $= 0.500 \,\cancel{\text{liter}_{\,2.0\ N}} \times \dfrac{2.0 \text{ eq } H_2SO_4}{\cancel{\text{liter}_{\,2.0\ N}}} = 1.00 \text{ eq } H_2SO_4$

Total volume $= 0.750$ liter   Total eq $H_2SO_4 = 1.0375$ eq H_2SO_4

? $N = \dfrac{\text{? eq } H_2SO_4}{\text{liter total soln.}}$

$$= \dfrac{1.0375 \text{ eq } H_2SO_4}{0.750 \text{ liter total soln.}}$$

$$= 1.4 \ N$$

15. (This problem is similar to the preceding one, except that you convert moles to equivalents.)

$$? \text{ eq } H_3PO_4 = 4.0 \text{ liters}_{6\ M} \times \frac{6.0 \text{ moles } H_3PO_4}{\text{liter}_{6\ M}}$$

$$\times \frac{2 \text{ eq } H_3PO_4}{\text{mole } H_3PO_4} = 48 \text{ eq } H_3PO_4$$

$$? \text{ eq } H_3PO_4 = 2.0 \text{ liters}_{1.5\ M} \times \frac{1.5 \text{ moles } H_3PO_4}{\text{liter}_{1.5\ M}}$$

$$\times \frac{2 \text{ eq } H_3PO_4}{\text{mole } H_3PO_4} = 6 \text{ eq } H_3PO_4$$

Total volume = 6.0 liters   Total eq H_3PO_4 = 54 eq H_3PO_4

$$? N = \frac{? \text{ eq. } H_3PO_4}{\text{liter total soln.}}$$

$$= \frac{54 \text{ eq } H_3PO_4}{6.0 \text{ liters total soln.}} \quad (H_3PO_4 \text{ loses } 2H^+ \text{ ions.})$$

$$= 9.0 \ N$$

16. $$? \text{ g HCOOH} = 100 \text{ g imp.} \times \frac{0.500 \text{ liter soln.}}{2.5 \text{ g imp.}} \times \frac{0.20 \text{ eq HCOOH}}{\text{liter soln.}}$$

$$\times \frac{\text{mole HCOOH}}{2 \text{ eq HCOOH}} \times \frac{46.0 \text{ g HCOOH}}{\text{mole HCOOH}}$$

$$= 92 \text{ g HCOOH} = 92\% \text{ pure}$$

(C goes from an oxidation number of +2 to +4.)

17. This is another one of those terrible problems where you will have to calculate the total volume of the solution by adding the weights of solvent and solute present. (Chapter 12, Section C.)

$$\text{Total weight of soln.} = 1000 \text{ g solvent} + 5.0 \text{ moles sucrose} \times \frac{342.0 \text{ g sucrose}}{\text{mole sucrose}}$$

$$= 1000 \text{ g} + 1710 \text{ g} = 2710 \text{ g soln.}$$

$$\text{Total volume soln.} = 2710 \text{ g soln.} \times \frac{\text{ml soln.}}{1.4 \text{ g soln.}} = 1936 \text{ ml soln.}$$

$$? N = \frac{? \text{ eq sucrose}}{\text{liter soln.}}$$

$$= \frac{5.0 \text{ moles sucrose}}{1.936 \text{ liters soln.}} \times \frac{24 \text{ eq sucrose}}{\text{mole sucrose}}$$

$$= 62 \ N$$

If you had trouble getting $\dfrac{24 \text{ eq sucrose}}{\text{mole sucrose}}$, consider

$$C_{12}H_{22}O_{11} \longrightarrow 6C_2H_5OH$$

⓪ $6 \times$ ④

+24 electrons

18. (This is like Problem 16, but backwards. It's a mess!)
    First determine the weight of one liter of solution.

$$? \text{ g soln.} = 1.0 \text{ liter soln.} \times \frac{\text{ml soln.}}{10^{-3} \text{ liter soln.}} \times \frac{1.05 \text{ g soln.}}{\text{ml soln.}}$$

$$= 1050 \text{ g soln.}$$

Then determine the number of grams $FeCl_3$ in one liter of solution.

$$? \text{ g } FeCl_3 = 1.0 \text{ liter soln.} \times \frac{0.10 \text{ eq } FeCl_3}{\text{liter soln.}} \times \frac{\text{mole } FeCl_3}{3 \text{ eq } FeCl_3}$$

$$\times \frac{162.3 \text{ g } FeCl_3}{\text{mole } FeCl_3}$$

$$= 5.4 \text{ g } FeCl_3$$

The weight of solvent is then gotten by subtracting the weight of solute, $FeCl_3$, from the total weight of solution.

$$? \text{ g solvent} = 1050 - 5.4 = 1044.6 \text{ g solvent}$$

You now know you have 1044.6 g solvent and 0.1 eq $FeCl_3$.

$$? \, m = \frac{? \text{ moles } FeCl_3}{\text{kg solvent}}$$

$$= \frac{0.10 \text{ eq } FeCl_3}{1.0446 \text{ kg solvent}} \times \frac{\text{mole } FeCl_3}{3 \text{ eq } FeCl_3}$$

$$= 3.2 \, m.$$

Chapter 15
Stoichiometry Involving Solutions

PRETEST

You will recall from the earlier chapter on stoichiometry that such problems were solved by converting the information given to a "bridge" expressing the molar relationships from the balanced equation. The same is true in solution stoichiometry. If you have done Chapters 13 and 14, you will have another "bridge" that you can use. This is the one-to-one relationship of equivalents of all substances in a reaction. Thus you can always use one equivalent A per one equivalent B regardless of what A and B are. Once you cross your "bridge" to the substance in the question asked, you simply convert it to the proper dimensions. Try the following questions. If you have not done Chapters 13 and 14, you will not be able to do 4 and 5 and will have to write balanced equations for the reactions in 3 and 6.

1. How many liters of a 0.30 M solution of HCl will you require for the complete reaction with 500 g $CaCO_3$ according to the reaction

    $2HCl + CaCO_3 = CaCl_2 + H_2O + CO_2$?

2. How many liters of 0.30 M solution of HCl will be required for the complete reaction with 5.0 liters of 0.10 M Na_2CO_3 according to the reaction

    $2HCl + Na_2CO_3 = 2NaCl + H_2O + CO_2$?

3. How many liters 0.30 N solution of H_2SO_4 will be required to react with 500 g $CaCO_3$? The products of the reaction are $CaSO_4$, CO_2, and H_2O. For those of you who have not done Chapter 14, the molarity of the H_2SO_4 is 0.15 M.

4. How many liters of 0.30 N solution of $KMnO_4$ will be required to react with 5.0 liters of 0.10 N $H_2C_2O_4$?

5. How many milliliters of a 6.0 N solution of $KMnO_4$ will be required to react with 1.0 liter of a 3.7 wt % solution of $Na_2C_2O_4$? The density of the $Na_2C_2O_4$ solution is 1.08 g/ml, and the product of its reaction is CO_2.

6. What is the percent purity of a Na_2CO_3 sample if 6.00 g of the sample titrates to an endpoint with 52.0 ml of 1.50 N HCl solution? The products of the reaction are NaCl, CO_2, and H_2O. For those who have not done Chapter 14, the HCl is 1.50 M.

7. An impure sample of $CaCO_3$ is reacted with an excess of 1.5 N HCl. The excess HCl is then back-titrated with an NaOH solution to the endpoint. The data is given below.

Weight impure $CaCO_3$ = 2.20 g

Volume HCl = 45.0 ml

Volume NaOH for back-titration = 14.0 ml

It was found that 30.0 ml of the HCl solution titrated to an end point with exactly 22.0 ml of the NaOH solution. If the product of reaction was $CaCl_2$, what is the percent purity of the $CaCO_3$ sample?

The equations for the reactions are

$2HCl + CaCO_3 = CaCl_2 + H_2O + CO_2$

$HCl + NaOH = NaCl + H_2O.$

PRETEST ANSWERS

If you got the correct answers to all the questions, skip the chapter and go directly to the problem set. Questions 1 and 2 are covered in Section A; questions 3 and 4 are covered in Section B; questions 5 and 6 are in Section C; and question 7 is in Section D. Guide your reading in the chapter accordingly.

If your instructor has suggested that you do not read Chapters 13 and 14, you will not be doing Sections B and C in this chapter.

1. 33 liters soln.

2. 3.3 liters soln.

3. 33 liters soln.

4. 1.7 liters $KMnO_4$ soln.

5. 99 ml $KMnO_4$ soln.

6. 68.9%

7. 88%

Most chemical reactions are run using solutions of the substances being reacted. There are several reasons for this: the reactions tend to be faster, handling the material is simpler, smaller amounts of the substance may be conveniently used, etc. Consequently, a great many stoichiometry problems for determining how much of one substance will react with or produce some other substance, are expressed in terms of the concentrations of the solutions of the substances.

As in the stoichiometry calculations of Chapter 8, the way to solve the problems is to multiply the information given by conversion factors until you have the substance in the information given expressed in moles. Next use the "bridge" conversion factor from the balanced equation to get to moles of the substance in the question asked. Then multiply this by conversion factors until you get to the dimensions of the substance asked for in the question asked.

SECTION A   Problems with Balanced Equations

When you have a balanced equation for the reaction, and one or more solutions whose concentration is expressed in molarity, the problems are quite simple. Consider question 1 on the pretest. How many liters of a 0.30 M solution of HCl will be required for the complete reaction with 500 g $CaCO_3$ according to the reaction

$$2HCl + CaCO_3 = CaCl_2 + H_2O + CO_2?$$

The question asked is "How many _____?" The

information given is _____, and the conversion factor given in the problem is the molarity, which you write as

_____. Since the problem has weight of $CaCO_3$ and you will need moles of $CaCO_3$, you will need the conversion factor

_____, the molecular weight. The two substances

referred to in the problem are HCl and _____, so the "bridge" you will need from the balanced equation is

$$\frac{_____ \text{ HCl}}{\text{mole } CaCO_3}.$$

You now have everything you will need to solve the problem. Starting in the usual way, then, you write

? _____ = _____.

To remove the unwanted dimension, _____, and get to moles $CaCO_3$ (so you can use the "bridge"), you multiply by the

conversion factor _____. Next you need the "bridge" to

liters solution

500 g $CaCO_3$

$\dfrac{0.30 \text{ mole HCl}}{\text{liter soln.}}$

$\dfrac{100.1 \text{ g } CaCO_3}{\text{mole } CaCO_3}$

$CaCO_3$

2 moles

liter soln.,   500 g $CaCO_3$

grams $CaCO_3$

$\dfrac{\text{mole } CaCO_3}{100.1 \text{ g } CaCO_3}$

$$\frac{\text{2 moles HCl}}{\text{mole CaCO}_3}$$

$$\frac{\text{liter soln.}}{\text{0.30 mole HCl}}$$

get to the substance in the question asked, HCl, so you multiply by _____. Now all you must do is to convert "mole HCl" to liter (HCl) soln. You can do this by using the conversion factor

_____. The only dimension left is the dimension of the answer, so the answer must be

$$? \text{ liters soln.} = 500 \text{ g } \cancel{\text{CaCO}_3} \times \frac{\text{mole } \cancel{\text{CaCO}_3}}{100.1 \text{ g } \cancel{\text{CaCO}_3}}$$

$$\times \frac{\text{2 moles } \cancel{\text{HCl}}}{\cancel{\text{mole CaCO}_3}} \times \frac{\text{liter soln.}}{0.30 \cancel{\text{ mole HCl}}}$$

$$= 33 \text{ liters soln.}$$

Often you will have two different solutions reacting. Such problems are no more difficult, but you must be very careful to write the dimensions so that you do not confuse the two different solutions. One way to do this is to write the substance contained in the solution (or an abbreviation for it) after the volume dimension. For example, 5.0 liters HCl would be 5.0 liters of the HCl solution. Or you might use an abbreviation like 5.0 liters MnO_4^- for 5.0 liters of a $KMnO_4$ solution.

Question 2 on the pretest is an example of this type of problem. How many liters of 0.30 *M* HCl will be required for the complete reaction with 5.0 liters of 0.10 *M* Na_2CO_3 solution according to the reaction

$$2HCl + Na_2CO_3 = 2NaCl + H_2O + CO_2?$$

liters HCl

liters Na_2CO_3

The question asked is "How many _____?" The information given is 5.0 _____. There are two conversion factors given in the problem, the two molarities,

$$\frac{\text{0.30 mole HCl}}{\text{liter HCl}} \quad \text{and} \quad \frac{\text{0.10 mole Na}_2\text{CO}_3}{\text{_____}},$$

liter Na_2CO_3

$$\frac{\text{2 moles HCl}}{\text{mole Na}_2\text{CO}_3}$$

and the "bridge" that you will need to relate the two substances is

_____. You will not need any conversion factor from the molecular weight, since there are no weights of the substances mentioned in the problem.

You now have everything you need, so, starting in the usual way, you write:

$$? \text{_____} = \text{_____}.$$

liter HCl, 5.0 liters Na_2CO_3

$$\frac{\text{0.10 mole Na}_2\text{CO}_3}{\text{liter Na}_2\text{CO}_3}$$

First you multiply by the conversion factor _____ to cancel the unwanted dimension and get you to moles so that you can use the "bridge." Now you can multiply by the "bridge,"

$$\frac{\text{2 moles HCl}}{\text{mole Na}_2\text{CO}_3}$$

_____. In order to get to the desired dimensions of HCl

you then multiply by _____. The only dimension remaining is the dimension of the answer, which must be

$$? \text{ liters HCl} = 5.0 \text{ liters Na}_2\text{CO}_3 \times \frac{0.10 \text{ mole Na}_2\text{CO}_3}{\text{liter Na}_2\text{CO}_3}$$

$$\times \frac{2 \text{ moles HCl}}{\text{mole Na}_2\text{CO}_3} \times \frac{\text{liter HCl}}{0.30 \text{ mole HCl}}$$

$$= 3.3 \text{ liters HCl.}$$

This type of problem can only become more complex if the volumes of the solutions are expressed in units other than liters or the weights in units other than grams. However, you are now expert at converting from one dimension to another, and you won't have any problems with these.

SECTION B   Problems without Balanced Equations

If your instructor has suggested that you omit Chapters 13 and 14, you should omit this section also.

In Chapter 13 we said that the only reason for using equivalency was to be able to take advantage of the extremely simple, unchanging "bridge" conversion factor of one equivalent of one substance per one equivalent of any other substance. The function of equivalency was to balance the equation for you. You are now going to see how this works.

We will start by doing question 3 on the pretest. How many liters of a 0.30 N solution of H_2SO_4 will be required to react with 500 g $CaCO_3$? The product of the reaction is $CaSO_4$.

Notice that you must know what the product of the reaction is in order to determine the number of equivalents per mole. In this case $CaCO_3$ goes to $CaSO_4$, which is a simple salt with no change in oxidation number. Since CO_3^{-2} can react with 2 H^+ ions (if you have forgotten this, refer back to Chapter 13, Section A), you have the conversion factor

$$\frac{\_\_ \text{ eq CaCO}_3}{\text{mole CaCO}_3}$$

There is one conversion factor given in the problem, the normality of the H_2SO_4 solution. This is

$$0.30 \_\_\_ H_2SO_4$$

_____

Since the $CaCO_3$ is expressed in grams, you will need the molecular weight conversion factor, _____. The question asked is "How many _____?" The information given is _____.

$$\frac{\text{liter HCl}}{0.30 \text{ mole HCl}}$$

2

eq

liter soln.

$$\frac{100.1 \text{ g CaCO}_3}{\text{mole CaCO}_3}$$

liters soln.,   500 g $CaCO_3$

Since you do not have a balanced equation, you are going to have to use as a "bridge" the unchanging conversion factor

1

1

$$\frac{\underline{\ \ } \text{ eq } H_2SO_4}{\underline{\ \ } \text{ eq } CaCO_3}.$$

You must therefore convert the dimensions of the information given to equivalents of the substance. Next you use the "bridge" to get to the substance in the question, and then you convert to the wanted dimensions of that substance. Therefore, starting in the usual way, you have

liters soln., 500 g CaCO₃

$$? \underline{\qquad\quad} = \underline{\qquad\quad}.$$

Now you must get grams CaCO₃ to equivalents CaCO₃. You can

$$\frac{\underline{\ \ } \text{mole } \cancel{CaCO_3}}{100.1 \text{ g } CaCO_3} \times \frac{2 \text{ eq } CaCO_3}{\underline{\ \ } \text{mole } \cancel{CaCO_3}}$$

do this with two conversion factors, $\underline{\qquad\qquad\qquad\qquad}$,

$$\frac{1 \text{ eq } H_2SO_4}{1 \text{ eq } CaCO_3}$$

and then you can multiply by the "bridge," $\underline{\qquad\qquad}$. The

equivalent H₂SO₄

unwanted dimension now is $\underline{\qquad\qquad\qquad}$. However, you can get to the dimensions of the answer by multiplying by

$$\frac{\text{liter soln.}}{0.30 \text{ eq } H_2SO_4}$$

$\underline{\qquad\qquad}$. The only dimension remaining is the dimension of the answer, which must be

$$? \text{ liters soln.} = 500 \times \frac{1}{100.1} \times 2 \times 1 \times \frac{1}{0.30}$$

$$= 33 \text{ liters soln.}$$

Doing this problem is really as simple as working with moles and molarity. The only point to remember is that the "bridge" will be in equivalents, and you must convert the information given into equivalents before you can use the "bridge." However, the real reason for expressing concentration in normality is that there is a terrific shortcut that you can use if you have two solutions reacting. First we will do a problem the long way and then we'll try the shortcut to do the same job. Question 4 on the pretest is an example of two solutions reacting together. Since there is more than one solution you will have to take care in writing the dimensions.

How many liters of 0.30 *N* KMnO₄ will be required to react with 5.0 liters of 0.10 *N* H₂C₂O₄? There are two conversion factors given in the problem, the two normalities.

eq

MnO₄⁻

eq

liter C₂O₄⁼

$$\frac{0.30 \underline{\ \ } \text{ KMnO}_4}{\text{liter} \underline{\qquad}}$$

$$\frac{0.10 \underline{\ \ } \text{ H}_2\text{C}_2\text{O}_4}{\underline{\qquad\qquad}}$$

You also know that the "bridge" you will use will be in equivalents

and will be _____; remember that there is always one
equivalent per one equivalent.

Now you start the problem in the usual way.

? liters MnO_4^- = _____

The unwanted dimension in the information given is _____.
You can go directly to equivalent $H_2C_2O_4$ by multiplying by the

normality conversion factor _____. Now that you are at

equivalents you can use the "bridge," _____. The unwanted

dimension now is _____, which can be converted to the
dimensions of the answer by multiplying by the other normality,

_____. The only dimension remaining is the dimension
of the answer, which must be

$$? \text{ liters } MnO_4^- = 5.0 \text{ liters } \overline{C_2O_4^=} \times \frac{0.10 \text{ eq } \overline{H_2C_2O_4}}{\overline{\text{liters } C_2O_4^=}}$$

$$\times \frac{1 \text{ eq } \overline{KMnO_4}}{1 \text{ eq } \overline{H_2C_2O_4}} \times \frac{\text{liter } MnO_4^=}{0.30 \text{ eq } \overline{KMnO_4}}$$

$$= 1.7 \text{ liters } MnO_4^-.$$

Now for the shortcut. When two solutions react and you have
their concentrations expressed in normality (and only in normality),
you can write

$$\boxed{V' \times N' = V'' \times N''}$$

where V' is the volume of one of the solutions and N' is its
normality, and V'' is the volume of the other solution and N'' is its
normality. Both volumes will be in the same units: that is, if V' is
liters, V'' is liters; if V' is milliliters, then V'' will be milliliters.

It is easy to see why this relationship holds true: multiplying the
volume times the normality gives you equivalents. Consider a 2 N
solution of any compound, X. If you have 3 liters of solution then
$V \times N$ is

$$3 \text{ liters} \times \frac{2 \text{ eq } X}{\text{liter}} = 6 \text{ eq } X.$$

The equation $V' \times N' = V'' \times N''$ means that the number of
equivalents of one substance equals the number of equivalents of
the other. That is exactly how we set up equivalency, and, of
course, it is true.

---

1 eq $KMnO_4$
─────────
1 eq $H_2C_2O_4$

5.0 liters $C_2O_4^=$

liter $C_2O_4^=$

0.10 eq $H_2C_2O_4$
─────────────
liter $C_2O_4^=$

eq $KMnO_4$
─────────
eq $H_2C_2O_4$

eq $KMnO_4$

liter MnO_4^-
─────────────
0.30 eq $KMnO_4$

Let's do the last example by this shortcut method.

V' = unknown volume of the $KMnO_4$ solution

N' = 0.30, the normality of the $KMnO_4$ solution

V'' = 5.0 liters of the $H_2C_2O_4$ solution

N'' = 0.10, the normality of the $H_2C_2O_4$ solution

Therefore,

$$V' \times 0.30 = 5.0 \times 0.10$$

$$V' = \frac{5.0 \times 0.10}{0.30} = 1.7 \text{ liters } MnO_4^-$$

If you look at the numbers that are multiplied together, you will see they are identical with those in the longer solution.

If you have forgotten how to solve this type of calculation by "cross-multiplication," check Section B in Chapter 10, which describes the method.

Here is another example of this. Suppose that it is found that it requires 25.32 ml of a 0.101 N solution of HCl to exactly neutralize 10.00 ml of an NaOH solution. What would be the normality of the NaOH solution?

You start with the shortcut equation,

$$V_{NaOH} \times N_{NaOH} = V_{HCl} \times N_{HCl}$$

and put in the values from the problem.

$$10.00 \times N_{NaOH} = 25.32 \times 0.101$$

Cross-multiplying gives

$$N_{NaOH} = \frac{25.32 \times 0.101}{10.00}$$

$$= 0.256 \, N_{NaOH}$$

You may find, in the future, that it will be much faster to solve a problem, even when the concentrations are given in molarity, if you shift the concentrations to normality and use the $V' \times N' = V'' \times N''$ relationship. Section A in Chapter 12 covers this method. Say you have a 3.0 M solution of H_2SO_4 and want to know its normality. If the product of the reaction is SO_4^{-2}, then there are __ equivalents per mole. To get the normality you have

2

$$? N = \frac{? \text{ eq } H_2SO_4}{\text{liter soln.}}$$

$$= \frac{3.0 \text{ moles } H_2SO_4}{\text{liter soln.}} \times \frac{2 \text{ eq } H_2SO_4}{\text{mole } H_2SO_4}$$

$$= 3.0 \times \frac{2 \text{ eq } H_2SO_4}{\text{liter soln.}}.$$

What you do, simply, is multiply the molarity by the number of equivalents per mole and you have the normality.

Here is another problem of this type. What is the normality of a solution of $FeCl_3$ if 35 ml of the solution react with 175 ml of a 1.0 M solution of H_3AsO_3? The reaction product is H_3AsO_4.

First you have to get the normality of the H_3AsO_3 solution.

Since there are __ equivalents per mole (As goes from $+3$ to $+5$

2

oxidation number), then the normality is __ $\times 1.0$ N. Then you can use

2

$$V' \times N' = V'' \times N''$$

$$35 \times N' = \underline{\hspace{2cm}}$$

175 $\times$ 2.0

$$N' = \underline{\hspace{0.8cm}}.$$

10

You can see just how fast this is.

Incidentally, there is a special vocabulary that goes along with reactions that are carried out with a liquid. When you slowly add known volumes of the liquid, you say you are "titrating" with the liquid. When exactly the amount of the liquid is added to completely react with the substance to which it is being added, you say you have reached the "endpoint" or sometimes the

"equivalence point." The reaction of an acid with a base is called
"neutralization" of the acid or base.

SECTION C More Complicated Problems with Solutions

These problems can become more complex when they have more
conversion factors given in the problem, do not have the
volumes in liters or the concentrations in molarity or normality, or
deal with substances that are not pure. However, you
should be able to handle these complications without too much
strain. Just remember to write down all the conversion factors
given in the problem before you start.

Question 5 in the pretest is a good example of this type of
problem. This problem was mentioned in the "To the Students" at
the beginning of the book.

How many milliliters of a 6.0 N solution of $KMnO_4$ will be
required to react with 1.0 liter of a 3.7 wt % solution of $Na_2C_2O_4$?
The density of the $Na_2C_2O_4$ solution is 1.08 g/ml, and the product
of its reaction is CO_2.

Before you start, write down all the conversion factors given in
the problem. To keep the two solutions straight you can label the
volume of $KMnO_4$ solution with "MnO_4^-" and the volume of
$Na_2C_2O_4$ solution with "$C_2O_4^=$." Then, when you are referring to
the compound itself, you will write the full formula. Your
abbreviation, when you are referring to the solution, is to just
write the ion.

The conversion factors given in the problem are

eq	$\dfrac{6.0 \;\_\_\_ \; KMnO_4}{\text{liter} \;_____}$
MnO_4^-	
$Na_2C_2O_4$	$\dfrac{3.7 \text{ g} \;_____}{\_\_\_ \text{ g } C_2O_4^=}$
100	
$C_2O_4^=$	$\dfrac{1.08 \text{ g } C_2O_4^=}{\text{ml} \;\_\_\_}$

Since a weight percent of $Na_2C_2O_4$ is given, you will probably also need the conversion factor from its _____ weight,

   $\dfrac{134.0 \text{ g } Na_2C_2O_4}{\rule{3cm}{0.4pt}}$.

Now you are ready to start.

   ? _____ = _____

The only conversion factor given in the problem which contains a

volume of the $C_2O_4^=$ solution is the _____ of the solution. But
this is given in ml $C_2O_4^=$. You must therefore multiply the unwanted

dimension in the information given by _____. Now you

can multiply by the density, _____, and the unwanted

dimension which remains is g $C_2O_4^=$. This appears in the conversion

factor from the weight _____, and so next you multiply by

   $\dfrac{3.7 \text{ g } Na_2C_2O_4}{100 \text{ g } C_2O_4^=}$.

You have finally gotten to weight of $Na_2C_2O_4$, and it will take
just two more conversion factors to get to equivalents $Na_2C_2O_4$,

which will be your "bridge." First you can get to _____
by using the inverted form of the molecular weight conversion

factor, _____. All you need now is a conversion

factor which will get you from moles $Na_2C_2O_4$ to _____
$Na_2C_2O_4$, that is, the number of equivalents per mole. You were
told in the problem that the product of the reaction was CO_2. This
is a redox reaction.

$$Na_2C_2O_4 \longrightarrow 2CO_2$$
$$\underset{\textstyle +6}{\bigcirc} \qquad\qquad 2\times\underset{\textstyle +4}{\bigcirc}$$
$$|\!\!\underline{\qquad 2 \text{ electrons} \qquad}\!\!\uparrow$$

(If you are still shaky with these, check back to Chapter 13,
Section C.)

molecular

mole $Na_2C_2O_4$

ml MnO_4^-,   1.0 liter $C_2O_4^=$

density

   $\dfrac{\text{ml } C_2O_4^=}{10^{-3} \text{ liter } C_2O_4^=}$

   $\dfrac{1.08 \text{ g } C_2O_4^=}{\text{ml } C_2O_4^=}$

percent

mole $Na_2C_2O_4$

   $\dfrac{\text{mole } Na_2C_2O_4}{134.0 \text{ g } Na_2C_2O_4}$
equivalents

You now can write the conversion factor that you needed,

2

$$\frac{\_\_ \text{ eq Na}_2\text{C}_2\text{O}_4}{\text{mole Na}_2\text{C}_2\text{O}_4},$$

and if you multiply this factor you reach the point where you can use your "bridge." Next, then, you multiply by the "bridge,"

1
1

$$\frac{\_\_ \text{ eq KMnO}_4}{\_\_ \text{ eq Na}_2\text{C}_2\text{O}_4}.$$

You have now reached the compound in the question asked. All

equivalents,   milliliters

you must do is to convert _____ KMnO$_4$ to _____ MnO$_4^-$, the dimension of the answer. The one remaining conversion factor given in the problem will do this for you. Thus, you multiply by

ml

$$\frac{10^3 \_\_ \text{ MnO}_4^-}{6.0 \text{ eq KMnO}_4}.$$

(We take a slight shortcut and express liter MnO$_4^-$ as 10^3 ml MnO$_4^-$.)

If you have been writing down your own setup as we have worked our way through this problem, the numbers you should have with all the dimensions cancelled are

$$? \text{ ml MnO}_4^- = 1.0 \times \frac{1}{10^{-3}} \times 1.08 \times 3.7 \times \frac{1}{134.0}$$

$$\times 2 \times \frac{1}{1} \times \frac{10^3}{6.0} \text{ ml MnO}_4^-$$

$$= 9.9 \times 10^3 \text{ ml MnO}_4^-.$$

From now on the entire setup will no longer be written out after the examples we work through in the body of the chapters. Unless there is something unusual, all you are going to have in the text to check against your setup is the numbers which remain after the dimensions are cancelled.

Frequently a chemist will analyze a material to determine its percent purity. Remember that when you are asked, "What is the percent purity?" you must reword the question so that it contains both the dimension of the question asked and the information

100

given. You ask, "How many grams of pure substance in _____ g of impure?" If you don't recall this, you can check back to Section C in Chapter 4. Thus you always start these problems with

100 g imp.

$$? \text{ g pure substance} = _____.$$

Question 6 in the pretest is a typical example. What is the percent purity of a Na$_2$CO$_3$ sample if 6.00 g of the sample titrates to an endpoint with 52.0 ml of 1.50 N HCl solution? The product of the reaction is NaCl.

One conversion factor given in the problem is the normality of

the HCl solution, _____. Another is that there are 6.00 g

_____ sample per _____.

$$\frac{6.00 \text{ g imp.}}{52.0 \text{ ml soln.}}$$

Since the HCl solution is given in normality and you have no
balanced equation, you will need the number of equivalents of
Na_2CO_3 per mole. Since the product is given as NaCl, there is no
redox reaction and you can simply say that the $CO_3^=$ ion can
combine with 2 H^+ ions so that the conversion factor will be

$$\frac{\_\_ \text{ eq } Na_2CO_3}{\text{mole } Na_2CO_3}.$$

Since the question asks about weight of Na_2CO_3, you will also

need the conversion factor from the _____ weight,

$$\frac{106.0 \text{ g } Na_2CO_3}{_____ \ Na_2CO_3}.$$

Now you can start by rewording the question to ask how many

grams _____ in 100 g _____. Thus you have

$$? \text{ g } Na_2CO_3 = _____.$$

To cancel the unwanted dimension, _____, you can

multiply by the conversion factor _____. Notice that the
volume of solution was expressed in liters rather than milliliters.
This saves one conversion factor, that's all. Remember that in these
stoichiometry problems you are heading toward your "bridge."
Since there is no balanced equation, you are heading toward

_____.

Multiplying by the conversion factor _____ will get you

there. Now you multiply by _____, the "bridge." To get
your answer all that you must do is to convert equivalents

Na_2CO_3 to _____ Na_2CO_3. You can do this with two conversion

factors: the first is _____ and the second is

_____/mole Na_2CO_3. The only dimension left is the
dimension of the question asked, so the answer must be

$$? \text{ g } Na_2CO_3 = 100 \times \frac{0.052}{6.00} \times 1.50 \times 1 \times \frac{1}{2}$$
$$\times 106.0 \text{ g } Na_2CO_3$$
$$= 68.9 \text{ g} = 68.9\%.$$

$$\frac{1.50 \text{ eq HCl}}{\text{liter soln.}}$$
impure,   52.0 ml soln.

2

molecular

mole

Na_2CO_3, imp.

100 g imp.

grams impure
$$\frac{0.052 \text{ liter soln.}}{6.00 \text{ g imp.}}$$

equivalents
$$\frac{1.50 \text{ eq HCl}}{\text{liter soln.}}$$
$$\frac{1 \text{ eq } Na_2CO_3}{1 \text{ eq HCl}}$$

grams
$$\frac{\text{mole } Na_2CO_3}{2 \text{ eq } Na_2CO_3}$$
$106.0 \text{ g } Na_2CO_3$

SECTION D Back-titration

When running a titration chemists may pass the endpoint. This may be either accidental or intentional. They may do so intentionally in order to approach the endpoint in a more convenient way (more easily recognized color change, for example) or to ensure a complete and rapid reaction. What they then do is to "back-titrate" the reaction mixture to determine how much excess has been added. They then must subtract this amount from the amount of titrant initially added. Then they use the difference to make the calculation. What chemists must know is the equivalency of the substance used in back-titrating to the titrant initially added. This equivalency will always be given in the problem.

Question 7 on the pretest is a good example of this type of problem. An impure sample of $CaCO_3$ is reacted with an excess of 1.5 N HCl. The excess HCl solution is then back-titrated with an NaOH solution to the endpoint. The data are given below.

Weight impure $CaCO_3$ = 2.20 g

Volume HCl = 45.0 ml

Volume NaOH for back-titration = 14.0 ml

And 30.0 ml of the HCl solution titrates with 22.0 ml of the NaOH solution. The product of the reaction is $CaCl_2$; what is the percent purity of the $CaCO_3$ sample?

First you must determine how much of the HCl solution was really required. You do this by determining how much was in excess, and then subtracting this from the amount put in initially.

22.0 ml NaOH soln.

$$\text{ml HCl soln. excess} = 14.0 \text{ ml NaOH soln.} \times \frac{30.0 \text{ ml HCl soln.}}{\underline{\hspace{2cm}}}$$

19.1

$$= \underline{\hspace{1cm}} \text{ml HCl soln.}$$

ml HCl soln. required = ml HCl soln. init. − ml HCl soln. excess

19.1

$$= 45.0 - \underline{\hspace{1cm}}$$

25.9 ml HCl soln.

$$= \underline{\hspace{3cm}}$$

Now you solve the problem in the same way as if you had the 25.9 ml HCl soln. given as the amount needed to reach the endpoint. Since this is a percent purity problem, you start with

$CaCO_3$, 100 g imp.

$$? \text{ g} \underline{\hspace{1.5cm}} = \underline{\hspace{2.5cm}}.$$

From the data given and calculated you can write two conversion factors,

$$\underline{\qquad} \text{ eq HCl}$$

$$\underline{\qquad\qquad}$$

$$\frac{0.0259 \ \underline{\qquad\qquad}}{2.2 \ \underline{\qquad}},$$

and since the question asked concerns weight of $CaCO_3$, you will have to use the conversion factor from the _____ weight,

$$\underline{\qquad} \ CaCO_3$$

$$\underline{\qquad} \ CaCO_3$$

Since there is no balanced equation, you will also need the number of _____ $CaCO_3$ per ____.

$$\underline{\qquad} \ CaCO_3$$

$$\underline{\qquad} \ CaCO_3$$

This is the same reaction as in the preceding example; $CaCl_2$ is formed.

Now you have all the conversion factors you will need except

the "bridge" between $CaCO_3$ and HCl, which will be _____. Finally, you set up the solution to the problem and then check your answer with what appears below.

$$? \text{ g } CaCO_3 = 100 \times \frac{0.0259}{2.2} \times 1.5 \times 1 \times \frac{1}{2} \times 100.1 \text{ g } CaCO_3$$

$$= 88 \text{ g } CaCO_3 = 88\% \text{ pure}$$

Sometimes you will not be given the relationship between titrant and back-titrant in milliliters per milliliter. You may get the molarity or normality of both solutions. In this case you can determine the number of equivalents of the solute that were really required by calculating the equivalents that were in excess and subtracting this from the number of equivalents that were put in initially. The following example will bring this out.

Commercially the acidity of fats and oils is expressed as the "Saponification Number." This is the milligram of KOH, which will react with 1.0 g of the fat or oil. In order to ensure complete reaction of the fat or oil, you always put in an excess of KOH

1.5

liter HCl soln.

liter HCl soln.

g imp.

molecular

100.1 g

mole

equivalents,   mole

2 eq

mole

$$\frac{1 \text{ eq } CaCO_3}{1 \text{ eq HCl}}$$

solution. Then you back-titrate with an HCl solution. The following data were obtained for an oil sample.

Weight of oil	3.60 g
Volume 0.35 *N* KOH	50.0 ml
Volume 0.55 *N* HCl for back-titration	7.1 ml

The reaction HCl + KOH = KCl + H_2O occurs.

The first thing you must do is calculate the number of equivalents of KOH which were in excess.

$$? \text{ eq KOH excess} = 0.0071 \text{ liter HCl soln}$$

$$\times \frac{0.55 \text{ eq HCl}}{\text{liter HCl soln.}} \times \frac{\text{eq KOH}}{\text{eq HCl}}$$

$$= 0.0039 \text{ eq KOH excess}$$

Now that you know the equivalents of KOH in excess, you must determine the number of equivalents of KOH put in initially.

$$\frac{0.35 \text{ eq KOH}}{\text{liter KOH soln.}}$$

$$0.0175$$

$$? \text{ eq KOH init} = 0.050 \text{ liter KOH soln.} \times \underline{\hspace{3cm}}$$

$$= \underline{\hspace{2cm}} \text{ eq KOH init.}$$

To get the number of equivalents of KOH actually needed you

subtract

initially

must _____ the equivalents in excess from the number of

equivalents _____ present.

0.0175, 0.0039

0.0136

$$\text{Equivalent KOH required} = \underline{\hspace{2cm}} - \underline{\hspace{2cm}}$$

$$= \underline{\hspace{2cm}} \text{ eq KOH}$$

Now you are ready to solve the problem since you know the number of equivalents of KOH required by 3.60 g oil.

0.0136

$$\frac{\underline{\hspace{2cm}} \text{ eq KOH}}{3.60 \text{ g oil}}$$

You also know that, since KOH has only one OH group to lose, it must have

1

$$\frac{\underline{\hspace{1cm}} \text{ eq KOH}}{\text{mole KOH}}.$$

and since the problem asks for weight of KOH, you will probably need the conversion factor

56.1 g

$$\frac{\underline{\hspace{2cm}} \text{ KOH}}{\text{mole KOH}}.$$

Since the problem asks for the "Saponification Number" of the oil, you reword the question to "How many milligrams of KOH are required by 1.0 g oil?"

mg KOH,   1.0 g oil

? _____ = _____.

To cancel the unwanted dimension of the information given you

$$\frac{0.0136 \text{ eq KOH}}{3.60 \text{ g oil}}$$

multiply by _____. Next, to cancel equivalents KOH and

$$\frac{\text{mole KOH}}{\text{eq KOH}}$$

move towards weight KOH you multiply first by _____ and

$$\frac{56.1 \text{ g KOH}}{\text{mole KOH}} \text{,   milligrams}$$

then by _____. The dimensions of the answer are _____

$$\frac{\text{mg KOH}}{10^{-3} \text{ g KOH}}$$

KOH, so finally you must multiply by _____.
   The answer is therefore

$$\text{Saponification Number} = 1.0 \times \frac{0.0136}{3.60} \times 1 \times 56.1 \times \frac{\text{mg KOH}}{10^{-3}}$$

$$= 212 \text{ mg KOH per gram oil.}$$

PROBLEM SET

This set of problems contains both problems which deal with molarity and moles and problems which deal with normality and equivalents. If you have not worked through Chapters 13 and 14, you will not be able to do those problems which are marked with an asterisk (*).

The first fourteen problems will be based on the balanced equations shown below. The pertinent equation will be indicated in each problem. Also, to save you some time, a listing of the molecular weights that you will need is given after the balanced equations.

(a) $CaO + 2HCl = CaCl_2 + H_2O$

(b) $KMnO_4 + 5FeCl_2 + 8HCl = MnCl_2 + 5FeCl_3 + 4H_2O + KCl$

(c) $K_2Cr_2O_7 + 3H_2C_2O_4 + 8HCl = 2CrCl_3 + 6CO_2 + 7H_2O + 2KCl$

(d) $H_3PO_4 + 3KOH = K_3PO_4 + 3H_2O$

(e) $2H_3BO_3 + Ca(OH)_2 = Ca(H_2BO_3)_2 + 2H_2O$

Molecular Weights

AgCl	143.4 g/mole	Fe	55.8 g/mole	$K_2Cr_2O_7$	294.2 g/mole
$Ba(OH)_2$	171.3 g/mole	$FeCl_2$	126.8 g/mole	KOH	56.1 g/mole
Br_2	160.0 g/mole	H_3AsO_4	141.9 g/mole	$KMnO_4$	158.0 g/mole
CH_3COOH	60.0 g/mole	$H_2C_2O_4$	90.0 g/mole	$Na_2C_2O_4$	134.0 g/mole
$CaCO_3$	100.1 g/mole	HCl	36.5 g/mole	NaOH	40.0 g/mole
$Ca(OH)_2$	74.1 g/mole	KCl	75.6 g/mole		

1. How many liters of a 0.50 *M* solution of HCl will be required to react completely with 0.56 g of CaO according to reaction (a)?

2. How many grams of $KMnO_4$ will react completely with 150 ml of a 0.30 *M* solution of $FeCl_2$ according to equation (b)?

3. According to reaction (c), 4.5 liters of a 0.10 *M* solution of $K_2Cr_2O_7$ can react with how many liters of a 0.50 *M* solution of $H_2C_2O_4$?

4. What is the molarity of an H_3PO_4 solution if 35 ml of the solution will react completely with 0.025 g of KOH according to equation (d)?

5. What is the molarity of an H_3BO_3 solution if 650 ml of the solution react with exactly 1.3 liters of a 0.025 *M* solution of $Ca(OH)_2$ according to equation (e)?

6. How many grams of pure $Ca(OH)_2$ are required to react with 0.75 liter of a 0.50 *M* solution of H_3BO_3 according to equation (e)?

7. How many moles of $K_2Cr_2O_7$ are needed to produce 5.0 liters of a 3.0 *M* solution of KCl according to reaction (c)?

8. How many liters of a 0.10 *M* solution of $FeCl_2$ will be required to react completely with 15.8 g of $KMnO_4$ according to reaction (b)?

*9. According to equation (e), 4.0 liters of a 0.50 *N* solution of H_3BO_3 will react with how many grams of $Ca(OH)_2$?

*10. How many milliliters of a 0.12 N solution of $KMnO_4$ will be required to react with 60 ml of a 0.30 M $FeCl_2$ solution according to equation (b)?

*11. What is the normality of an $H_2C_2O_4$ solution if 35.0 ml of the solution are required to react with 0.580 g $K_2Cr_2O_7$ according to equation (c)?

*12. How many liters of a 0.10 N solution of HCl are required to react with 4.0 moles of CaO according to reaction (a)?

*13. What is the normality of an H_3PO_4 solution according to equation (d), if 35.0 ml of it will react with 50.0 ml of a 2.5 M KOH solution?

*14. How many liters of a 0.15 N solution of H_3BO_3 will react with 3.0 liters of a 0.020 N $Ca(OH)_2$ solution according to equation (e)?

The following eleven problems do not have balanced equations given, but can be worked in terms of equivalents. If you have not done Chapters 13 and 14, you will not be able to do them unless you write a balanced equation.

*15. How many liters of a 0.10 N $KMnO_4$ solution will react with 13.4 g of $Na_2C_2O_4$? The $Na_2C_2O_4$ reacts to produce CO_2.

*16. How many liters of a 0.060 N $SnCl_4$ solution will react with 1.2 liters of a 0.030 M $FeCl_3$ solution? The $FeCl_3$ reacts to produce Fe metal.

*17. How many grams of H_3AsO_4 are required to react with 150 ml of a 0.20 N solution of $Na_2C_2O_4$? The H_3AsO_4 goes to H_3AsO_3.

18. How many milliliters of a 0.10 M H_2SO_4 solution are required to react with 0.30 liter of 0.0035 M $Ba(OH)_2$ solution? The products $BaSO_4$ and H_2O are formed.

19. How many milliliters of a 0.10 M H_3PO_4 are required to react with 30 ml of 0.20 N NaOH? The products are Na_2HPO_4 and H_2O.

20. How many milliliters of a 6.0 M H_2SO_4 solution are needed to react with 75 ml of a 2.0 M $Fe(NO_3)_3$ solution? The products are $Fe_2(SO_4)_3$ and HNO_3.

*21. What is the normality of a $Na_2Cr_2O_7$ solution if 25.0 ml of the solution are required to react with 0.126 g $FeCl_2$? The $FeCl_2$ goes to $FeCl_3$.

*22. What is the normality of a $KMnO_4$ solution if 65.0 ml of the solution are needed to react with 130 ml of a 2.00 M $H_2C_2O_4$ solution? The $H_2C_2O_4$ reacts to give CO_2.

*23. What is the normality of an HCl solution if 25.0 ml are required to react with 50.0 ml of a 0.10 N solution of NH_4OH?

*24. How many equivalents per mole are there for an acid if 30.0 ml of a 0.50 M solution of the acid react with exactly 150 ml of a 0.200 N solution of NaOH?

*25. What volume of 2.0 N $Na_2Cr_2O_7$ will be required to react with 36.0 ml of a 3.0 N solution of anything, provided the normality has been based on the correct reaction?

These remaining problems are more complicated than those which have preceded, but you should be able to set them up. None is impossibly difficult, and all represent calculations which you may be faced with later in chemistry. A word to the wise: write down all conversion factors given in the problems before you start.

26. What is the molarity of an H_2SO_4 solution if 2.0 liters of the solution titrate 300 g of an 80% pure NaOH sample to an endpoint? The reaction is

    $H_2SO_4 + 2NaOH = Na_2SO_4 + 2H_2O$.

*27. What is the percent purity of a sample of Fe ore if 2.00 g of the ore titrate to an endpoint with 100 ml of 0.600 N $Na_2Cr_2O_7$? The product is Fe^{+3}.

*28. It was found that 0.43 g of AgCl precipitated out when an excess of NaCl was added to 75 ml of an $AgNO_3$ solution. Write the equation for the reaction and then determine the molarity of the $AgNO_3$ solution. How many equivalents per mole are there for $AgNO_3$ in the reaction? What, then, is the normality of the $AgNO_3$ solution?

29. How many milliliters of 0.20 M NaOH solution will be needed to exactly neutralize (bring to an endpoint) 5.0 ml of a 30 wt % solution of CH_3COOH whose density is 1.04 g/ml? The products are CH_3COONa and H_2O.

*30. What is the normality of a KOH solution if it requires 25.0 ml of a 20.2 wt % HCl solution (density = 1.1 g/ml) to reach an endpoint with 40.4 ml of the KOH solution?

31. If 75.0 ml of a 0.20 M HCl solution is added to 25.0 ml of a 0.50 M NaOH solution, what is the molarity of the HCl solution remaining after the following reaction has taken place? (Don't forget to add the volumes.)

    $HCl + NaOH = NaCl + H_2O$

*32. The "Bromine Number" of a lubricating oil is a measure of the number of double bonds in the oil. It is expressed as "milligrams Br_2 per gram oil." To ensure complete reaction an excess of Br_2 is added, and when the reaction is completed the unused Br_2 is titrated to an endpoint with $Na_2S_2O_3$ of known normality. In the reaction with $Na_2S_2O_3$, the Br_2 is converted to Br^-. What is the "Bromine Number" of an oil if 16.0 g of the oil are reacted with 50.0 ml of a 0.100 M Br_2 solution, and then the excess Br_2 requires 8.00 ml of a 0.100 N $Na_2S_2O_3$ solution for back-titration?

33. What is the percent purity of a $Ba(OH)_2$ sample if it is titrated with an HCl solution that has been previously standardized by titration against a pure $CaCO_3$ sample? The following data were obtained.

    Standardization of the HCl solution:

    45.0 ml of HCl titrated against 0.240 g $CaCO_3$

    Relationship of HCl solution to NaOH solution used in back-titration:

    1.20 ml HCl titrated against 1.30 ml NaOH solution

Titration of impure $Ba(OH)_2$ (since excess HCl is present, reaction is complete):

Weight impure $Ba(OH)_2$ = 0.500 g

Volume HCl solution = 37.0 ml

Volume NaOH used in back-titrating excess HCl = 1.8 ml.

The equations for the reactions which are occurring are

$$2HCl + CaCO_3 = CaCl_2 + H_2O + CO_2$$

$$2HCl + Ba(OH)_2 = BaCl_2 + 2H_2O.$$

As you have probably noticed, the problem sets are getting longer and harder. However, by this point you should be really expert at setting up simple problems, so now we will try to really stir up your brains.

PROBLEM SET ANSWERS

If you have done Chapters 12, 13, and 14, you should be able to do twenty-five out of the thirty-three problems. If, however, you have not worked through Chapters 13 and 14, you will not be able to work the problems having to do with equivalents or normality. Of the fourteen problems which only deal with moles and molarity, you should be able to get the correct answer for twelve.

1. ? liters soln. = $0.56 \text{ g CaO} \times \dfrac{\text{mole CaO}}{56.1 \text{ g CaO}}$

$\times \dfrac{2 \text{ moles HCl}}{\text{mole CaO}} \times \dfrac{\text{liter soln.}}{0.50 \text{ mole HCl}}$

$= 4.0 \times 10^{-2}$ liter soln.

2. ? g $KMnO_4$ = $0.150 \text{ liter soln.} \times \dfrac{0.30 \text{ mole FeCl}_2}{\text{liter soln.}}$

$\times \dfrac{\text{mole KMnO}_4}{5 \text{ moles FeCl}_2} \times \dfrac{158.0 \text{ g KMnO}_4}{\text{mole KMnO}_4}$

$= 1.4$ g $KMnO_4$

3. ? liters $C_2O_4^=$ = $4.5 \text{ liters Cr}_2O_7^= \times \dfrac{0.10 \text{ mole K}_2\text{Cr}_2\text{O}_7}{\text{liter Cr}_2\text{O}_7^=}$

$\times \dfrac{3 \text{ moles H}_2\text{C}_2\text{O}_4}{\text{mole K}_2\text{Cr}_2\text{O}_7} \times \dfrac{\text{liter C}_2\text{O}_4^=}{0.50 \text{ mole H}_2\text{C}_2\text{O}_4}$

$= 2.7$ liters $C_2O_4^=$

(Notice how the two solutions were labelled so that they didn't get confused in the problem.)

4. $? M = \dfrac{? \text{ moles } H_3PO_4}{\text{liter soln.}}$

   $= \dfrac{0.025 \text{ g KOH}}{0.035 \text{ liter soln.}} \times \dfrac{\text{mole KOH}}{56.1 \text{ g KOH}} \times \dfrac{\text{mole } H_3PO_4}{3 \text{ moles KOH}}$

   $= 4.2 \times 10^{-3} \ M$

5. $? M = \dfrac{? \text{ moles } H_3BO_3}{\text{liter } H_3BO_3 \text{ soln.}}$

   $= \dfrac{1.3 \text{ liters } Ca(OH)_2 \text{ soln.}}{0.650 \text{ liter } H_3BO_3 \text{ soln.}} \times \dfrac{0.025 \text{ mole } Ca(OH)_2}{\text{liter } Ca(OH)_2 \text{ soln.}}$

   $\times \dfrac{2 \text{ moles } H_3BO_3}{\text{mole } Ca(OH)_2}$

   $= 0.10 \ M$

6. $? \text{ g } Ca(OH)_2 = 0.75 \text{ liter soln.} \times \dfrac{0.50 \text{ mole } H_3BO_3}{\text{liter soln.}}$

   $\times \dfrac{\text{mole } Ca(OH)_2}{2 \text{ moles } H_3BO_3} \times \dfrac{74.1 \text{ g } Ca(OH)_2}{\text{mole } Ca(OH)_2}$

   $= 14 \text{ g } Ca(OH)_2$

7. $? \text{ moles } K_2Cr_2O_7 = 5.0 \text{ liters soln.} \times \dfrac{3 \text{ moles KCl}}{\text{liter soln.}} \times \dfrac{\text{mole } K_2Cr_2O_7}{2 \text{ moles KCl}}$

   $= 7.5 \text{ moles } K_2Cr_2O_7$

8. $? \text{ liters soln.} = 15.8 \text{ g } KMnO_4 \times \dfrac{\text{mole } KMnO_4}{158.0 \text{ g } KMnO_4}$

   $\times \dfrac{5 \text{ moles } FeCl_2}{\text{mole } KMnO_4} \times \dfrac{\text{liter soln.}}{0.10 \text{ mole } FeCl_2}$

   $= 5.0 \text{ liters soln.}$

9. $? \text{ g } Ca(OH)_2 = 4.0 \text{ liters soln.} \times \dfrac{0.50 \text{ eq } H_3BO_3}{\text{liter soln.}} \times \dfrac{\text{mole } H_3BO_3}{1 \text{ eq } H_3BO_3}$

   $\times \dfrac{\text{mole } Ca(OH)_2}{2 \text{ moles } H_3BO_3} \times \dfrac{74.1 \text{ g } Ca(OH)_2}{\text{mole } Ca(OH)_2}$

   $= 74 \text{ g } Ca(OH)_2$

   (There is 1 eq H_3BO_3/mole H_3BO_3 since only one H^+ reacts.)

10. $? \text{ ml } MnO_4^- = 0.060 \text{ liter } Fe^{++} \times \dfrac{0.30 \text{ mole } FeCl_2}{\text{liter } Fe^{++}} \times \dfrac{\text{mole } KMnO_4}{5 \text{ moles } FeCl_2}$

   $\times \dfrac{5 \text{ eq } KMnO_4}{\text{mole } KMnO_4} \times \dfrac{10^3 \text{ ml } MnO_4^-}{0.12 \text{ eq } KMnO_4}$

   $= 1.5 \times 10^2 \text{ ml } MnO_4^-$

   (There are several ways to do this problem. You could change the 0.30 M $FeCl_2$ to 0.30 N $FeCl_2$, since there is one equivalent per mole, and then solve as $V'N' = V''N''$.)

11. $? N = \dfrac{? \text{ eq } H_2C_2O_4}{\text{liter soln.}}$

$= \dfrac{0.580 \text{ g } K_2Cr_2O_7}{0.0350 \text{ liter soln.}} \times \dfrac{\text{mole } K_2Cr_2O_7}{294.2 \text{ g } K_2Cr_2O_7}$

$\times \dfrac{6 \text{ eq } K_2Cr_2O_7}{\text{mole } K_2Cr_2O_7} \times \dfrac{\text{eq } H_2C_2O_4}{\text{eq } K_2Cr_2O_7}$

$= 0.338 \ N$

(Two Cr's go from an oxidation number of $+12$ to $+6$.)

12. $? \text{ liters soln.} = 4.0 \text{ moles } CaO \times \dfrac{2 \text{ moles } HCl}{\text{mole } CaO}$

$\times \dfrac{\text{eq } HCl}{\text{mole } HCl} \times \dfrac{\text{liter soln.}}{0.10 \text{ eq } HCl}$

$= 80 \text{ liters soln.}$

(You could also solve this by noting that there are 2 eq CaO per mole CaO. Then convert using the equivalent "bridge.")

13. We will solve this one by converting the KOH solution into normality.

$? N \text{ KOH} = \dfrac{? \text{ eq KOH}}{\text{liter soln.}}$

$= \dfrac{2.5 \text{ moles } KOH}{\text{liter soln.}} \times \dfrac{1 \text{ eq KOH}}{\text{mole } KOH}$

$= 2.5 \ N$

Now you can use $V'N' = V''N''$

$35.0 \ N' = 50.0 \times 2.5$

$N' = 3.6 \ N.$

14. This is a straight $V'N' = V''N''$ problem

$V' \times 0.15 = 3.0 \times 0.020$

$V' = 0.40 \text{ liter}$

15. $? \text{ liters soln.} = 13.4 \text{ g } Na_2C_2O_4 \times \dfrac{\text{mole } Na_2C_2O_4}{134.0 \text{ g } Na_2C_2O_4} \times \dfrac{2.0 \text{ eq } Na_2C_2O_4}{\text{mole } Na_2C_2O_4}$

$\times \dfrac{\text{eq } KMnO_4}{\text{eq } Na_2C_2O_4} \times \dfrac{\text{liter soln.}}{0.10 \text{ eq } KMnO_4}$

$= 2.0 \text{ liters soln.}$

(Two C's go from an oxidation number of $+6$ to $+8$.)

16. $\text{? liters Sn}^{+4} = 1.2 \text{ liters Fe}^{+3} \times \dfrac{0.030 \text{ mole FeCl}_3}{\text{liter Fe}^{+3}} \times \dfrac{3 \text{ eq FeCl}_3}{\text{mole FeCl}_3}$

$$\times \dfrac{\text{eq SnCl}_4}{\text{eq FeCl}_3} \times \dfrac{\text{liter Sn}^{+4}}{0.060 \text{ eq SnCl}_4}$$

$$= 1.8 \text{ liters Sn}^{+4}$$

(Fe goes from an oxidation number of +3 to 0.)

17. $\text{? g H}_3\text{AsO}_4 = 0.150 \text{ liter soln.} \times \dfrac{0.20 \text{ eq Na}_2\text{C}_2\text{O}_4}{\text{liter soln.}}$

$$\times \dfrac{\text{eq H}_3\text{AsO}_4}{\text{eq Na}_2\text{C}_2\text{O}_4} \times \dfrac{\text{mole H}_3\text{AsO}_4}{2 \text{ eq H}_3\text{AsO}_4} \times \dfrac{141.9 \text{ g H}_3\text{AsO}_4}{\text{mole H}_3\text{AsO}_4}$$

$$= 2.1 \text{ g H}_3\text{AsO}_4$$

18. You can convert both solutions from molarity to normality and use $V'N' = V''N''$. This is probably fastest. Since the H_2SO_4 loses 2 H's and the $Ba(OH)_2$ loses both OH's, both have 2 eq per mole. Therefore,

$$N \text{ H}_2\text{SO}_4 = (2 \times 0.10) = 0.20 \ N$$

$$N \text{ Ba(OH)}_2 = (2 \times 0.0035) = 0.0070 \ N$$

$$V' \times 0.20 = 300 \times 0.0070$$

$$V' = 10 \text{ ml H}_2\text{SO}_4 \text{ soln.}$$

19. This problem could be done in the same way as 18, but we will do it the somewhat longer way.

$$\text{? ml H}_3\text{PO}_4 = 0.030 \text{ liter NaOH} \times \dfrac{0.20 \text{ eq NaOH}}{\text{liter NaOH}} \times \dfrac{\text{eq H}_3\text{PO}_4}{\text{eq NaOH}}$$

$$\times \dfrac{\text{mole H}_3\text{PO}_4}{2 \text{ eq H}_3\text{PO}_4} \times \dfrac{10^3 \text{ ml H}_3\text{PO}_4}{0.10 \text{ mole H}_3\text{PO}_4}$$

$$= 30 \text{ ml H}_3\text{PO}_4$$

20. This is the same type of problem as the above two. We will do it the shorter way, converting both solutions from molarity to normality.
The H_2SO_4 loses both H's, so it has 2 equivalents per mole.
The $Fe(NO_3)_3$ is a salt with 3 $(NO_3)^-$ groups, each of which is equivalent to 1 H, so it has 3 equivalents per mole.

$$N \text{ H}_2\text{SO}_4 = (2 \times 6.0) = 12 \ N$$

$$N \text{ Fe(NO}_3)_3 = (3 \times 2.0) = 6.0 \ N$$

$$V'N' = V''N''$$

$$V' \times 12.0 = 75 \times 6.0$$

$$V' = 38 \text{ ml H}_2\text{SO}_4 \text{ soln.}$$

21. $? N = \dfrac{? \text{ eq } Na_2Cr_2O_7}{\text{liter soln.}} = \dfrac{0.126 \text{ g } \cancel{FeCl_2}}{0.0250 \text{ liter soln.}} \times \dfrac{\text{mole } \cancel{FeCl_2}}{126.8 \text{ g } \cancel{FeCl_2}}$

$\times \dfrac{1 \text{ eq } \cancel{FeCl_2}}{\cancel{\text{mole } FeCl_2}} \times \dfrac{\text{eq } Na_2Cr_2O_7}{\cancel{\text{eq } FeCl_2}}$

$= 0.0397 \ N$

(Fe goes from an oxidation number of $+2$ to $+3$.)

22. $? N = \dfrac{? \text{ eq } KMnO_4}{\text{liter } MnO_4^-}$

$= \dfrac{0.130 \text{ liter } \cancel{C_2O_4^=}}{0.0650 \text{ liter } MnO_4^-} \times \dfrac{2.00 \text{ moles } H_2C_2O_4}{\cancel{\text{liter } C_2O_4^=}}$

$\times \dfrac{2 \text{ eq } \cancel{H_2C_2O_4}}{\cancel{\text{mole } H_2C_2O_4}} \times \dfrac{\text{eq } KMnO_4}{\cancel{\text{eq } H_2C_2O_4}}$

$= 8.00 \ N$

(This problem can also be solved using $V'N' = V''N''$ if you first convert the molarity of the $H_2C_2O_4$ solution to normality. The number of equivalents per mole is found in the same way as in Problem 15.)

23. $V'N' = V''N''$

$25.0 \ N' = 50.0 \times 0.10$

$N' = 0.20 \ N$

24. $? \text{ eq acid} = 1 \ \cancel{\text{mole acid}} \times \dfrac{\cancel{\text{liter acid}}}{0.50 \ \cancel{\text{mole acid}}} \times \dfrac{0.150 \ \cancel{\text{liter NaOH}}}{0.0300 \ \cancel{\text{liter acid}}}$

$\times \dfrac{0.200 \ \cancel{\text{eq NaOH}}}{\cancel{\text{liter NaOH}}} \times \dfrac{\text{eq acid}}{\cancel{\text{eq NaOH}}}$

$= 2 \text{ eq acid per mole acid}$

(This must be a whole number.)

25. $V'N' = V''N''$

$V' \times 2.0 = 36.0 \times 3.0$

$V' = 54 \text{ ml of anything}$

26. $? M = \dfrac{? \text{ moles } H_2SO_4}{\text{liter soln.}}$

$= \dfrac{300 \ \cancel{\text{g imp.}}}{2.0 \text{ liters soln.}} \times \dfrac{80 \text{ g } \cancel{NaOH}}{100 \ \cancel{\text{g imp.}}}$

$\times \dfrac{\cancel{\text{mole NaOH}}}{40.0 \ \cancel{\text{g NaOH}}} \times \dfrac{\text{mole } H_2SO_4}{2 \ \cancel{\text{moles NaOH}}}$

$= 1.5 \ M$

27. $? \text{ g Fe} = 100 \text{ g ore} \times \dfrac{0.100 \text{ liter soln.}}{2.00 \text{ g ore}} \times \dfrac{0.600 \text{ eq Na}_2\text{Cr}_2\text{O}_7}{\text{liter soln.}}$

$\times \dfrac{\text{eq Fe}}{\text{eq Na}_2\text{Cr}_2\text{O}_7} \times \dfrac{\text{mole Fe}}{3 \text{ eq Fe}} \times \dfrac{55.8 \text{ g Fe}}{\text{mole Fe}}$

$= 55.8 \text{ g Fe or } 55.8\%$

28. $\text{AgNO}_3 + \text{NaCl} = \text{AgCl} + \text{NaNO}_3$

$? M = \dfrac{? \text{ moles AgNO}_3}{\text{liter soln.}}$

$= \dfrac{0.43 \text{ g AgCl}}{0.075 \text{ liter soln.}} \times \dfrac{\text{mole AgCl}}{143.4 \text{ g AgCl}} \times \dfrac{\text{mole AgNO}_3}{\text{mole AgCl}}$

$= 0.040 \ M$

There is one equivalent of AgNO_3 per mole AgNO_3 since this is a simple salt reaction with no change in oxidation number. One $(\text{NO}_3)^-$ is equivalent to one H^+. The normality is the same as the molarity $= 0.040 \ N$.

29. $? \text{ ml NaOH soln.} = 5.0 \text{ ml CH}_3\text{COOH soln.} \times \dfrac{1.04 \text{ g CH}_3\text{COOH soln.}}{\text{ml CH}_3\text{COOH soln.}}$

$\times \dfrac{30 \text{ g CH}_3\text{COOH}}{100 \text{ g CH}_3\text{COOH soln.}} \times \dfrac{\text{mole CH}_3\text{COOH}}{60.0 \text{ g CH}_3\text{COOH}}$

$\times \dfrac{\text{eq CH}_3\text{COOH}}{\text{mole CH}_3\text{COOH}} \times \dfrac{\text{eq NaOH}}{\text{eq CH}_3\text{COOH}} \times \dfrac{10^3 \text{ ml NaOH soln.}}{0.20 \text{ eq NaOH}}$

$= 1.3 \times 10^2 \text{ ml NaOH soln.}$

Notice the extreme care you have to take in labeling the dimensions so that the two solutions are not confused with their contents. Here we have chosen to write ions to indicate which solution is being referred to. You may prefer to use some other way. But you must keep them straight!

30. $? N = \dfrac{? \text{ eq KOH}}{\text{liter KOH soln.}}$

$= \dfrac{25.0 \text{ ml HCl soln.}}{0.0404 \text{ liter KOH soln.}} \times \dfrac{1.1 \text{ g HCl soln.}}{\text{ml HCl soln.}} \times \dfrac{20.2 \text{ g HCl}}{100 \text{ g HCl soln.}}$

$\times \dfrac{\text{mole HCl}}{36.5 \text{ g HCl}} \times \dfrac{\text{eq HCl}}{\text{mole HCl}} \times \dfrac{\text{eq KOH}}{\text{eq HCl}}$

$= 3.8 \ N$

(Here we have used a slightly different way of labeling the solutions to differentiate between them. You might prefer this.)

31. First determine how many moles of HCl are consumed by the NaOH.

? moles HCl = 0.0250 ~~liter NaOH soln.~~ × $\dfrac{0.50 \text{ mole NaOH}}{\text{liter NaOH soln.}}$

$\times \dfrac{\text{mole HCl}}{\text{mole NaOH}}$

= 0.0125 mole HCl consumed

Then calculate the number of moles of HCl present initially.

? moles HCl = 0.0750 ~~liter HCl soln.~~ × $\dfrac{0.20 \text{ mole HCl}}{\text{liter HCl soln.}}$

= 0.0150 mole HCl initially

Then get the remaining unconsumed HCl by subtracting.

? moles HCl = 0.0150 − 0.0125 = 0.0025 mole HCl remaining

The total volume of the system is the sum of the volumes of the HCl solution and the NaOH solution.

? liters total = 0.075 + 0.025 = 0.100 liter total soln.

Now the molarity of the remaining HCl is easily calculated.

$? M = \dfrac{? \text{ moles HCl}}{\text{liter total soln.}} = \dfrac{0.0025 \text{ mole HCl}}{0.100 \text{ liter total soln.}}$

= 0.025 *M*

32. This is a typical back-titration problem. First you must calculate the amount of Br_2 actually consumed. You do this by calculating the initial number of equivalents of Br_2 put in the reaction.

? eq Br_2 init. = 0.0500 ~~liter Br₂ soln.~~ × $\dfrac{0.100 \text{ mole Br}_2}{\text{liter Br}_2 \text{ soln.}}$

$\times \dfrac{2 \text{ eq Br}_2}{\text{mole Br}_2}$

= 0.0100 eq Br_2 init.

(Two Br's go from a total oxidation number of 0 to −2.)

$Br_2 \longrightarrow 2Br^-$
⓪　　　$2 \times$ Ⓝ(−1)

Then you calculate the number of equivalents of Br_2 remaining (from the back-titration data).

? eq Br_2 remaining = 0.00800 ~~liter S₂O₃~~ × $\dfrac{0.100 \text{ eq Na}_2S_2O_3}{\text{liter S}_2O_3^=}$

$\times \dfrac{\text{eq Br}_2}{\text{eq Na}_2S_2O_3}$

= 0.000800 eq Br_2 remaining

Then you get the number of equivalents of Br_2 actually used in the reaction by subtracting.

? eq Br_2 used = $0.0100 - 0.000800 = 0.0092$ eq Br_2 used

(Did you have the correct number of significant figures?)
Now you can calculate the "Bromine Number.")

$$? \text{ mg } Br_2 = 1 \text{ g oil} \times \frac{0.0092 \text{ eq } Br_2}{16.0 \text{ g oil}} \times \frac{\text{mole } Br_2}{2 \text{ eq } Br_2}$$

$$\times \frac{160.0 \text{ g } Br_2}{\text{mole } Br_2} \times \frac{\text{mg } Br_2}{10^{-3} \text{ g } Br_2}$$

$$= 46 \text{ mg } Br_2/1.0 \text{ g oil}$$

33. This is another back-titration problem, but you are given a relationship between the titrant and back-titrant in milliliters. First, then, you calculate the number of milliliters of HCl actually used. You do this by calculating the excess HCl equivalent to the ml NaOH used in back-titration.

$$? \text{ ml HCl excess} = 1.80 \text{ ml NaOH} \times \frac{1.20 \text{ ml HCl}}{1.30 \text{ ml NaOH}}$$

$$= 1.7 \text{ ml HCl excess}$$

Then you subtract this from the amount of HCl put in initially.

? ml HCl used = $37.0 - 1.7 = 35.3$ ml HCl used

Now you can solve the problem.

$$? \text{ g } Ba(OH)_2 = 100 \text{ g imp.} \times \frac{35.3 \text{ ml HCl}}{0.500 \text{ g imp.}} \times \frac{0.240 \text{ g } CaCO_3}{45.0 \text{ ml HCl}}$$

$$\times \frac{\text{mole } CaCO_3}{100.1 \text{ g } CaCO_3} \times \frac{2 \text{ eq } CaCO_3}{\text{mole } CaCO_3} \times \frac{\text{eq } Ba(OH)_2}{\text{eq } CaCO_3}$$

$$\times \frac{\text{mole } Ba(OH)_2}{2 \text{ eq } Ba(OH)_2} \times \frac{171.3 \text{ g } Ba(OH)_2}{\text{mole } Ba(OH)_2}$$

$$= 64.4 \text{ g } Ba(OH)_2 = 64.4\%$$

Chapter 16
Colligative Properties
of Solutions

Boiling Point, Freezing
Point, and Osmotic
Pressure

There will be no pretest on this chapter. The material is completely different from what has come before. As a matter of fact, we will have to start with a straight discussion of what will happen to certain properties of a solvent when you dissolve a nonvolatile solute in it.

SECTION A   Colligative Properties

There are several properties of a solution which depend on the relative numbers of solute and solvent particles. Among these are

1.   the lowering of the vapor pressure,

2.   the lowering of the freezing point,

3.   the raising of the boiling point,

4.   the raising of the osmotic pressure.

These are called the "colligative properties."
   The changes in the freezing and boiling points are both due to the lowering of the vapor pressure. It is easy to show how this occurs. Below is a "phase diagram" which shows the vapor pressure versus temperature for a pure solvent (————) and for a solution (— — — —). The lines represent equilibrium between liquid-gas, liquid-solid, and solid-gas.
   The boiling point is defined as the temperature at which the liquid is in equilibrium with gas at 1.0 atmosphere pressure. (To be very exact, this is called the *standard boiling point*. We will consider no other.) The freezing point is the temperature at which liquid and solid are at equilibrium under 1.0 atm pressure. On the following diagram, they will be the temperatures at which the two equilibrium lines cross 1.0 atm pressure.

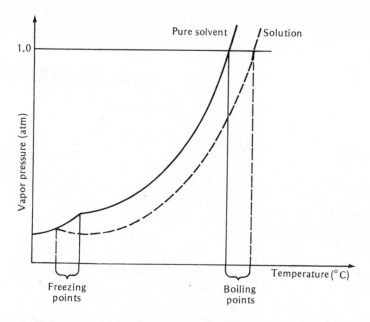

As you see, because the solution's vapor pressure line is lower, the crossing of the 1.0 atm line will be at a higher temperature for the boiling points and at a lower temperature for the freezing point. The number of degrees centigrade it will change is determined by three factors:

1. the number of solute particles,
2. the number of solvent particles,
3. the nature of the solvent.

The amount the vapor pressure is lowered is independent of what the solute is so long as the vapor pressure of the solute is much, much lower than the vapor pressure of the solvent.

It has been found to be most convenient to express the number of solute particles in moles and always to refer to the number of solvent particles present in 1000 g of solvent. This concentration unit,

$$\frac{\text{mole solute}}{1000 \text{ g solvent}} \quad \text{or} \quad \frac{\text{mole solute}}{\text{kg solvent}} \, ,$$

molality

should look familiar to you. It is called _____ and abbreviated,

m

___. In Chapter 12 we said that this concentration unit was useful in only one type of problem. These problems are that type.

Now we are going to set up a conversion factor which is the number of °C that the boiling point will be raised or the freezing point lowered by a 1 molal solution. Each solvent will have its own numerical value for this conversion factor. Thus, in much the same way that we can say 18.0 g H_2O per mole of H_2O,

$$\frac{18.0 \text{ g } H_2O}{\text{mole } H_2O} \, ,$$

or 72.0 g of C_6H_6 per mole C_6H_6,

$$\frac{72.0 \text{ g } C_6H_6}{\underline{\hspace{1cm}} C_6H_6},$$

mole

we can say that the boiling point elevation is 0.51°C per molal solution in water,

$$\frac{0.51°C}{m \text{ H}_2O},$$

and 2.5°C per molal solution in benzene (C_6H_6),

$$\frac{2.5°C}{m \text{ } C_6H_6}.$$

It is customary to write the number of °C that the boiling point is raised as ΔT_B and the number of °C the freezing point is lowered as ΔT_F. They are always positive numbers and their units are °C. We can therefore write our conversion factor as $(\Delta T)°C/m$. The numerical value of ΔT will depend on the solvent. Of course, molality (m) is itself a conversion factor,

$$m = \frac{\text{mole solute}}{\text{kg} \underline{\hspace{1cm}}}.$$

solvent

Thus, rewriting our conversion factor for the boiling point elevation or the freezing point lowering, we have

$$\frac{\Delta T°C}{\dfrac{\text{mole solute}}{\text{kg solvent}}} \quad \text{or} \quad \frac{(\Delta T°C) \text{ (kg solvent)}}{\text{mole solute}}.$$

The second conversion factor is simply an algebraic transposition of the first. As a matter of fact, later on you may find it more convenient to write these conversion factors as

$$\frac{(\Delta T°C) \text{ (10}^3 \text{ g solvent)}}{\text{mole solute}}.$$

These conversion factors have a name. If you are referring to the number of degrees centigrade that 1.0 mole of solute raises the boiling point of 1.0 kg of solvent, it is called the *molal boiling point elevation constant* and abbreviated K_B. If you are talking about the number of degrees centigrade that the freezing point is lowered by the presence of 1.0 moles of solute in 1.0 kg of solvent, it is called the *molal freezing point depression constant*, K_F. The important thing to remember is that K_B and K_F are conversion factors with the very strange units

$$\frac{(°C) \text{ (kg solvent)}}{\text{mole solute}}.$$

This is the first time that you have seen a conversion factor with three different dimensions in it. However, if you recall how we initially set up a conversion factor from an equality (in Chapter 2), there is no reason at all why the equality could not have more than two dimensions in it Perhaps you have heard of "man-hours," a term which is usually used in reference to work; for example, "It took 1500 man-hours to build a tunnel." You could have one person working 1500 hours or 1500 persons working just one hour; the product of the number of persons times the number of hours is still the same. Similarly, our dimension, (°C) (kg solvent), is just such a double dimension.

When you were doing gas laws in Chapter 10, you had a conversion factor with two of these double dimensions, a total of four different units in one conversion factor. The important thing to remember is simply to make sure when you solve a problem that all the unwanted dimensions cancel.

SECTION B Determining K_F or K_B

All you need to know in order to experimentally determine K_F or K_B is the freezing point of a solution of known amount of solute and known amount of solvent, and the freezing point of pure solvent (or for K_B the boiling points respectively). Here is a simple example.

What is the molal boiling point elevation constant, K_B, for the solvent cyclohexane, if it was found that dissolving 0.20 moles of a solute in 4.0 kg of cyclohexane increased the boiling point 0.42°C over that of pure cyclohexane?

K_B

The question asked is, "What is ___ for cyclohexane?" You must, of course, put down the dimensions in order to solve the problem.

°C, kg cyclohexane
———————————
mole

$$? K_B = \frac{(? \underline{\quad})(\underline{\qquad\qquad})}{\underline{\qquad}\text{ solute}}$$

There is one conversion factor given in the problem,

0.20 mole

$$\frac{\underline{\qquad}\text{ solute}}{\underline{\qquad}},$$

4.0 kg cyclohexane

0.42°C

and the information given in the problem is ΔT_B, which is ———. You have everything you need to get the dimensions of the answer.

$$? K_B = \frac{(?\ °C)\ (\text{kg cyclohexane})}{\text{mole solute}}$$

$$= 0.42°C \times \frac{4.0\ \text{kg cyclohexane}}{0.20\ \text{mole solute}}$$

$$= \frac{(8.4°C)\ (\text{kg cyclohexane})}{\text{mole solute}}$$

Here is a more complicated example which has the amount of solute given in grams not moles. A solution of 50.0 g $C_2H_6O_2$ in 700 g of water freezes at $-2.14°C$. Pure water freezes at $0.00°C$.

What is K_F for water? Since the question asked is "What is _____ for water?", you put this in dimensions and have

$$? \, K_F = \frac{(? \, \underline{\quad}) \, (\underline{\qquad})}{\underline{\quad} \, C_2H_6O_2} .$$

The information given is the lowering of the freezing point of water.

$$(\Delta T_F)°C = 0.00°C - (-2.14°C) = \underline{\qquad}$$

Note: ΔT is always a $+$ number.
There is one conversion factor given in the problem.

$$\frac{\underline{\qquad} \, C_2H_6O_2}{0.700 \, \underline{\qquad}}$$

(The grams water was changed immediately to kilograms water to save one step.)

Since the answer will have mole $C_2H_6O_2$ and the conversion factor has grams $C_2H_6O_2$, you will need a conversion factor which relates these two. It is _____, the molecular weight.
You now have everything you will need to solve the problem.

$$? \, K_F = \frac{(? \, °C) \, (kg \, water)}{mole \, C_2H_6O_2} = \underline{\qquad}$$

You put in the information given first and then multiply by the conversion factor given in the problem, _____. You now have two of the dimensions of the answer, $°C$ and kilograms water; but you still have the unwanted dimension _____ and you miss the dimension _____ $C_2H_6O_2$. If, however, you multiply by the conversion factor _____, you then have only the dimensions of the answer left. The answer is therefore

$$? \, K_F = \frac{(? \, °C) \, (kg \, water)}{mole \, C_2H_6O_2}$$

$$= 2.14°C \times \frac{0.700 \, kg \, water}{50.0 \, g \, C_2H_6O_2} \times \frac{62.0 \, g \, C_2H_6O_2}{mole \, C_2H_6O_2}$$

$$= \frac{1.86°C \, (kg \, water)}{mole \, solute} .$$

Actually, chemists rarely calculate the values for K_B and K_F. They are well known and you look them up in a text or in the

---

Right column annotations:

K_F

$°C$, kg water

mole

$2.14°C$

50.0 g

kg water

$\dfrac{62.0 \, g \, C_2H_6O_2}{mole \, C_2H_6O_2}$

$2.14°C$

$\dfrac{0.700 \, kg \, water}{50.0 \, g \, C_2H_6O_2}$

grams $C_2H_6O_2$

mole

$\dfrac{62.0 \, g \, C_2H_6O_2}{mole \, C_2H_6O_2}$

Handbook of Chemistry and Physics [Ohio: The Chemical Rubber Co.]. Below is a table which shows the values of K_B and K_F for a few solvents to just two significant figures.

Molal Freezing and Boiling Point Constants

Solvent	Boiling point	K_B	Freezing point	K_F
Acetic acid	118.5°C	3.1°/m	16.6°C	3.9°/m
Benzene	80.1°C	2.5°/m	5.5°C	5.1°/m
Carbon disulfide	46.3°C	2.3°/m		
Carbon tetrachloride	76.8°C	5.0°/m		
Naphthalene			80.2°C	6.8°/m
Camphor			178.4°C	38.0°/m
Water	100°C	0.51°/m	0.0°C	1.86°/m

SECTION C Determining Molecular Weight

A chemist will frequently want to know the molecular weight of an unknown compound. The easiest way to find out is to determine either the freezing point lowering or boiling point elevation caused by a known weight of the compound in a known weight of some solvent. You will always have one conversion factor given in the problem (the relative weights of unknown and solvent) and one conversion factor, the molal freezing point or the boiling point constant, which you look up in the table. You will also have ΔT given in some form.

Here is an example. What is the molecular weight of an unknown compound, X, if a solution prepared by dissolving 1.0 g of X in 10.0 g benzene freezes at 2.3°C? Pure benzene freezes at 5.5°C.

The conversion factor given in the problem is

10.0 g benzene

$$\frac{1.0 \text{ g } X}{\rule{3cm}{0.4pt}}.$$

The molal freezing depression constant for benzene listed in the table is

kg benzene

mole X

$$\frac{(5.1°C) (\rule{2cm}{0.4pt})}{\rule{2cm}{0.4pt}}.$$

You must calculate the value for ΔT_F, the number of degrees that the freezing point of pure benzene has been lowered.

2.3

3.2

$$\Delta T_F = 5.5°C - \rule{1cm}{0.4pt}°C$$

$$= \rule{1cm}{0.4pt}°C$$

You now have everything you need. The question asked is, "What is the molecular weight of X?" You must reword this using

mole

units, so you ask "How many grams of X are in a _____ of X?"

You start, then, with the question

$$\frac{? \text{ g } X}{\text{mole } X}.$$

If you now put in the conversion factor given in the problem, you get grams X in the numerator where you want it. The unwanted

dimension is in the denominator and is _____.

The conversion factor, K_F, has kilograms benzene in it. You may save yourself a step by writing the conversion factor in units of grams benzene.

$$K_F = \frac{(5.1°C) \, (10^3 \text{ g benzene})}{\text{mole } X}$$

If you multiply by this conversion factor next, you find that you have grams X in the numerator and mole X in the denominator, which is what you want for the answer; but there is one unwanted

dimension left, ___, in the numerator. You can cancel this unit by

multiplying by _____, which is the reciprocal of the information

given, $1/$_____. This is the first time you have done this, and it may seem a little uncomfortable. However, if you were willing to multiply by the information given to cancel unwanted dimensions, why not divide? So the answer is

$$\frac{? \text{ g } X}{\text{mole } X} = \frac{1.0 \text{ g } X}{10.0 \text{ g benzene}} \times \frac{(5.1\cancel{°C}) \, (10^3 \cancel{\text{g benzene}})}{\text{mole } X} \times \frac{1}{3.2\cancel{°C}}$$

$$= \frac{1.6 \times 10^2 \text{ g } X}{\text{mole } X}.$$

These problems are really very simple if you always express K_F or K_B as a conversion factor with all its dimensions. Sometimes, however, the relation between weight of solute and weight of solvent will be a little more complicated, as in the next example.

What is the molecular weight of an unknown compound, X, if a 10 wt % solution of X in acetic acid boils at 122.5°C? Here the weight of solute per weight of solvent is in the 10 wt %. If you

take 100 g of solution it will contain ___ g of X and ___ g of acetic acid. So you can write the conversion factor

$$\frac{\_\_ \text{ g } X}{\_\_ \text{ g acetic acid}}$$

The other conversion factor, given in the table, is the K_B for acetic acid.

$$\frac{(3.1°C) \, (10^3 \, \text{_____})}{\text{_____}}$$

---

grams benzene

°C

$\dfrac{1}{3.2°C}$

ΔT_F

10, 90

10

90

g acetic acid

mole X

Finally, you must calculate the number of °C that the boiling point of pure benzene has been raised.

122.5

$$\Delta T_B = \underline{\hspace{1cm}}°C - 118.5°C$$

4.0

$$= \underline{\hspace{0.5cm}}°C$$

You now have all the information you will need. To get molecular weight you ask the question

grams

$$\frac{? \underline{\hspace{1cm}} X}{\underline{\hspace{1cm}} X} =$$

mole

You start with the conversion factor given in the problem,

$$\frac{10\ g\ X}{90\ g\ \text{acetic acid}}$$

$$\frac{(3.1°C)\ (10^3\ g\ \text{acetic acid})}{\text{mole}\ X}$$

$$\frac{1}{4.0°C}$$

_____. Next you multiply by K_B,

_____, to remove the unwanted dimension,

grams acetic acid, and, finally, you multiply by _____ to cancel

the unwanted dimension, °C. Once again you multiply by the reciprocal of the information given. The answer will then be

$$\frac{?\ g\ X}{\text{mole}\ X} = \frac{10\ g\ X}{90} \times \frac{(3.1)\ (10^3)}{\text{mole}\ X} \times \frac{1}{4}$$

$$= \frac{86\ g\ X}{\text{mole}\ X}.$$

SECTION D Determining Freezing and Boiling Points of Solutions

Often chemists will be interested in knowing at what temperature a solution of a substance of known molecular weight will boil or freeze. What they must do is determine ΔT_B or ΔT_F and then add or subtract this number of degrees from the boiling point or freezing point of pure solvent.

Here is a very simple example. What will the boiling point be for a 6.0 *m* solution of hexane in benzene? The information given in the problem is that the solution is 6.0 *m*. Written as a conversion factor this is

mole

$$\frac{6.0\ \underline{\hspace{1cm}}\ \text{hexane}}{\underline{\hspace{0.5cm}}\ \text{benzene}}.$$

kg

From the table you know another conversion factor, the molal

$$\frac{(2.5°C)\ (\text{kg benzene})}{\text{mole hexane}}$$

boiling point elevation constant, which is _____.

The question asked is really "How many °C is the boiling point of pure benzene raised?" You start, then, by writing

$$?\ °C =$$

The information given is the conversion factor that the solution is 6.0 *m*, so you can start there.

$$? \,°C = \frac{6.0 \text{ moles hexane}}{\text{kg benzene}}$$

If you now multiply by K_B,

$$? \,°C = \frac{6.0 \cancel{\text{ moles hexane}}}{\cancel{\text{kg benzene}}} \times \frac{(2.5°C) \cancel{(\text{kg benzene})}}{\cancel{\text{mole hexane}}},$$

the only dimension left is the dimension of the answer.

$$? \,°C = 6.0 \times 2.5°C$$

$$= 15.°C = \Delta T_B$$

To get the boiling point of the solution you will have to add ΔT_B to the boiling point of pure benzene.

$$\text{Boiling point } °C = 80.1°C + 15°C = 95°C$$

(Notice how you add to get the correct number of significant figures.)

The last example was very easy because you were given the concentration of solute in molality. Such problems are not always so straightforward, but with all your experience at converting from one dimension to another, a problem like the following shouldn't be too difficult.

At what temperature will a solution of ethyl alcohol in water freeze if 200 ml of ethyl alcohol, C_2H_5OH, (density 0.79 g/ml) is added to 1000 ml of water (density 1.00 g/ml) to prepare the solution?

There are three conversion factors given in the problem.

$$\frac{200 \text{ ml ethyl alcohol}}{\underline{\hspace{2cm}}} \qquad \frac{1000 \text{ ml water}}{\text{ethyl alcohol}}$$

$$\frac{0.79 \text{ g} \underline{\hspace{1cm}}}{\underline{\hspace{2cm}}} \qquad \frac{\text{ml ethyl alcohol}}{\text{water}}$$

$$\frac{1.00 \text{ g} \underline{\hspace{1cm}}}{\underline{\hspace{2cm}}} \qquad \frac{\text{water}}{\text{ml water}}$$

From the table you can get the molal freezing point depression constant for water, which is

$$\frac{(1.86°C) (10^3 \underline{\hspace{1cm}})}{\text{mole ethyl alcohol}} \qquad \text{g water} \,.$$

Since you have mole ethyl alcohol in one conversion factor and grams ethyl alcohol in another, you will probably need the

conversion factor $\underline{\hspace{4cm}}$, which will get you from one to the other. Now you have all the conversion factors you need.

$$\frac{46.0 \text{ g ethyl alcohol}}{\text{mole ethyl alcohol}}$$

You start, then, with the question

°C

$$? \underline{\hspace{1cm}} =$$

First you have the conversion factor given in the problem,

$$\frac{200 \text{ ml ethyl alcohol}}{1000 \text{ ml water}}$$

$$\frac{0.79 \text{ g ethyl alcohol}}{\text{ml ethyl alcohol}}$$

_____. To cancel the unwanted dimension milliliters ethyl alcohol you can multiply by _____.

$$\frac{\text{ml water}}{1.00 \text{ g water}}$$

To cancel the unwanted dimension milliliters water you can multiply by _____. You can cancel the unwanted dimension grams water by using the conversion factor from K_F,

$$\frac{(1.86°C)(10^3 \text{ g water})}{\text{mole ethyl alcohol}}$$

_____. Finally, to cancel both the unwanted dimensions mole ethyl alcohol and grams ethyl alcohol you can

$$\frac{\text{mole ethyl alcohol}}{46.0 \text{ g ethyl alcohol}}$$

°C

multiply by _____. The only dimension left now is

___, and so the number of degrees the freezing point has been lowered must be

$$? \, \Delta T_F = ? \, °C = \frac{200}{1000} \times 0.79 \times \frac{1}{1.00} \times 1.86°C \times 10^3 \times \frac{1}{46.0}$$

$$= 6.4°C.$$

subtract

To get the freezing point of the solution you now must _____ ΔT_F from the freezing point of pure water.

$$\text{Freezing point} = 0.0°C - 6.4°C = -6.4°C$$

SECTION E Preparing Solutions with Known Freezing and Boiling Points

These problems are simply the problems in Section D turned the other way around. For example, how many grams of $CHCl_3$ would you have to add to 100 g of carbon tetrachloride solvent to have a solution which will boil at 78.3°C?

grams $CHCl_3$

The question asked is "How many _____?" The information given is the number of °C that the boiling point must be raised, ΔT_B. You must calculate this.

78.3

$$(\Delta T_B)°C = \underline{\hspace{0.5cm}}°C - 76.8°C$$

1.5

$$= \underline{\hspace{0.5cm}}°C$$

100 g

Also given in the problem is the information that there are _____ carbon tetrachloride. You can start with this information; it is best to leave the temperature until last, as it may go in as the reciprocal.

The molal boiling point elevation constant for carbon tetrachloride is a conversion factor available from the table. Using it, you start in the normal way.

g $CHCl_3$

$$? \, \underline{\hspace{1cm}} = 100 \text{ g carbon tetrachloride}$$

Multiplying by the inverted K_B cancels the unwanted dimension, grams carbon tetrachloride.

$$? \text{ g CHCl}_3 = 100 \underline{\text{ g carbon tetrachloride}}$$

$$\times \frac{\text{mole CHCl}_3}{(5.0°C) (10^3 \underline{\text{ g carbon tetrachloride}})}$$

Next you can cancel the unwanted dimension, mole CHCl$_3$, and get directly to the dimensions of the answer by multiplying by

$$\frac{119.5 \text{ g CHCl}_3}{\text{mole CHCl}_3}$$

_____. You now have the dimensions of the answer, but there remains one unwanted dimension, ___. Consequently, you

°C

multiply by the other piece of information given, _____, which is ΔT_B. The answer will therefore be

1.5°C

$$? \text{ g CHCl}_3 = 100 \times \frac{1}{(5.0)\ (10^3)} \times 119.5 \text{ g CHCl}_3 \times 1.5$$

$$= 3.6 \text{ g CHCl}_3.$$

The problem is a little strange because there are two pieces of "information given." You could have put them both in initially, but it is better to wait with °C until the end. The next example will show why. It is the same sort of problem which asks for the solvent instead of the solute. What's more, just to make your life miserable, the units are not metric.

How many gallons of water could you add to 1.0 quart of ethylene glycol antifreeze ($C_2H_6O_2$; density = 0.80 g/ml) in order to have a solution in your radiator that will not freeze until the temperature drops below 23.0°F? The density of water is 1.0 g/ml.

First you should determine ΔT_F in °C. To do this, determine the temperature in °C below which it will freeze.

$$°C = \frac{5}{9} (\underline{})$$

23.0°F − 32.0

$$= \underline{}$$

−5.0°C

(If you have forgotten how to do this, check the last section of Chapter 3.)

Now you calculate ΔT_F.

$$\Delta T_F = 0.0°C - (\underline{})$$

−5.0°C

$$= \underline{}$$

5.0°C

The amount of $C_2H_6O_2$ is the information given in the problem;

it is _____. We will pick up the other conversion factors given in the problem as we go along. You also will need the conversion factor K_F for water from the table. But first let's start the problem.

1.0 qt $C_2H_6O_2$

$$? \underline{} =$$

gal water

1.0 qt $C_2H_6O_2$

1.0 qt $C_2H_6O_2$

946

ml

$\dfrac{\text{mole } C_2H_6O_2}{62.0 \text{ g } C_2H_6O_2}$, mole $C_2H_6O_2$

10^3 g water

mole

~~qt water~~

1.0 g water

°C

$\dfrac{1}{5.0°C}$

The information given is _____, so, as usual, you start there.

? gal water = _____

You can now cancel the unwanted dimension and work your way toward moles $C_2H_6O_2$ by using the following conversion factors.

$$\frac{\underline{\quad} \text{ ml } C_2H_6O_2}{\text{qt } C_2H_6O_2} \times \frac{0.80 \text{ g } C_2H_6O_2}{\underline{\quad} C_2H_6O_2}$$

To get from grams $C_2H_6O_2$ to moles $C_2H_6O_2$ you multiply by

_____. The unwanted dimension now is _____, but this is in K_F for water,

$$\frac{(1.86°C) \,(\underline{\qquad\qquad})}{\underline{\qquad} C_2H_6O_2} \, .$$

You can convert from grams water to gallons water in three steps.

$$\frac{\underline{\quad} \text{ ml water}}{\underline{\qquad}} \times \frac{\underline{\qquad}}{946 \text{ ml water}} \times \frac{\text{gal water}}{4 \text{ qt water}}$$

You now have the dimensions of the answer, but still have not

cancelled ___. However, you have a value in °C for ΔT_F that you

can use, so you multiply by _____. Notice that once again you have used the reciprocal. The answer, therefore, is

$$? \text{ gal water} = 1.0 \times 946 \times 0.80 \times \frac{1}{62.0} \times (1.86)\,(10^3)$$

$$\times \frac{1}{1.0} \times \frac{1}{946} \times \frac{\text{gal water}}{4} \times \frac{1}{5.0}$$

$$= 1.2 \text{ gal water.}$$

That problem was especially complicated. If you were able to do it, congratulate yourself; if not, don't knock yourself out.

SECTION F What Happens when the Solute Breaks Up into Several Particles

(Do not do this complex section unless instructed to do so.)
You may have noticed all along that the molecules of solute have not been mentioned, while the term "particles of solute" has been used. The reason for this distinction is that some solute molecules will dissociate (break up) into several charged particles (ions) when they are dissolved in a very polar solvent like water. The solutes which will do this are either inorganic salts or acids or bases.

If these compounds dissociated 100% (that is, if every molecule broke up into its ions), then the determination of the colligative properties wouldn't be that much more difficult. Thus, if you had a 1 molal NaCl solution and you knew that the NaCl broke up into Na^+ and Cl^- ions, then the solution would be 2 molal in particle concentration.

In general, dilute solutions (less than 0.1 m) of salts and strong acids and bases will be 100% dissociated. As the concentration of the solute increases, the percentage dissociation falls off. Weak acids and bases will, even in the most dilute solutions, only dissociate very slightly. The discussion of which acids and bases fall into which category falls beyond the scope of this book, but by the time you are faced with problems which require that you know this, you will have other texts to give you the information you will need.

Since even compounds which are capable of complete dissociation are not 100% dissociated at all concentrations, we will figure out how to count the total number of particles, both ions and molecules, present in the compound.

To see how you count particles when you know the percentage dissociation, let's try a sample problem. What is the total number of particles present if you have 5.0 moles of NaCl, which is only 90% dissociated into Na^+ and Cl^- ions?

$$NaCl \longrightarrow Na^+ + Cl^-$$

By inspection you can see that 4.5 moles are dissociated, and each mole dissociated gives 2 moles of ions (1 Na^+ and 1 Cl^-). Thus you will have 9.0 moles of mixed ions. Ten percent of the initial NaCl will still be molecules, so you have 0.5 moles of molecules. Thus, the total number of moles of particles will be 9.5.

Now let's see if we can make this a little more systematic. Instead of talking about percentage dissociation (the number per 100 total molecules), it is more convenient to talk about "fraction" dissociated (the number per 1 molecule). This fraction, f, equals the percentage dissociation divided by 100. Thus, if a salt is 80% dissociated, then the fraction dissociated, f, equals _____. If an acid is 0.15% dissociated, then the value for f is _____.

0.80

0.0015

Therefore, if you have 5.0 moles of NaCl initially and it is 90% dissociated, then the number of particles can be counted this way.

$$(5.0) (f) = \text{Number dissociated} \quad = 5.0 \times 0.90$$
$$= 4.5$$
$$(5.0) (1-f) = \text{Number undissociated} = 5.0 \times 0.10$$
$$= 0.5$$

Now we can count particles.

$$\text{Moles } Na^+ = (5.0) (f) \quad = \text{Number dissociated}$$
$$\text{Moles } Cl^- = (5.0) (f) \quad = \text{Number dissociated}$$
$$\text{Moles } NaCl = (5.0) (1-f) = \text{Number undissociated}$$

Adding these all up, you have

$$\text{Moles particles} = (5.0)(f) + (5.0)(f) + (5.0)(1-f)$$
$$= (5.0)(1+f).$$

Since f in this case is 0.90,

$$\text{Moles particles} = (5.0)(1.90) = 9.5.$$

We can make this still more general if we let the number of moles of molecules initially be n. Then in the case where the molecule breaks up into two ions, the total number of particles can be expressed as moles particles $= n(1+f)$.

Now you try one. How many total moles of particles will be present in a KI solution in which 5 moles of KI have dissolved but 98% of the KI has dissociated into K^+ and I^- ions?

0.98

$$f = \underline{}$$

5, +0.98

$$\text{Particles} = \underline{}\,(1\underline{})$$

9.9

$$= \underline{} \text{ moles particles}$$

Try an example involving weights of solute rather than moles. How many moles of particles are present in a sample containing 18 g of CH_3COOH which is 2% dissociated into H^+ and CH_3COO^- ions?

0.02

$$f = \underline{}$$

$$\text{Moles } CH_3COOH \text{ init.} = n = 18 \text{ g } CH_3COOH$$

$\dfrac{\text{mole } CH_3COOH}{60.0 \text{ g } CH_3COOH}$

$$\times \underline{}$$

0.30 mole

$$= \underline{}$$

n

$$\text{Moles particles} = \underline{}\,(1+f)$$

0.306

$$= \underline{}$$

Remember, n must be expressed in moles of solute initially put in the solution.

What will happen if you have a molecule which dissociates into 3 ions, as, for example, $CaCl_2$?

$$CaCl_2 \longrightarrow Ca^{+2} + 2Cl^-$$

We will count as we did before. Say you start with n moles of $CaCl_2$ and f represents the fraction which is dissociated.

$$\text{Number moles dissociated} = n(f)$$
$$\text{Number moles undissociated} = n(1-f)$$
$$\text{Number moles } Ca^{+2} = n(f)$$
$$\text{Number moles } Cl^- = 2n(f)$$
$$\text{Number moles } CaCl_2 = n(1-f)$$

Adding these up, you have

Total number moles $= n(f) + 2n(f) + n(1-f)$

$$= n(1+2f).$$

Here is another problem. What will the total number of moles of particles be if 8 moles of $CaCl_2$ are 94% dissociated?

$f =$ _____ | 0.94

$n =$ _____ | 8 moles

Total number of particles $= n($ _____ $)$ | $1 + 2f$

$= 8($ _____ $)$ | 2.88

$= 23$ moles particles

Now that you can count the number of moles of particles, you will be able to do colligative property problems like the example which follows.

What will the boiling point be for a solution prepared by dissolving 11.1 g $CaCl_2$ in 250 g water? The $CaCl_2$ is 97% dissociated at this concentration. First you must calculate the number of moles of particles, Ca^{+2}, Cl^-, and $CaCl_2$. To do this you need n, the number of moles of $CaCl_2$ put in initially.

$n = ?$ init. moles $CaCl_2 = 11.1$ g $CaCl_2 \times$ _____ | $\dfrac{\text{mole } CaCl_2}{111.1 \text{ g } CaCl_2}$

$=$ _____ mole $CaCl_2$ | 0.100

Since $f =$ _____ and $CaCl_2$ dissociates into _____ ions, the total | 0.97, 3

number of particles present will be _____ mole. | 0.294

Now you can solve the problem exactly as you did in Section D. The conversion factor given in the problem has become

$$\frac{0.294 \text{ mole particles}}{\text{\_\_\_\_\_ kg water}}.$$ | 0.250

The question asked is, "How many degrees above the boiling point of pure water will the solution boil?" or, "What is the value

for _____?" Starting with the conversion factor | ΔT_B

$(? \Delta T_B)°C =$ _____ , | $\dfrac{0.294 \text{ mole particle}}{0.250 \text{ kg water}}$

You next multiply by the conversion factor, K_B,

_____ . Everything except the dimension of the | $\dfrac{(0.51°C)(\text{kg water})}{\text{mole particles}}$

answer cancels, so the answer is

$\Delta T_B = 0.60°C.$

To get the boiling point of the solution you must _____ ΔT_B to the | add
boiling point of pure water. The boiling point of the solution

therefore will be _____. | 100.60

It is also possible to use an experimentally determined value of ΔT to get a percentage dissociation for a molecule. What you do is determine the number of moles of particles present in the first step. Using this information, you can then calculate the value for f.

Here is an example of this type of problem. What is the percentage dissociation of HIO_4 in a solution which contains 3.84 g HIO_4 in 250 g water if the solution freezes at $-0.15°C$? First you calculate ΔT_F. Since water freezes at $0.00°C$, it will be

0.15°C

$$\Delta T_F = \underline{\hspace{2cm}}.$$

Then you solve for the total number of moles of particles present.

$$\frac{\text{mole particle}}{(1.86°C)\,(\text{kg water})}$$

$$? \text{ moles particles} = 0.250 \text{ kg water} \times \underline{\hspace{3cm}}$$

0.15°C

$$\times \underline{\hspace{1.5cm}}$$

0.0202

$$= \underline{\hspace{1.5cm}} \text{ mole particles}$$

2

Now the acid HIO_4 will dissociate into __ ions, H^+ and IO_4^- so you know that the total number of particles is represented by the equation

$n(1+f)$

$$\text{Number of particles} = \underline{\hspace{2cm}}.$$

You can calculate n from the initial weight of HIO_4 used.

3.84 g

$$n = \text{moles } HIO_4 = \underline{\hspace{1.5cm}} HIO_4 \times \frac{\text{mole } HIO_4}{191.9 \text{ g } HIO_4}$$

0.020 mole NaIO$_4$

$$= \underline{\hspace{3cm}}$$

Now you solve for f.

$$0.0202 = 0.020\,(1+f)$$

0.01

$$f = \underline{\hspace{1.5cm}}$$

Since f is a fraction and the question asks for percent, you must

100

multiply f by __ to get percent. Therefore, the percentage

1%

dissociation is __.

Recall:
If the molecule splits into 2 ions,

$$\text{Total particles} = n\,(1+f).$$

If the molecule splits into 3 ions,

$$\text{Total particles} = n\,(1+2f).$$

If the molecule splits into 4 ions,

3

$$\text{Total particles} = n\,(1+\underline{\hspace{0.5cm}}f).$$

And (although there are not many) if it splits into 5 ions,

$(1+4f)$

$$\text{Total particles} = n\,(\underline{\hspace{2cm}}).$$

SECTION G   Osmotic Pressure

This last colligative property is observed when you have a solution separated from pure solvent by a membrane which allows the solvent molecules to go through but doesn't allow passage of the solute particles. This type of membrane is called "semipermeable." The liquids on the two sides of the membrane try to reach the same vapor pressure; that is, they try to have the same concentration of solute on both sides of the membrane. Since the solute particles cannot cross the membrane to increase the concentration of solute in pure solvent, the solvent molecules leave the pure solvent and attempt to dilute the solution until the concentration of solute becomes zero. This, of course, is not possible.

The osmotic pressure is defined as the pressure that must be applied to stop the flow of solvent molecules into the solution. It has been found to be dependent not only on the concentration of solute in the solution but also on the temperature. The relationship is

$$\pi = MRT, \quad \text{where} \quad \pi = \text{osmotic pressure}$$

$$M = \text{molarity}$$

$$R = \text{gas constant}$$

$$T = \text{temperature in °K}$$

This equation looks very much like the Ideal Gas Law, and you know that if

π is expressed in atmospheres (atm),

M is expressed in moles solute per liter solution, and

T is expressed in °K, which equals (°C + 273), then

$$R \text{ is } 0.0820 \, \frac{\text{(liter) (atm)}}{\text{(mole solute) (°K)}}.$$

(It has been found that at higher concentrations it is best to express the concentration in molality. However, almost all osmotic pressure measurements are done at very low concentrations.)

A major reason for determining osmotic pressure is to get the molecular weights of very large molecules. Its advantage is that it is simple to read very low pressures. A pressure of 10^{-2} atm is shown by a column of mercury 7.6 mm high, which is easily read. To read a temperature difference of 10^{-2}°C for a freezing point lowering is quite a problem in the lab. You can get good results with osmotic pressure even at very low molal concentrations of solute.

The expression you use for molecular weights is easily calculated from the expression

$$\pi = MRT.$$

We will do it by dimensional analysis.

We can start right off with an example. What is the molecular weight of polyethylene plastic if the osmotic pressure of a solution prepared by dissolving 1.00 g of the polyethylene in 500 ml of tetralin solvent was found to be 2.2×10^{-2} atm at 27°C?

First you must convert the temperature from °C to °K.

300°K

$$°K = (273 + °C) = \underline{\hspace{3em}}$$

Since the solution is so diluted, you can assume that the volume of the solution is equal to the volume of the solvent. You therefore can write the conversion factor given in the problem as

$$\frac{1.00 \text{ g polyethylene}}{\underline{\hspace{3em}} \text{ liter soln.}} \cdot$$

0.500

What is called the *gas constant*, R, is a conversion factor with four dimensions, shown above to be

$$\frac{0.0820 \text{ (liter) (atm)}}{\text{(mole solute) (°K)}} \cdot$$

There are two pieces of information given in the problem. The

2.2×10^{-2} atm

osmotic pressure is \underline{\hspace{4em}}, and the temperature is 300°K.

Now we are ready to start. You want to know the molecular weight of polyethylene which is a conversion factor having the dimensions g/mole. Consequently, you ask

$$\frac{? \text{ g polyethylene}}{\text{mole polyethylene}} =$$

You may start with any conversion factor or information given which has either of these dimensions. Let's start with

$$\frac{? \text{ g polyethylene}}{\text{mole polyethylene}} = \frac{1.00 \text{ g polyethylene}}{0.500 \text{ liter soln.}}$$

liter solution

The unwanted dimension left is \underline{\hspace{4em}}, and this appears in R. You therefore multiply by R.

$$\frac{? \text{ g polyethylene}}{\text{mole polyethylene}} = \frac{1.00 \text{ g polyethylene}}{0.500 \cancel{\text{ liter soln.}}}$$

$$\times \frac{0.0820 \cancel{\text{(liter soln.)}} \text{ (atm)}}{\text{(mole polyethylene) (°K)}} \cdot$$

You now have the dimensions of your answer, but you still have

atmosphere

two unwanted dimensions, \underline{\hspace{3em}} in the numerator and °K

denominator

in the \underline{\hspace{3em}}. You can cancel these by multiplying by

reciprocal

temperature, 300°K, and by the \underline{\hspace{3em}} of the osmotic pressure, 2.2×10^{-2} atm. Then you will have

$$\frac{? \text{ g polyethylene}}{\text{mole polyethylene}} = \frac{1.00 \text{ g polyethylene}}{0.500 \text{ liter soln.}}$$

$$\times \frac{0.0820 \text{ (liter soln.) (atm)}}{\text{(mole polyethylene) (}^\circ\text{K)}}$$

$$\times \frac{300^\circ \text{K}}{2.2 \times 10^{-2} \text{ atm}}$$

$$= \frac{2.24 \times 10^3 \text{ g polyethylene}}{\text{mole polyethylene}} .$$

PROBLEM SET

The questions below are divided into sections corresponding to the section in the chapter which pertains to them. Problems with an asterisk are optional.

Section B

1. The boiling point of pure tetralin is 207.3°C. When 8.0 g of hexane, C_6H_{14}, are dissolved in 225 g of tetralin, the solution boils at 210.1°C. What is the value of K_B for tetralin?

2. A 10.0 wt % alloy of Cu in Ag melts at 876°C. Pure Ag melts at 961°C. What is the molal freezing point depression constant for Ag?

Section C

3. What is the molecular weight of an unknown substance if a solution which contains 10.0 g unknown in 150 g carbon disulfide boils at 48.3°C?

4. The freezing point of a solution containing 2.0 g of an unknown compound in 20 g benzene is 3.4°C lower than the freezing point of pure benzene. What is the molecular weight of the unknown?

Section D

5. What is the freezing point of a solution which contains 3.42 g sugar (molecular weight = 342 g/mole) in 10.0 g water?

6. What is the boiling point of a solution which contains 1.5 g $CHCl_3$ in 10.0 g carbon tetrachloride?

*7. What is the boiling point of a 0.600 *M* solution of sugar in benzene? The molecular weight of sugar is 342 g/mole, and the density of the solution is 0.90 g/ml.

Section E

8. How many grams of ethyl alcohol, C_2H_5OH, must be added to 250 g of water to have a solution which will not freeze until the temperature goes below −3.72°C?

9. How many grams of octane, C_8H_{18}, must you add to 500 ml carbon tetrachloride (density = 1.59 g/ml) to increase the boiling point to 82.8°C?

*10. How many quarts of ethylene glycol, $C_2H_6O_2$, must you add to 3.0 gal of water (density = 1.00 g/ml) so that the solution will not freeze until the temperature gets below 14°F? There are 946 ml/qt, 4 qt/gal, and the density of ethylene glycol is 1.1 g/ml.

* Section F (Optional)

11. What is the freezing point of a 0.10 *m* solution $CaCl_2$ in water if the $CaCl_2$ is 100% dissociated at this concentration into Ca^{+2} and Cl^- ions?

12. What is the boiling point of a solution prepared by dissolving 5.85 g NaCl in 100 g water if the NaCl is 90% dissociated at this concentration?

13. What is the freezing point of an 18.75 wt % solution of $Cu(NO_3)_2$ in water? The salt is 80% dissociated into Cu^{+2} and (NO_3^-) ions.

14. What is the percentage dissociation into CH_3COO^- and H^+ ions if a solution prepared by dissolving 0.50 moles CH_3COOH in 1.0 kg water freezes at $-0.95°C$?

15. What is the percentage dissociation if a solution prepared by dissolving 292.5 g NaCl in 1.0 kg water boils at 103.5°C?

Section G

16. What is the molecular weight of a polypropylene sample if a solution prepared by dissolving 1.0 g of the polypropylene in 1.25 liters of hexane has an osmotic pressure, π, of 0.00020 atm at 350°K?

17. What is the molecular weight of a protein fraction if a solution containing 2.00 g protein in 200 g water has an osmotic pressure, π, of 7.6 mm Hg at 27°C? There are 760 mm Hg/1.0 atm, and the density of water is 1.0 g/ml.

PROBLEM SET ANSWERS

In order to satisfactorily complete this chapter, you should be able to do Problems 1–6, 8, 9, and either 16 or 17. If you run into trouble, check the section in the chapter which pertains to the problems you cannot do.

1. $? K_B = \dfrac{(? °C)(\text{kg tetralin})}{\text{mole hexane}}$

$= \dfrac{225 \text{ g tetralin}}{8.0 \text{ g hexane}} \times \dfrac{86.0 \text{ g hexane}}{\text{mole hexane}}$

$\times \dfrac{\text{kg tetralin}}{10^3 \text{ g tetralin}} \times 2.8°C$

$= \dfrac{6.8°C \,(\text{kg tetralin})}{\text{mole solute}} \quad (\Delta T = 210.1 - 207.3 = 2.8°C)$

2. $? K_F = \dfrac{(? °C)(\text{kg Ag})}{\text{mole Cu}}$

$= \dfrac{90.0 \text{ g Ag}}{10.0 \text{ g Cu}} \times \dfrac{\text{kg Ag}}{10^3 \text{ g Ag}} \times \dfrac{63.5 \text{ g Cu}}{\text{mole Cu}} \times 85°C$

$= \dfrac{49°C \,(\text{kg Ag})}{\text{mole solute}} \quad (\Delta T = 961 - 876 = 85°C)$

3. $\dfrac{? \text{ g } X}{\text{mole } X} = \dfrac{10.0 \text{ g } X}{0.150 \text{ kg carbon disulfide}}$

$\times \dfrac{2.3°\text{C (kg carbon disulfide)}}{\text{mole } X} \times \dfrac{1}{2.0°\text{C}}$

$= \dfrac{77 \text{ g } X}{\text{mole } X} \quad (\Delta T = 48.3 - 46.3 = 2.0°\text{C})$

4. $\dfrac{? \text{ g } X}{\text{mole } X} = \dfrac{2.0 \text{ g } X}{20 \text{ g benzene}} \times \dfrac{5.1°\text{C (10}^3 \text{ g benzene)}}{\text{mole } X} \times \dfrac{1}{3.4°\text{C}}$

$= \dfrac{1.5 \times 10^2 \text{ g } X}{\text{mole } X}$ (Notice the correct way to give the number of significant figures.)

5. $? \Delta T_F °\text{C} = \dfrac{3.42 \text{ g sugar}}{10.0 \text{ g water}} \times \dfrac{1.86°\text{C (10}^3 \text{ g water)}}{\text{mole sugar}} \times \dfrac{\text{mole sugar}}{342 \text{ g sugar}}$

$= 1.86°\text{C}$

Freezing point $= (0.00 - 1.86)°\text{C} = -1.86°\text{C}$

6. $? \Delta T_B °\text{C} = \dfrac{1.5 \text{ g CHCl}_3}{10.0 \text{ g carbon tetrachloride}}$

$\times \dfrac{5.0°\text{C (10}^3 \text{ g carbon tetrachloride)}}{\text{mole CHCl}_3} \times \dfrac{\text{mole CHCl}_3}{119.5 \text{ g CHCl}_3}$

$= 6.3°\text{C}$

Boiling point $= 76.8°\text{C} + 6.3°\text{C} = 83.1°\text{C}$

7. First calculate the weight of 1.0 liter of solution.

$? \text{ g soln.} = 1.0 \text{ liter soln.} \times \dfrac{\text{ml soln.}}{10^{-3} \text{ liter soln.}} \times \dfrac{0.90 \text{ g soln.}}{\text{ml soln.}}$

$= 900 \text{ g soln.}$

Then calculate the weight of sugar in 1.0 liter of solution.

$? \text{ g sugar} = 1.0 \text{ liter soln.} \times \dfrac{0.600 \text{ mole sugar}}{\text{liter soln.}} \times \dfrac{342 \text{ g sugar}}{\text{mole sugar}}$

$= 205 \text{ g sugar}$

You then get the weight of solvent by subtracting

$? \text{ g benzene} = 900 \text{ g} - 205 \text{ g} = 695 \text{ g benzene.}$

Now you know that you have 0.600 mole sugar/695 g benzene.

$? \Delta T_B °\text{C} = \dfrac{0.600 \text{ mole sugar}}{0.695 \text{ kg benzene}} \times \dfrac{2.5°\text{C (kg benzene)}}{\text{mole sugar}}$

$= 2.2°\text{C}$

Boiling point $= 80.1°\text{C} + 2.2°\text{C} = 82.3°\text{C}$

8. $\Delta T_F = 0.00°C - (-3.72°C) = 3.72°C$

   $\text{? g C}_2\text{H}_5\text{OH} = 0.250 \text{ kg water} \times \dfrac{\text{mole C}_2\text{H}_5\text{OH}}{1.86°C \text{ (kg water)}}$

   $\times \dfrac{46.0 \text{ g C}_2\text{H}_5\text{OH}}{\text{mole C}_2\text{H}_5\text{OH}} \times 3.72°C$

   $= 23.0 \text{ g C}_2\text{H}_5\text{OH}$

9. $\Delta T_B = 82.8°C - 76.8°C = 6.0°C$

   $\text{? g C}_8\text{H}_{18} = 500 \text{ ml carbon tetrachloride} \times \dfrac{1.59 \text{ g carbon tetrachloride}}{\text{ml carbon tetrachloride}}$

   $\times \dfrac{\text{mole C}_8\text{H}_{18}}{5.0°C \text{ (10}^3 \text{ g carbon tetrachloride)}}$

   $\times \dfrac{114 \text{ g C}_8\text{H}_{18}}{\text{mole C}_8\text{H}_{18}} \times 6.00°C$

   $= 1.1 \times 10^2 \text{ g C}_8\text{H}_{18}$

10. First determine the freezing point in °C.

    $°C = \dfrac{5}{9}(°F - 32) = \dfrac{5}{9}(-18) = -10°C$

    $\Delta T_F = 0.0°C - (-10°C) = 10°C$

    $\text{? qt eth. glyc.} = 3.0 \text{ gal water} \times \dfrac{4 \text{ qt water}}{\text{gal water}} \times \dfrac{946 \text{ ml water}}{\text{qt water}} \times \dfrac{1.00 \text{ g water}}{\text{ml water}}$

    $\times \dfrac{\text{mole ethylene glycol}}{1.86°C \text{ (10}^3 \text{ g water)}} \times \dfrac{62.0 \text{ g ethylene glycol}}{\text{mole ethylene glycol}}$

    $\times \dfrac{\text{ml ethylene glycol}}{1.1 \text{ g ethylene glycol}} \times \dfrac{\text{qt ethylene glycol}}{946 \text{ ml ethylene glycol}} \times 10°C$

    $= 3.6 \text{ qt ethylene glycol}$

11. Since the salt is 100% dissociated and a molecule of $CaCl_2$ will produce three ions, one Ca^{++} and two Cl^-'s, the solution will be $3 \times 0.10 \, m$ in total particles.

    $\text{? } \Delta T_F °C = \dfrac{0.30 \text{ mole particles}}{\text{kg water}} \times \dfrac{1.86°C \text{ (kg water)}}{\text{mole particle}}$

    $= 0.56°C$

    Freezing point $= 0.00°C - 0.56°C = -0.56°C$

12. Mole particles $= n(1+f) = 5.85 \text{ g NaCl} \times \dfrac{\text{mole NaCl}}{58.5 \text{ g NaCl}}(1+0.90)$

$= 0.19$ mole particles

$? \Delta T_B °C = \dfrac{0.19 \text{ mole particles}}{0.100 \text{ kg water}} \times \dfrac{0.51°C \text{ (kg water)}}{\text{mole particle}}$

$= 0.97°C$

Boiling point $= 100.00°C + 0.97°C = 100.97°C$

13. Mole particles $= n(1+2f) = 18.75 \text{ g Cu(NO}_3)_2$

$\times \dfrac{\text{mole Cu(NO}_3)_2}{187.5 \text{ g Cu(NO}_3)_2}(1+1.6)$

$= 0.26$ mole particles

Weight of water $= 100 \text{ g} - 18.75 \text{ g} = 81.25 \text{ g water}$

$? \Delta T_F °C = \dfrac{0.26 \text{ mole particles}}{0.08125 \text{ kg water}} \times \dfrac{1.86°C \text{ (kg water)}}{\text{mole particle}}$

$= 6.0°C$

Freezing point $= 0.00°C - 6.0°C = -6.0°C$

14. First calculate the number of moles of particles from the freezing point lowering, $\Delta T_F = 0.95°C$.

$? \text{ mole particles} = 1.0 \text{ kg water} \times \dfrac{\text{mole particle}}{1.86°C \text{ (kg water)}} \times 0.95°C$

$= 0.51$ mole particles

Then note that the moles CH_3COOH initially $= 0.50$ mole $= n$. Then calculate f.

Mole particles $= n(1+f)$

$0.51 = 0.50(1+f)$

$f = 0.02$

Percent dissociation $= 100(f) = 2\%$ dissociated

15. This problem is the same as the one above.

$? \Delta T_B °C = 103.5°C - 100.0°C = 3.5°C$

$? \text{ moles particles} = 1.0 \text{ kg water} \times \dfrac{\text{mole particle}}{0.51°C \text{ (kg water)}} \times 3.5°C$

$= 6.9$ moles particles

$? \text{ moles NaCl init.} = 292.5 \text{ g NaCl} \times \dfrac{\text{mole NaCl}}{58.5 \text{ g NaCl}}$

$= 5.00$ moles NaCl $= n$

Mole particles $= n(1+f)$

$$6.9 = 5.00(1+f)$$

$$f = 0.38$$

Percent dissociation $= 100(f) = 38\%$ dissociated
(This is not a very realistic example; it is only to show you how to do these calculations.)

16. Since the solution is so diluted, you can assume that the volume of the solution equals the volume of the solvent.

$$\frac{?\ g\ polypropylene}{mole\ polypropylene} = \frac{1.0\ g\ polypropylene}{1.25\ liters\ soln.} \times \frac{0.0820\ (liter\ soln.)\ (atm)}{(mole\ polypropylene)\ (°K)}$$

$$\times \frac{350°K}{0.00020\ atm}$$

$$= \frac{1.1 \times 10^5\ g\ polypropylene}{mole\ polypropylene}$$

17. This problem is the same as the one above, except that you will have to convert the pressure to atmosphere, the temperature to °K, and the volume of solution to liters.

$$?\ atm = 7.6\ mm\ Hg \times \frac{atm}{760\ mm\ Hg} = 0.010\ atm$$

$$?\ °K = 27°C + 273 = 300°K$$

$$?\ liters\ soln. = 200\ g\ water \times \frac{ml\ water}{1.0\ g\ water} \times \frac{10^{-3}\ liter\ water}{ml\ water} \times \frac{liter\ soln.}{liter\ water}$$

$$= 0.20\ liter\ soln.$$

(Once again we have assumed that the volume of solution equals volume solvent.)

$$\frac{?\ g\ protein}{mole\ protein} = \frac{2.00\ g\ protein}{0.20\ liter\ soln.} \times \frac{0.0820\ (liter\ soln.)\ (atm)}{(mole\ protein)\ (°K)} \times \frac{300°K}{0.010\ atm}$$

$$= \frac{2.5 \times 10^4\ g\ protein}{mole\ protein}$$

Chapter 17
Calculations Involving Heat

PRETEST

1. How many calories are needed to raise the temperature of 50.0 g of methyl alcohol 15.0°C? The specific heat of methyl alcohol is 0.600 cal/g°C.

2. How many kilocalories are needed to heat 11.0 moles of water from 20.0°C to 100.0°C?

3. What is the specific heat of lead metal if 14.4 calories were given off when a 25.0 g sample of lead cooled from 98.0°C to 80.0°C?

4. How many calories are required to vaporize 35.0 g of methyl alcohol? The heat of vaporization for the alcohol is 263 cal/g.

5. How many calories are needed to convert 45.0 ml of ether at 0.0°C to its vapor at its boiling point (34.6°C)? Ether's density is 0.714 g/ml. The specific heat of ether is 0.547 cal/g°C, and its heat of vaporization is 89.3 cal/g.

6. How much heat can you get by burning 4.80 kg of carbon with all substances at 25°C? The reaction is $C(s) + O_2(g) \rightarrow CO_2(g)$. The standard heat of reaction is -94.1 kcal/mole.

7. What is the heat of reaction for the process

   $FeS(s) + 2\ HCl(aq) = FeCl_2(aq) + H_2S(g)$?

   The notation (aq) means in dilute solution in water.
   The standard heats of formation are

$FeS(s)$	$\Delta H_f^0 = -22.7$ kcal/mole
$HCl(aq)$	$\Delta H_f^0 = -39.8$ kcal/mole
$FeCl_2(aq)$	$\Delta H_f^0 = -100.4$ kcal/mole
$H_2S(g)$	$\Delta H_f^0 = -4.8$ kcal/mole

8. A 10.0-g piece of aluminum is heated to 102.25°C and placed in a calorimeter containing 25.0 g of water at 20.50°C. The temperature of the water rises to 27.25°C. What is the specific heat of aluminum? Assume that the body of the calorimeter does not absorb any heat.

9. Supose that you put 25.0 g of ice at 0.00°C into a calorimeter that contains 300 g of water at 21.16°C. The ice melts and the temperature of the water in the calorimeter finally drops to 13.40°C. What is the heat of fusion for water, in kcal/mole? The heat capacity K_c of the calorimeter equals 0.80 cal/°C.

PRETEST ANSWERS

All the questions are worked out as examples in the body of the chapter. If you were able to do questions 1, 2, and 3, you can skip Section A in this chapter. If you were able to do question 4 but not 5, you need to cover only the last half of Section B. If you missed both, do all of Section B. You should be able to do both question 6 and question 7. If you missed either, do Section C. You should be able to do question 8. You can find it worked out in the first part of Section D. Question 9 is rough. It is covered at the end of Section D. Your instructor will tell you whether you need to be able to do this type of calculation.

1. 450 cal

2. 15.8 kcal

3. 0.0320 cal/g°C

4. 9.20×10^3 cal

5. 3.48×10^3 cal

6. -3.76×10^4 kcal

7. -2.9 kcal/mole

8. 0.225 cal/g°C

9. 1.44 kcal

In this chapter, we will be examining the types of calculations that you use when you are dealing with heat. We will only consider the following three types of processes that can either consume or give off heat.

1. Changing the temperature of a substance

2. Changing the phase of a substance

3. Having the substance undergo a chemical reaction

Finally, we will consider how this heat is measured in the laboratory. This is called *calorimetry*.

As we go through these calculations, you will see that occasionally a simple, one-step dimensional analysis solution is not possible. You may have to add or subtract several solutions to arrive at the final answer. However, this shouldn't pose any real difficulty for you because the methods are very straightforward.

SECTION A   Specific Heat

To raise the temperature of a substance, you must put heat into the substance. When the substance cools down, it will give off heat to its surroundings. The amount of heat it gives off depends on two things: the amount of the substance and the change in its temperature. When you express this heat per gram of substance and also per °C change, you have a conversion factor that has the units

$$\frac{\text{cal}}{(\text{g})\,(^\circ\text{C})}$$

where cal is the abbreviation for calories, the commonly used unit of heat. This conversion factor is called the *specific heat* of the substance. Every substance has its own specific heat which varies somewhat with the temperature at which you consider the substance. This is similar to the boiling point. Each substance has its own boiling point which varies with the pressure at which you consider the substance. There are tables of the specific heats of various substances at various temperatures. For example, the specific heat of methyl alcohol at 0°C is

$$\frac{0.566 \text{ cal}}{(\text{g alcohol})\,(^\circ\text{C})}$$

but at 20°C it is

$$\frac{0.600 \text{ cal}}{(\text{g alcohol})\,(^\circ\text{C})}$$

For copper metal at 0°C the specific heat is

$$\frac{0.0910 \text{ cal}}{(\text{g Cu}) (\underline{})}$$

°C

and at 20°C it is

$$\frac{0.0921 \underline{}}{(\text{g Cu}) (\text{°C})}$$

cal

If you look up the specific heat of NaCl at 0°C in a table, you will find that it is

$$\frac{0.204 \underline{}}{(\underline{}) (\underline{})}$$

cal

g NaCl,   °C

The definition of one calorie is based on the amount of heat required to raise the temperature of exactly one gram of water from 14.5°C to 15.5°C. This means that the specific heat of water is

$$\frac{1 \text{ cal}}{(\text{g water}) (\text{°C})}$$

Since calorie is a defined term, it does not enter into any considerations of significant figures. If you have forgotten about the difference between defined and measured quantities, check the beginning of Section C, Chapter 1. Although the specific heat of water varies with temperature, we are going to use the value

$$\frac{1 \text{ cal}}{(\text{g water}) (\underline{})}$$

°C

for all our calculations. The variation is so slight that it will not affect your results significantly.

There is another, more fundamental, unit of heat, the joule. It is used by physicists, and has now replaced the calorie in the new SI units. We will not be using the joule, but it can be related to calories by the conversion factor

$$\frac{4.184 \text{ joules}}{\text{cal}}$$

You will frequently use the unit of heat, the kilocalorie. The prefix kilo- follows the normal metric system usage, so there are

$$\frac{10^3 \text{ cal}}{\text{kcal}}$$

Now you are ready to see the types of problems that you may have to solve which involve specific heat. Here is question 1 on the pretest. How many calories are needed to raise the temperature of 50.0g of methyl alcohol 15.0°C? The specific heat of the alcohol is 0.600 cal/g°C.

calories

The question asked is "How many _____?" There are two pieces of information given. It may be best, in general, to start with the amount of the substance. So you write

cal

$$? \_\_\_\_ = 50.0 \text{ g alcohol}$$

Now you can use the conversion factor, the specific heat, to cancel the unwanted dimension, g alcohol.

$$\dfrac{0.600 \text{ cal}}{\text{(g alcohol) (°C)}}$$

$$? \text{ cal} = 50.0 \text{ g alcohol} \times _____$$

°C

The unwanted dimension that remains is ____.
This can be cancelled by multiplying by the other piece of

15.0°C

information given in the problem, _____. Thus you have as the complete solution

$$? \text{ cal} = 50.0 \text{ g alcohol} \times \dfrac{0.600 \text{ cal}}{\text{(g alcohol) (°C)}} \times 15°C$$

$$= 450 \text{ cal} \quad (\text{or } 4.50 \times 10^2 \text{ cal in scientific notation})$$

In some problems, you may not be given the weight of the substance in grams. Then you have to convert whatever units it may have into grams. Here is an example of this type. How many calories are given off when 25.0 ml of mercury cool from 25.5°C to 20.4°C? The density of mercury is 13.5 g/ml, and its specific heat is 0.0332 cal/g°C.

Starting in the usual way, you write

cal

$$? \_\_\_\_ =$$

25.0 ml

The information given is the volume of Hg, _____, and its

weight

initial and final temperatures. What you must know is the _____ of Hg and its *change* in temperature. The density gets you from the volume to the weight of Hg. You can get the change in

subtracting

temperature by _____ the initial temperature from the final temperature. This change is abbreviated as ΔT.

$$\Delta T = T_{\text{final}} - T_{\text{initial}}$$

20.4, 25.5

$$= \_\_\_\_°C - \_\_\_\_°C$$
$$= -5.1°C$$

Notice that since the temperature dropped, the sign on ΔT is negative.
Now we can set up the problem. The amount of Hg given in the

25.0

problem was _____ ml Hg.

$$? \text{ cal} = 25.0 \text{ ml Hg}$$

and you can get from ml Hg to g Hg by using the conversion

factor _____

$$? \text{ cal} = 25.0 \,\cancel{\text{ml Hg}} \times \frac{13.5 \text{ g Hg}}{\cancel{\text{ml Hg}}}$$

$\dfrac{13.5 \text{ g Hg}}{\text{ml Hg}}$

Now you can use the _____ , $\dfrac{0.0332 \text{ cal}}{(\text{g Hg}) (°\text{C})}$, to cancel the

specific heat

unwanted dimension g Hg

$$? \text{ cal} = 25.0 \,\cancel{\text{ml Hg}} \times \frac{13.5 \,\cancel{\text{g Hg}}}{\cancel{\text{ml Hg}}} \times \frac{0.0332 \text{ cal}}{(\cancel{\text{g Hg}}) (°\text{C})}$$

The only remaining unwanted dimension is _____. You can cancel

°C

this by multiplying by _____.

ΔT

$$? \text{ cal} = 25.0 \,\cancel{\text{ml Hg}} \times \frac{13.5 \,\cancel{\text{g Hg}}}{\cancel{\text{ml Hg}}} \times \frac{0.0332 \text{ cal}}{(\cancel{\text{g Hg}}) (°\text{C})} \times \text{\_\_\_\_\_}$$

−5.1°C

$$= -57 \text{ cal}$$

Notice that the number of calories has a negative sign. This is
the conventional way of showing that the process has given off
heat. Such a process is called *exothermic*. A process that takes up
heat is called *endothermic*, and the sign of the heat is positive.
As you know, + signs are generally not written; they are understood.
   You may run across problems that ask you to find the heat
required for a temperature change and give you the amount of the

substance in moles. You can easily convert moles to grams by

using as a conversion factor the _____ of the
substance. Here is question 2 on the pretest. How many
kilocalories are needed to heat 11.0 moles of water from 20.0°C
to 100°C?

molecular weight

The _____ of water will get you from moles of
water to grams of water. To get the temperature change, you must

molecular weight

subtract the _____ temperature from the _____ temperature.

initial, final

$$\Delta T = T\text{\_\_\_\_\_} - T\text{\_\_\_\_\_}$$

final,   initial

$$= \text{\_\_\_\_\_}°\text{C} - \text{\_\_\_\_\_}°\text{C}$$

100.0, 20.0

$$= \text{\_\_\_\_\_}°\text{C}$$

80.0

Notice that the sign on ΔT is positive because this is an _____
process. Heat must be put in.
   Starting in the usual way, you write

endothermic

$$? \text{ kcal} = 11.0 \text{ moles H}_2\text{O}$$

Using the molecular weight of water to get to grams of water, you have

$$? \text{ kcal} = 11.0 \text{ moles } H_2O \times \underline{\hspace{2cm}}$$

18.0 g water
----
mole water

specific heat

Now you can use the conversion factor, the _____ of water.

$$? \text{ kcal} = 11.0 \text{ \underline{moles water}} \times \frac{18.0 \text{ g water}}{\text{\underline{mole water}}} \times \underline{\hspace{2cm}}$$

$$\frac{1 \text{ cal}}{(\text{g water}) (°C)}$$

You can cancel the unwanted °C by multiplying by _____, which equals _____ °C.

$$\Delta T$$

80.0

$$? \text{ kcal} = 11.0 \text{ \underline{moles water}} \times \frac{18.0 \text{ \underline{g water}}}{\text{\underline{mole water}}} \times \frac{1 \text{ cal}}{(\text{\underline{g water}}) (\text{\underline{°C}})}$$

$$\times 80.0°\underline{C}$$

The only dimension remaining is cal, but the dimension of the answer must be _____. Therefore you multiply by the conversion factor _____

kcal

$$\frac{\text{kcal}}{10^3 \text{ cal}}$$

$$? \text{ kcal} = 11.0 \text{ \underline{moles water}} \times \frac{18.0 \text{ \underline{g water}}}{\text{\underline{mole water}}} \times \frac{1 \text{ \underline{cal}}}{(\text{\underline{g water}}) (\text{\underline{°C}})}$$

$$\times 80°\underline{C} \times \frac{\text{kcal}}{10^3 \text{ \underline{cal}}}$$

$$= 15.8 \text{ kcal}$$

Calculating the specific heat of a substance is very simple if you know the number of calories involved in changing the temperature of a given weight by a known number of degrees Celsius. Since you want to calculate a conversion factor, the specific heat, you could reword the problem as we have done before for determining density, percent, molal freezing point depression constants, and so on. However, you may find it much easier just to take the values for calories, for ΔT, and for the weight, and then set them into a fraction that gives you all the correct dimensions in their correct places. Question 3 on the pretest is a typical example of this type of calculation. What is the specific heat of lead metal, if 14.4 calories were given off when a 25.0 g sample of lead cooled from 98.0°C to 80.0°C?

First you must determine ΔT.

final, initial

80.0, 98.0

−18.0

$$\Delta T = T\underline{\hspace{1.5cm}} - T\underline{\hspace{1.5cm}}$$

$$= \underline{\hspace{1cm}}°C - \underline{\hspace{1cm}}°C$$

$$= \underline{\hspace{1cm}}°C$$

The problem tells you that 14.4 cal were given off. Since heat is

given off, this is an ＿＿＿＿＿ process and the sign on the   exothermic

number of calories is ＿＿＿＿＿.   − (negative)

You are also told that this heat was given off by a 25.0 g sample.
You now have all the data that you need to set up the specific

heat conversion factor. The units of this factor are ＿＿＿＿. When $\dfrac{cal}{g°C}$
you put in the values −18.0°C, −14.4 cal, and 25.0 g so that
the dimensions are as desired for specific heat, you have

$$\text{specific heat} = \frac{\underline{\hspace{2cm}}}{(\underline{\hspace{1.5cm}})\,(\underline{\hspace{1.5cm}})}$$

−14.4 cal

25.0 g, −18.0°C

Dividing the numbers in the numerator by the numbers in the
denominator then gives you

$$\text{specific heat} = \frac{\underline{\hspace{2cm}}}{g°C}$$

0.0320 cal

Notice that the two negative signs have cancelled. The specific
heat always has a + sign.

You may prefer to reword problems in which the answer is itself
a conversion factor so that the question asked and the information
given both come from the desired conversion factor. If so, you can
reword specific-heat problems so that the question is "How many
calories?" and the information given is "1.0 g substance" and
"1.0°C." Both the 1.0's are defined terms and do not affect the
number of significant figures. In the example we have just done, it
would go like this.

$$? \text{ cal} = 1.0 \text{ g lead} \times 1.0°C$$

The remaining information given in the problem is that −14.4 cal
are required per 25.0 g lead per −18.0°C. Using this information
to cancel the unwanted dimensions gives you

$$? \text{ cal} = 1.0 \text{ g lead} \times 1.0°C \times \frac{\underline{\hspace{2.5cm}}}{(\underline{\hspace{1.5cm}})\,(-18.0°C)}$$

−14.4 cal

25.0 g lead

0.0320 cal

$$= \underline{\hspace{2cm}}$$

Since this is the number of calories needed to change the

temperature of 1 g of lead 1°C, it is, of course, the ＿＿＿＿＿＿＿.   specific heat

Another type of problem that doesn't come up too frequently
involves determining the change in temperature that results from
the gain or loss of a given number of calories. With dimensional
analysis, it is no more difficult to determine the temperature
change when you know the number of calories than it is to
determine the number of calories when you know the temperature
change. In both cases, you use the specific heat of the substance

as your main conversion factor. Here is an example of this type of problem.

If you add 750 calories to a 50.0 g sample of water at 11.0°C, what does the temperature become? First you must realize that you cannot calculate the final temperature directly. The specific heat

change

tells you only about the _____ in temperature. You can calculate ΔT. Once you know ΔT, you can get the final temperature by

adding

_____ ΔT to the initial temperature.

+

$$T_{final} = T_{init} \underline{\quad} \Delta T$$

Thus the question is "How many °C is ΔT?" and the information given is $+750$ cal and 50.0 g water. The calories have a plus sign

endothermic

because heat is being added. This is an _____ process.

You will find it best to start with 750 cal as the information given

$$? \,°C = 750 \text{ cal}$$

You can cancel the unwanted dimension calories by using the

$\dfrac{1 \text{ cal}}{(\text{g water}) (°C)}$

specific heat of water, _____, as your conversion factor.

$$? \,°C = 750 \, \cancel{\text{cal}} \times \frac{(\text{g water}) (°C)}{1 \, \cancel{\text{cal}}}$$

inverted

Notice that you had to use the specific heat in its _____ form to cancel the unwanted cal.

g water

The unwanted dimension that remains is _____, and you have this in the information given. But be careful. You must put it in as $1/\text{g water}$ in order to cancel the unwanted dimension. This may look a little weird to you, but we did the same thing in Chapter 16 when we were dealing with colligative properties. If you can multiply by the information given, there is no reason that you cannot divide by it, since division is the same as multiplying by one over something. So you finish the problem off by writing

$\dfrac{1}{50.0 \text{ g water}}$, 15.0°C

$$? \,°C = 750 \, \cancel{\text{cal}} \times \frac{(\text{g water}) (°C)}{1 \, \cancel{\text{cal}}} \times \underline{\qquad} = \underline{\qquad}$$

This 15.0°C represents ΔT. To find the final temperature of the

add

water, you must _____ ΔT to the starting temperature of the water.

ΔT

$$T_{final} = T_{initial} + \underline{\quad}$$

11.0

$$= \underline{\quad}°C + 15.0°C$$

26.0

$$= \underline{\quad}°C$$

SECTION B   Heat Involved in Changes of Phase

When ice melts, its temperature remains constant at 0°C throughout the melting process. Nevertheless, heat must be added to make the melting continue. Similarly, when water is boiling at standard pressure, its temperature remains at 100°C. Nonetheless, heat must be added in order to convert liquid water at 100°C to gaseous water at 100°C. The heat required to change the phase of a substance without changing its temperature is called the *latent heat*. Generally a more exact name is used to indicate which change of phase is occurring. Thus you speak of the *heat of vaporization* when you are having a phase change between liquid and gas. You speak about the *heat of fusion* when the phase change is between solid and liquid. If the phase change is directly from solid to gas, you use the term *heat of sublimation*.

These latent heats are all conversion factors, since the more substance you have, the more heat is involved. However, since the temperature does not change when the phase changes, °C doesn't appear as it did in specific heat. So the dimensions of these latent heats are

$$\frac{cal}{g}$$

Thus the heat of vaporization of water is given as

$$\frac{540 \ cal}{g \ water}$$

and the heat of fusion of water is given as

$$\frac{79.7 \ cal}{g \ water} .$$

The heat of fusion for Cu metal is given as

$$49.0 \ \frac{cal}{g \ Cu}$$

Latent heats for these changes in phase are frequently given with the dimensions

$$\frac{kcal}{mole}$$

and you can easily calculate the value in one set of dimensions from the value in the other. Here is an example of that calculation. What is the heat of vaporization for water in kcal/mole if its value is 540 cal/g?

In this case, it is best to reword the question as "How many kcal?" and have as the information given "1 mole." Once again, the mole is a defined number and doesn't affect the significant figures.

1 mole

? kcal = _____

To convert the unwanted "mole water" to "g water," you can

molecular weight

$$\frac{18.0 \text{ g water}}{\text{mole water}}$$

use the _____ of water as your conversion factor.

? kcal = 1 mole water × _____

Now you may cancel the unwanted dimension "g water" with the

vaporization

conversion factor given in the problem, the heat of _____ of water.

$$\frac{540 \text{ cal}}{\text{g water}}$$

$$? \text{ kcal} = 1 \text{ } \cancel{\text{mole water}} \times \frac{18.0 \text{ g water}}{\cancel{\text{mole water}}} \times \text{_____}$$

kcal

Now all you must do is convert cal to _____.

$$\frac{\text{kcal}}{10^3 \text{ cal}}$$

$$? \text{ kcal} = 1 \text{ } \cancel{\text{mole water}} \times \frac{18.0 \text{ } \cancel{\text{g water}}}{\cancel{\text{mole water}}} \times \frac{540 \text{ cal}}{\cancel{\text{g water}}} \times \text{_____}$$

$$= 1 \times 18.0 \times 540 \times \frac{\text{kcal}}{10^3}$$

9.72 kcal

$$= \text{_____}$$

Thus the vaporization of 1 mole of water at 100°C requires 9.72 kcal. This is the latent heat of vaporization in kcal/mole. Just as specific heats vary slightly with the temperature at which they are considered, so do heats of vaporization. We won't consider this problem. Instead, we will always consider the heat of vaporization at the standard boiling point.

Question 4 on the pretest is a good example of the type of problem that you will face. How many calories are required to vaporize 35.0 g of methyl alcohol? The heat of vaporization for the alcohol is 263 cal/g.

calories

The question asked is "How many _____?" and the

35.0 g alcohol

information given is "_____." So start by writing

cal

? _____ = 35.0 g alcohol

You can cancel the unwanted dimension by multiplying by the

heat of vaporization

conversion factor given in the problem, the _____ of alcohol.

$$\frac{263 \text{ cal}}{\text{g alcohol}}, 9205$$

? cal = 35.0 g alcohol × _____ = _____ cal

This is a large number, and it has an incorrect number of significant figures, so it would be best to express it in kcal.

$$\frac{\text{kcal}}{10^3 \text{ cal}}$$

? kcal = 9205 cal × _____ = 9.20 kcal

Since converting a liquid to a gas or converting a solid to a liquid always requires heat, they are _____ processes and

the sign on the number of calories is always _____. On the other hand, converting a gas to a liquid or converting a liquid to a

solid gives off heat. These are _____ processes, and the

number of calories is _____.

endothermic

+ (positive)

exothermic

− (negative)

We can write the processes as a reaction

      liquid + heat → gas
      solid + heat → liquid

and

      gas − heat → liquid
      liquid − heat → solid

If you consider the same amount of substance, then the same number of calories is consumed in changing a liquid to the gas as is given off by changing the gas to liquid. And the same number of calories is needed to change the solid to the liquid as is given off

by changing the _____ to the _____. This is always true. The same amount of heat is involved in both a process and its reverse. Only the sign on the heat is opposite. You know that it requires 9.20 kcal to convert one mole of water to one mole of steam. The heat you get when one mole of steam condenses to liquid water

is _____. It has a _____ sign because it is an _____ process.

liquid,   solid

−9.20 kcal,   −,   exothermic

Problems with latent heats are almost always very simple. The possible hang-ups occur only when the amount of substance is not given in grams and you have to convert to grams, or when the substance is not at the temperature at which the phase change occurs. In this last case, you must calculate the amount of heat needed to get the substance to the constant temperature of the phase change, and then add this amount to the heat needed for the phase change. Question 5 on the pretest has both these complications in it.

How many calories are needed to convert 45.0 ml of ether at 0.0°C to its vapor at its boiling point, 34.6°C? Ether's density is 0.714 g/ml. The specific heat of ether is 0.547 cal/g°C, and its heat of vaporization is 89.3 cal/g.

The first thing that you must determine is the heat needed to bring the 45.0 ml of ether from 0.0°C to its boiling point, 34.6°C.

The ΔT for this process is _____°C. Proceeding as you did in Section A for specific heat, you write

      ? _____ = _____

Then you must convert ml to _____ so that you can use the

34.6

cal,   45.0 ml ether

g

specific-heat conversion factor. You can do this by multiplying
by the _____.

density

$$\frac{0.714 \text{ g ether}}{\text{ml ether}}$$

$$? \text{ cal} = 45.0 \text{ ml ether} \times \underline{\hspace{3cm}}$$

Then you can use the specific heat.

$$\frac{0.547 \text{ cal}}{(\text{g ether}) (°C)}$$

$$? \text{ cal} = 45.0 \text{ ml ether} \times \frac{0.714 \text{ g ether}}{\text{ml ether}} \times \underline{\hspace{2cm}}$$

°C

The remaining unwanted dimension is _____, and you can cancel this

34.6°C

by multiplying ΔT, which is _____.

$$? \text{ cal} = 45.0 \text{ ml ether} \times \frac{0.714 \text{ g ether}}{\text{ml ether}} \times \frac{0.547 \text{ cal}}{(\text{g ether}) (°C)}$$

$$\times 34.6°C$$

$$= 608 \text{ cal}$$

This is the amount of heat needed to get the ether to the
temperature at which it changes, to gas. Now you must determine
the heat needed for the phase change that occurs at the
boiling point.

$$? \text{ cal} = 45.0 \text{ ml ether}$$

Once again, you must get to grams to use the heat of vaporization.

$$\frac{0.714 \text{ g ether}}{\text{ml ether}}$$

$$? \text{ cal} = 45.0 \text{ ml ether} \times \underline{\hspace{3cm}}$$

Now you can multiply by the heat of vaporization,

$$\frac{89.3 \text{ cal}}{\text{g ether}}$$

$$? \text{ cal} = 45.0 \text{ ml ether} \times \frac{0.714 \text{ g ether}}{\text{ml ether}} \times \underline{\hspace{2cm}}$$

and the only dimension left is the desired dimension, calories.

$$? \text{ cal} = 45.0 \text{ ml ether} \times \frac{0.714 \text{ g ether}}{\text{ml ether}} \times \frac{89.3 \text{ cal}}{\text{g ether}}$$

$$= 2869 \text{ cal}$$

The total heat needed to warm the ether and to vaporize it is thus

608 cal, 2869 cal

$$\underline{\hspace{2cm}} + \underline{\hspace{2cm}} = 3477 \text{ cal}$$

Written to the correct number of significant figures, this is

3.48×10^3, 3.48

$$\underline{\hspace{3cm}} \text{ cal or } \underline{\hspace{2cm}} \text{ kcal.}$$

You can see that in this example, we took a process that occurs in
several steps, and added the heats of the individual steps to get the
total heat of the process. If we write down the steps, we have

Step 1: $\text{ether}_{(\text{liq. 0°C})} + \text{heat}_1 = \text{ether}_{(\text{liq. 34.6°C})}$

Step 2: $\text{ether}_{(\text{liq. 34.6°C})} + \text{heat}_2 = \text{ether}_{(\text{gas 34.6°C})}$

If you add these two equations by adding all the things on the left side of the equals sign and setting these equal to the sum of all the things on the right side of the equals sign, you have

ether$_{(liq.0°C)}$ + heat$_1$ + ~~ether$_{(liq.34.6°C)}$~~ + heat$_2$

= ~~ether$_{(liq.34.6°C)}$~~ + ether$_{(gas 34.6°)}$

Since ether$_{(liq. 34.6°C)}$ appears on both sides of the equal sign, it cancels and the sum of the two steps is really

ether$_{(liq. 0°C)}$ + heat$_1$ + heat$_2$ = ether$_{(gas 34.6°C)}$

The fact that you can determine the heat of a process by adding the heats of the steps that get you from the starting conditions to the final conditions is called *Hess's law*. Your process may not actually go through these steps, but if you can write a series of steps that, by addition, result in the process in which you are interested, then you can simply add the heats of these steps.

Here is another example of this. Say that you want to know the heat needed to convert one mole of ice at 0°C to water vapor at the same temperature. The process of going from the solid directly to the gas is called *sublimation*. You can write this process as two steps.

Step 1:   ice + heat$_1$ = water

Step 2:   water + heat$_2$ = water vapor

Adding both steps, you have

ice + heat$_1$ + ~~water~~ + heat$_2$ = ~~water~~ + water vapor

Since water appears on both sides of the equal sign, it cancels and you have for the total process

ice + _____ + _____ = water vapor   heat$_1$,   heat$_2$

This is the sublimation of ice. The values for heat$_1$ and heat$_2$ can be found in tables. They are 1.43 kcal/mole and 10.72 kcal/mole, respectively. Therefore you can write

ice + _____ + _____ = water vapor   1.43 kcal,   10.72 kcal

ice + _____ kcal = water vapor   12.15

You need 12.15 kcal to sublime one mole of water.

The use of *Hess's law* is going to be much more important in the next section, in which we consider the heat involved in chemical reactions.

SECTION C   Heat Involved in Chemical Reactions

Practically every chemical reaction either needs heat to occur or gives off heat as it proceeds. The amount of heat depends on how much material is reacting. It is most convenient to express this amount of heat as a conversion factor that has the dimensions

$$\frac{kcal}{mole}$$

endothermic

+ (positive)

exothermic, − (negative)

If the reaction requires heat, it is an _____ reaction and the sign on the heat is _____. If the reaction gives off heat, it is an _____ reaction and the sign on the heat is _____.

For example, the burning of coal is an exothermic reaction that gives off 94.1 kcal heat per mole of carbon burned. You would write its heat of reaction as

$$\Delta H^0 = \frac{94.1 \text{ kcal}}{\text{mole C}}$$

The symbol Δ is used because the heat involved in a reaction is a measure of the differences between the heats of the products and the heats of the reactants. It is a change in heat.

The ^{0} that follows the *H* indicates that the reaction is occurring at 25°C and that if gases are present, they are at 1.0 atm pressure. When you write the equation for the reaction, it is important that you note the phases, solid (*s*), liquid (*l*) or gas (*g*), since the heat is different if a change of phase is required. Thus, when you write the equation for the burning of coal (carbon) with oxygen, you have

$$C(s) + O_2(g) = CO_2(g), \quad \Delta H^0 = -94.1 \text{ kcal/mole}$$

You can actually write the heat in the equation as if it were a reactant or a product.

$$C(s) + O_2(g) = CO_2(g) + 94.1 \text{ kcal}$$

This shows that 94.1 kcal are produced. However, the convention is that you always show the heat term on the reactant side of the equation. You can get it there by subtracting 94.1 kcal from both sides of the equation.

$$C(s) + O_2(g) - 94.1 \text{ kcal} = CO_2(g) + 94.1 \text{ kcal} - 94.1 \text{ kcal}$$

or

$$C(s) + O_2(g) - 94.1 \text{ kcal} = CO_2(g)$$

This convention determines that the heat term in an exothermic

− (negative)

+ (positive)

process always has a _____ sign and the heat in an endothermic process has a _____ sign.

By writing the heat of reaction in the equation, you can easily solve problems that ask you to find the amount of heat involved in a reaction. You consider the heat as if it were one of the substances in the reaction, and you set up a "bridge" conversion factor relating heat to moles of the other components. Thus, in the example above, you can write

$$\frac{-94.1 \text{ kcal}}{\text{mole C}} \quad \text{or} \quad \frac{-94.1 \text{ kcal}}{\underline{\hspace{1cm}} \text{ } O_2} \quad \text{or} \quad \frac{\underline{\hspace{2cm}}}{\text{mole } CO_2}$$

94.1 kcal

mole

Question 6 on the pretest is a typical example of the type of problem that you may face. How much heat can you get from burning 4.80 kg of carbon The reaction is $C(s) + O_2(g) \rightarrow CO_2(g)$, with all substances at 25°C. The standard heat of reaction is

−94.1 kcal/mole.

As with normal stoichiometry problems, the first thing that you

must have is a _____ equation.

balanced

$$C(s) + O_2(g) - 94.1 \text{ kcal} = CO_2(g)$$

The question asked is "How many kcal (the most convenient unit

for heat)?" and the information given is "_____."

4.80 kg C

$$? \text{ kcal} = \underline{\hspace{2cm}}$$

4.80 kg C

To use the "bridge," you must convert kg C to _____. You can do this in two steps.

mole C

$$? \text{ kcal} = 4.80 \text{ kg C} \times \underline{\hspace{1.5cm}} \times \underline{\hspace{1.5cm}}$$

$$\frac{10^3 \text{ g C}}{\text{kg C}}, \frac{\text{mole C}}{12.0 \text{ g C}}$$

Now you can use the "bridge" conversion factor that relates the heat to moles C.

$$? \text{ kcal} = 4.80 \,\cancel{\text{kg C}} \times \frac{10^3 \,\cancel{\text{g C}}}{\cancel{\text{kg C}}} \times \frac{\text{mole C}}{12.0 \,\cancel{\text{g C}}} \times \underline{\hspace{1.5cm}}$$

$$\frac{-94.1 \text{ kcal}}{\text{mole C}}$$

$$= \underline{\hspace{2cm}}$$

−37,640 kcal

Written in scientific notation to the correct number of significant figures, this is

$$= \underline{\hspace{2cm}}$$

-3.76×10^4 kcal

You may have a problem in which this entire procedure is reversed. You know the amount of heat involved when a given weight of a reactant is used, and you are asked to determine ΔH^0.

Since ΔH^0 is a conversion factor, _____, you reword and ask

$\dfrac{\text{kcal}}{\text{mole}}$

"How many _____?" and take as your information given "_____."
This is our usual way to solve for conversion factors. Here is an example of this type of problem.

kcal,   1 mole

What is ΔH^0 for the reaction

$$PCl_5(g) = PCl_3(g) + Cl_2(g)$$

if 463 cal are needed to decompose 4.17 g of PCl_5? Since the problem states that the heat is "required," the sign on the heat

+ (positive)

+0.463 kcal

kcal, 1 mole PCl$_5$

g

$$\frac{208.5 \text{ g PCl}_5}{\text{mole PCl}_5}$$

$$\frac{+0.463 \text{ kcal}}{4.17 \text{ g PCl}_5}$$

+23.2 kcal

is _____. Also, since it is more convenient to express the heat

in kcal immediately, the heat would be _____.

   So, starting with the reworded question and the information given, you have

   ? _____ = _____

The problem tells you that you need 0.463 kcal for 4.17 g PCl$_5$.

Therefore you must convert mole PCl$_5$ to _____ PCl$_5$ by using the molecular weight, 208.5 g PCl$_5$ per mole PCl$_5$

   ? kcal = 1 mole PCl$_5$ × _____

Now you can use the conversion factor given in the problem, the number of kcal per g PCl$_5$

   ? kcal = 1 mole PCl$_5$ × $\dfrac{208.5 \text{ g PCl}_5}{\text{mole PCl}_5}$ × _____

   = _____ (to the correct number of significant figures)

And this is ΔH° for the reaction.

Adding Heats of Stepwise Reactions

In Section B, we introduced Hess's law for summing the heats of the steps that lead from starting conditions to final conditions. All the cases shown there were for only a single substance that changed temperature or phase. But Hess's law can be applied equally well to chemical reactions. In a way, this is quite the same thing as doing stoichiometry for consecutive reactions. If you don't remember this, refer to Section E, Chapter 9. There we covered two methods of doing this. In the first, we used two separate "bridges" from the two equations. In the second, we added the two consecutive reactions and then got the "bridge" from this composite equation. Only this last method works for thermodynamic equations because the heats are additive and are not multiplied together.

   Here is an example that will make this clear. Suppose that you wanted to find the amount of heat that you get from burning 4.80 kg of carbon in oxygen, to yield CO_2, but you know only the heats of reaction for the two reactions

$$C(s) + \tfrac{1}{2}O_2(g) - 26.4 \text{ kcal} = CO(g)$$

$$CO(g) + \tfrac{1}{2}O_2(g) - 67.7 \text{ kcal} = CO_2(g)$$

You can add these two steps by adding all the terms on the left side of the equals sign and setting this equal to the sum of everything on the right side of the equals sign. This gives you

$$C(s) + \tfrac{1}{2}O_2(g) - 26.4 \text{ kcal} + \cancel{CO(g)} + \tfrac{1}{2}O_2(g) - 67.7 \text{ kcal}$$

$$= \cancel{CO(g)} + CO_2(g)$$

Since $CO(g)$ appears on both sides of the equals sign, it cancels, and two $\tfrac{1}{2}O_2(g)$ is the same as just one $O_2(g)$, so you have as the sum

$$C(s) + O_2(g) - 94.1 \text{ kcal} = CO_2(g)$$

This is exactly the same thermodynamic equation that you had before when you were given the heat of reaction. The sum of the steps that make up a reaction gives you the heat according to

_____ law. Now you can solve the problem exactly as before.

$$? \underline{\quad} = 4.80 \text{ kg C}$$

$$= 4.80 \text{ kg C} \times \underline{\quad} \times \underline{\quad}$$

$$= 4.80 \cancel{\text{ kg C}} \times \frac{10^3 \cancel{\text{ g C}}}{\cancel{\text{kg C}}} \times \frac{\text{mole C}}{12.0 \cancel{\text{ g C}}} \times \underline{\quad}$$

$$= \underline{\quad\quad\quad}$$

Hess's

kcal

$\dfrac{10^3 \text{ g C}}{\text{kg C}}, \ \dfrac{\text{mole C}}{12.0 \text{ g C}}$

$\dfrac{-94.1 \text{ kcal}}{\text{mole C}}$

$-3.76 \times 10^4 \text{ kcal}$

It may look a little strange to you to have the balanced equations with $\tfrac{1}{2}O_2$. Up to this time, we have never used fractional coefficients to balance chemical equations. However, in equations that show the heat (thermodynamic equations), it is much more convenient to have the heat expressed per one mole of the substance in which you are interested. For example, if you are interested in the amount of heat that is given off when $HCl(g)$ is formed from its elements, you write the thermodynamic equation as

$$\tfrac{1}{2}H_2(g) + \tfrac{1}{2}Cl_2(g) - 22.1 \text{ kcal} = HCl(g)$$

The value -22.1 kcal represents the molar heat of formation for $HCl(g)$ at 25°C. You use the symbol ΔH_f^0 for molar heats of formation. We will soon see how very useful the values for ΔH_f^0 can be.

Sometimes when you use a series of steps to get the total reaction in which you are interested, you may find that you have to write one of the steps in reverse. For example, say that you are interested in the reaction.

$$2Al(s) + Fe_2O_3(s) = 2Fe(s) + Al_2O_3(s)$$

The steps whose heats of reaction you know are:

Step 1: $2Al(s) + \tfrac{3}{2}O_2(g) - 388 \text{ kcal} = Al_2O_3(s)$

Step 2: $2Fe(s) + \tfrac{3}{2}O_2(g) - 196 \text{ kcal} = Fe_2O_3(s)$

These two steps have all the substances that appear in the equation for the reaction in which you are interested. But notice that the $Fe_2O_3(s)$ appears on the left side in the equation of interest, and on the right side in Step 2. If you are going to add Step 1 to Step 2 to get the equation that you want, you have to write Step 2 in reverse. Once you have done this, you can add them together.

Step 1: $2Al(s) + \frac{3}{2}O_2(g) - 388 \text{ kcal} = Al_2O_3(s)$

Step 2: $Fe_2O_3(s) = -196 \text{ kcal} + \frac{3}{2}O_2(g) + 2Fe(s)$

Adding up everything on the left, you have

$$2Al(s) + \frac{3}{2}O_2(g) - 388 \text{ kcal} + Fe_2O_3(s)$$

and adding up everything on the right, you have

$$Al_2O_3(s) - 196 \text{ kcal} + \frac{3}{2}O_2(g) + 2Fe(s)$$

Since the $\frac{3}{2}O_2(g)$ appears on both the left and the right sides, it cancels, and you can write

$$2Al(s) - 388 \text{ kcal} + Fe_2O_3(s) = Al_2O_3(s) - 196 \text{ kcal} + Fe(s)$$

To get all the kcal on the left, you add $+196$ kcal to both sides, and you have

$$2Al(s) + Fe_2O_3(s) - 388 \text{ kcal} + 196 \text{ kcal} = Al_2O_3(s) + Fe(s)$$

If you algebraically add the -388 kcal and the $+196$ kcal, you have the final thermodynamic equation

-192

$$2Al(s) + Fe_2O_3(s) \underline{\hspace{1cm}} \text{kcal} = Al_2O_3(s) + 2Fe(s)$$

You now know the heat of the reaction for the initial reaction in which you were interested.

Let's try another one of these with the reactions of C with O_2. This should make the whole process very clear. Say that you want to know the heat of reaction for

$$C(s) + \frac{1}{2}O_2(g) = CO(g)$$

and the steps for which you have values are:

Step 1: $C(s) + O_2(g) - 94.1 \text{ kcal} = CO_2(g)$

Step 2: $CO(g) + \frac{1}{2}O_2(g) - 26.4 \text{ kcal} = CO_2(g)$

In the reaction in which you are interested, the $CO(g)$ is on

right

the \underline{\hspace{1cm}} side of the equal sign. In Step 2, the $CO(g)$ is on

left

the \underline{\hspace{1cm}} side of the equal sign. Therefore, before you add the

reverse

steps, you must write Step 2 in its \underline{\hspace{1cm}} form.

Step 1: $C(s) + O_2(g) - 94.1 \text{ kcal} = CO_2(g)$

Step 2: $CO_2(g) = -26.4 \text{ kcal} + \frac{1}{2}O_2(g) + CO(g)$

Now you can add these steps

$$C(s) + O_2(g) - 94.1 \text{ kcal} + CO_2(g)$$

$$= CO_2(g) - 26.4 \text{ kcal} + \underline{\hspace{2cm}} + CO(g)$$ $\frac{1}{2}O_2(g)$

The $CO_2(g)$ appears on both sides, and so it $\underline{\hspace{1.5cm}}$. If you   cancels

subtract $\frac{1}{2}O_2(g)$ from both sides and add $\underline{\hspace{2cm}}$ to both sides,   26.4 kcal
what remains is

$$C(s) + \frac{1}{2}O_2(g) \underline{\hspace{2cm}} = CO(g)$$ -67.7 kcal

This is the thermodynamic equation in which you are interested.
    From all these considerations, you can see that to get the
thermodynamic equations that allow you to solve a numerical
problem, you may have to add several steps whose thermodynamic
equations you know. And you can reverse any steps to get to
the reaction of interest. You may even have to multiply some of the
steps to give you the correct balanced equation of interest. You
may even need more than just two steps. Consider the following
example.

    The equation of interest is the burning of CH_4, methane.

$$CH_4(g) + 2O_2(g) = CO_2(g) + 2H_2O(l)$$

    The only thermodynamic equations that you have are the heats
of formation of the various compounds from their elements. Recall
that we said earlier that these ΔH_f^0 values would be very useful.

Step 1: $C(s) + 2H_2(g) - 17.9 \text{ kcal} = CH_4(g)$

Step 2: $C(s) + O_2(g) - 94.1 \text{ kcal} = CO_2(g)$

Step 3: $H_2(g) + \frac{1}{2}O_2(g) - 68.3 \text{ kcal} = H_2O(l)$

When you compare these steps with the equation of interest, you
see that the CH_4 is on the left side of the equal sign in the

equation of interest, but on the $\underline{\hspace{1.5cm}}$ side in Step 1. Therefore,   right

when you add the steps, you must $\underline{\hspace{1.5cm}}$ the form of Step 1.   reverse

Also, in the balanced equation of interest there are $\underline{\hspace{1cm}}$ H_2O's,   two
but in Step 3, there is only one. Consequently, you must multiply

Step 3 by $\underline{\hspace{0.7cm}}$ all the way across. This means the heat term will   2

be $\underline{\hspace{2cm}}$ kcal for two times Step 3. Now you can write -136.6
the steps in the form that adds up to the equation of interest.

Step 1: $CH_4(g) = \underline{\hspace{2cm}} + 2H_2(g) + C(s)$ -17.9 kcal

Step 2: $C(s) + O_2(g) - 94.1 \text{ kcal} = CO_2(g)$

Step 3: $2H_2(g) + O_2(g) \underline{\hspace{1.5cm}} \text{kcal} = 2H_2O(l)$ -136.6

Now you add all terms on the left side of the equals sign.

Step 1 : $CH_4(g) =$

-94.1	Step 2 : $C(s) + O_2(g)$ _____ kcal =
-136.6	Step 3 : $2H_2(g) + O_2(g)$ _____ kcal =

$$CH_4(g) + C(s) + 2H_2(g) + 2O_2(g) \text{ _____ kcal} =$$

Next you add all the terms on the right side of the equal sign.

-230.7	
-17.9	Step 1 : = _____ kcal + $2H_2(g) + C(s)$
	Step 2 : = $CO_2(g)$
$2H_2O(l)$	Step 3 : = _____

$$= -17.6 \text{ kcal} + 2H_2(g) + C(s) + CO_2(g) + \text{_____}$$

$2H_2O(l)$	As you can see, you have one $C(s)$ and two $H_2(g)$ on the left
cancel	and on the right side. Therefore they _____. Thus the total equation is

$$CH_4(g) + 2O_2(g) - 230.7 \text{ kcal}$$

-17.9	$= \text{_____ kcal} + CO_2(g) + 2H_2O(l)$

Adding $+17.9$ kcal to both sides of the equation brings all the kcal terms together. So as the final equation you have

-212.8	$CH_4(g) + 2O_2(g) \text{ _____ kcal} = CO_2(g) + 2H_2O(l)$

This is your equation of interest.

Using Heats of Formation, ΔH_f^0

As mentioned at the beginning of the last problem, each of the steps used to build up the total equation are simply the heats of formation of the compounds from their elements. These standard heats of formation have the abbreviation ΔH_f^0. You used the thermodynamic equation for the formation of the compounds on the left side of the equals sign, the *reactants*, as it stands. But for the compounds on the right side of the equals sign, the *products*, you had to reverse the equation for the heat of formation. The mathematical effect of reversing the thermodynamic equation is to change the sign of the heat. Also, when the compound in Step 3 appeared with a coefficient of 2 in the equation of interest [it was $2H_2O(l)$], you had to double the thermodynamic equation. This resulted in the heat being doubled for that step.

What all this means is that a very neat short cut is available if you know the heats of formation of all the compounds in an equation. All you have to do is write the equation and place the heats of formation beneath the compound. If there is a coefficient on a substance, you multiply the heat of formation by that coefficient. Then you add all the heats of formation of the products on the right side of the equals sign, and you subtract from this the sum of all the heats on the left side of the equals sign.

Let's look at this last example with the heats of formation written below each substance.

$$CH_4(g) \quad + O_2(g) = \quad CO_2(g) \quad + \quad 2H_2O(l)$$
$$(-17.9 \text{ kcal}) \quad (O) \quad (-94.1 \text{ kcal}) \quad 2(-68.3 \text{ kcal})$$

Notice that the element oxygen has a heat of formation of zero. This entire procedure is based on the fact that the standard heat of formation of all elements is zero.

Adding all the heats on the products side of the equation algebraically, you have

$$-94.1 \text{ kcal} + (\underline{\hspace{1cm}}) \text{ kcal} = \underline{\hspace{1cm}} \text{ kcal}$$

−136.6, −230.7

On the reactant side of the equation, you have simply

$$-17.9 \text{ kcal} + (\underline{\hspace{1cm}}) \text{ kcal} = \underline{\hspace{1cm}} \text{ kcal}$$

0, −17.9

If you subtract the reactant sum from the product sum, you have

$$-230.7 \text{ kcal} - (-17.9 \text{ kcal}) = \underline{\hspace{1cm}} \text{ kcal}$$

−212.8

which is exactly the same value that we got before.

To give you a little more practice, let's look again at an earlier example.

$$2Al(s) + Fe_2O_3(s) = 2Fe(s) + Al_2O_3(s)$$

The heats of formation that we used were

$$Fe_2O_3(s) \quad \Delta H_f^0 = -196 \text{ kcal/mole}$$

$$Al_2O_3(s) \quad \Delta H_f^0 = -388 \text{ kcal/mole}$$

You don't consider $Al(s)$ or $Fe(s)$ because they are _____ and

elements

their heats of formation are _____.

zero

Writing these values beneath the compounds in the equation, you have

$$2Al(s) \quad + \quad Fe_2O_3(s) \quad = \quad 2Fe(s) \quad + \quad Al_2O_3(s)$$
$$2(O) \text{ kcal} \quad (-196 \text{ kcal}) \quad 2(O) \text{ kcal} \quad (-388 \text{ kcal})$$

If you subtract the sum of the heats of formation of the reactants from that of the products, you have

$$\underline{\hspace{1cm}} \text{ kcal} - (\underline{\hspace{1cm}}) \text{ kcal} = -192 \text{ kcal}$$

−388, −196

This is exactly the same value that we got before. When you write the thermochemical equation, be very careful to put the heat on

the reactant, _____, side.

left

Here is question 7 on the pretest. What is the heat of reaction for the process

$$FeS(s) + 2HCl(aq) = FeCl_2(aq) + H_2S(g)$$

The notation (aq) is an abbreviation for aqueous, which means as a dilute solution in water. The heats of formation are $FeS(s) = -22.7$ kcal/mole, $HCl(aq) = -39.8$ kcal/mole, $FeCl_2(aq) = -100.4$ kcal/mole, and $H_2S(g) = -4.8$ kcal/mole.

To get the heat of reaction for the process, you write the various

balanced

heats of formation underneath the compounds in the _____ equation.

$$FeS(s) \quad + \quad 2HCl(aq) \quad = \quad FeCl_2(aq) \quad + \quad H_2S(g)$$

2(−39.8 kcal)

$$(-22.7 \text{ kcal}) \quad \underline{\hspace{2cm}} \quad (-100.4 \text{ kcal}) \quad (-4.8 \text{ kcal})$$

add

Then you _____ the heats of formation on the right (product) side

$$-100.4 \text{ kcal} + (-4.8 \text{ kcal}) = \underline{\hspace{2cm}}$$

−105.2 kcal

Next you add the heats of formation of the reactants on the

left

_____ side of the equals sign:

$$-22.7 \text{ kcal} + (\underline{\hspace{1cm}}) \text{ kcal} = \underline{\hspace{1cm}} \text{ kcal}$$

−79.6, −102.3

subtract

Then you must _____ the sum of the heats of the reactants

products

from the sum of the heats of the _____.

$$-105.2 \text{ kcal} - (-102.3 \text{ kcal}) = \underline{\hspace{1cm}} \text{ kcal}$$

−2.9

You now know that the heat of this reaction is −2.9 kcal

There are tables available that give the standard heats of formation of many compounds. The following relatively short table contains all the values that you will need for any problems in this text.

Standard Heats of Formation in Kcal/Mole at 25°C and 1 Atm

Any element in standard state	0.0	$C_4H_{10}(g)$	−29.8	$Na^+(aq)$	−57.3
		$C_6H_6(l)$	11.7	$NH_3(g)$	−11.0
		$C_6H_6(g)$	19.8	$NH_4^+(aq)$	−31.7
$Ag^+(aq)$	25.3	$Cu^{2+}(aq)$	15.5	$NO(g)$	21.6
$AgCl(s)$	−30.4	$F^-(aq)$	−78.7	$NO_2(g)$	8.1
$Al_2O_3(s)$	−388.	$FeO(s)$	−63.7	$NO_3^-(aq)$	−49.4
$Br^-(aq)$	−28.9	$Fe_2O_3(s)$	−196.5	$N_2O(g)$	19.5
$Ca^{2+}(aq)$	−127.8	$H^+(aq)$	0.0	$OH^-(aq)$	−55.0
$Cl^-(aq)$	−40.0	$HCl(g)$	−22.1	$PCl_3(g)$	−73.2
$CO(g)$	−26.4	$HCl(aq)$	−39.8	$PCl_5(g)$	−95.4
$CO_2(g)$	−94.1	$HI(g)$	6.2	$S^{2-}(aq)$	10.0
$CH_4(g)$	−17.9	$H_2O(l)$	−68.3	$SO_2(g)$	−71.0
$CH_3OH(l)$	−57.0	$H_2O(g)$	−57.8	$SO_3(g)$	−95.1
$CH_3OH(g)$	−48.1	$H_2S(g)$	−4.8	$SO_4^{2-}(aq)$	−216.9
$C_2H_5OH(l)$	−66.4	$I^-(aq)$	−13.4	$Zn^{2+}(aq)$	−36.4
$C_2H_5OH(g)$	−56.2	$K^+(aq)$	−60.0		

SECTION D Calorimetry

So far we have been talking about the heats that are given off and the heats that are consumed, but we never mentioned just how these heats were measured in the laboratory. Actually, it is quite simple. You just have to have the process in which you are interested occur in something whose specific heat is known. Water is used most often for this, but other liquids can be used just as well. If you know the amount of water and can measure how much the temperature changes as a result of the process occurring in the water, you can calculate the heat change of the water. Since heat is conserved in the same way that matter is conserved, the heat that the water has gained or lost must have come from your process having lost or gained an equivalent amount of heat. Thus:

Heat gained by water = heat lost by process

Heat lost by water = heat _____ by process   gained

The apparatus for measuring this change in temperature is called a *calorimeter*. It is important for the calorimeter body to be well insulated so that very little heat escapes from the calorimeter. There are many types of calorimeter, but the simplest version and the one that you are most likely to encounter is just a styrofoam cup in which a thermometer is suspended.

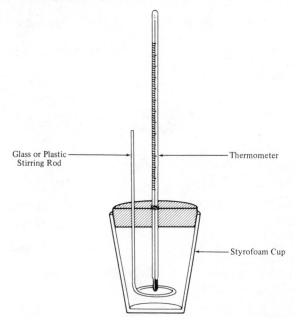

Glass or Plastic Stirring Rod — Thermometer — Styrofoam Cup

To measure the amount of heat gained or lost, you place a weighed amount (or a known volume) of water in the calorimeter and record its temperature. Then you allow the process in which you are interested to take place in the water. After the process is complete (indicated by no further change in the temperature of the water), you record the final temperature of the water. Subtracting the _____ temperature of the water from the _____ temperature *initial, final* gives you ΔT. Knowing the weight of the water, its specific heat, and ΔT lets you calculate the amount of heat either lost or gained by the calorimeter. We covered this in Section A of this chapter.

The symbol Q is frequently used to represent the heat. The idea of balancing the change in the heat of the process with that in the calorimeter can be shown in an equation,

$$Q_{calor.} = -Q_{process}$$

where $Q_{calor.}$ is the heat change in the calorimeter and $Q_{process}$ is the heat change of the process. $Q_{process}$ has a negative sign because, if the process loses heat, the calorimeter _____ an *gains* equivalent amount of heat.

Question 8 on the pretest is a good example of how this works. What is the specific heat of aluminum if, when a 10.0 g sample of the metal is heated to 102.25°C and placed in a calorimeter containing 25.0 g water at 20.50°C, the temperature of the water rises to 27.25°C? Assume that the calorimeter body does not absorb any heat.

To determine the heat change in the calorimeter, you must know three things: the weight of water, the specific heat of water, and ΔT, the temperature change of the water. The weight of water is given in the problem as _____. The specific heat of water *25.0 g* is known to be _____. So all you must do is calculate $\dfrac{1\ cal}{(g\ water)\,(°C)}$ ΔT. You do this by _____ $T_{initial}$ from T_{final}. *subtracting*

$$\Delta T = T\rule{2cm}{0.4pt} - T\rule{2cm}{0.4pt}$$ *final, initial*

In this problem, the calculation is

$$\Delta T = \rule{1.5cm}{0.4pt}°C - \rule{1.5cm}{0.4pt}°C$$ *27.25, 20.50*

$$= \rule{1.5cm}{0.4pt}°C$$ *6.75*

If you follow the methods you used in Section A of this chapter for determining the heat involved in a temperature change, you start by asking "How many _____?" and take as the *calories* information given the _____ of the substance. So in this problem *weight* you would start with

$$?\ cal = \rule{2.5cm}{0.4pt}$$ *25.0 g water*

To cancel the unwanted dimension "g water", you use the _____ of water. *specific heat*

$$?\ cal = 25.0\ g\ water \times \rule{2.5cm}{0.4pt}$$ $\dfrac{1\ cal}{(g\ water)\,(°C)}$

Finally, to cancel the unwanted °C, you multiply by _____, which ΔT
is 6.75°C.

$$? \text{ cal} = 25.0 \,\cancel{\text{g water}} \times \frac{1 \text{ cal}}{(\cancel{\text{g water}}) \,(°C)} \times \underline{\qquad}$$ 6.75°C

$$= \underline{\qquad}$$ 169 cal

The heat gained in the calorimeter, $Q_{calor.}$, must equal the heat
lost by the aluminum, $Q_{process}$. But since the calorimeter gains

heat, its calories have a ____ sign, and the aluminum must lose   +

heat, so its calories will have a ____ sign.   −

$$Q_{calor.} = \underline{\qquad} Q_{process}$$ −

So from the temperature change in the calorimeter, you have
determined that the number of calories the aluminum has lost is

_____. To answer the question "What is the _____ heat   −169 cal,   specific
of Al?" you must know three things: the number of calories, the

_____ of the aluminum, and its ____. As in the preceding   weight, ΔT

example, you can get ΔT by subtracting the _____ temperature   initial

from the _____ temperature. The initial temperature of the   final

aluminum is given in the problem as _____ °C, and its final   102.25

temperature is the same as the water in the calorimeter, _____ °C.   27.25

$$\Delta T = T\underline{\qquad} - T\underline{\qquad}$$ final,   initial

$$= \underline{\qquad}°C - \underline{\qquad}°C$$ 27.25,   102.25

$$= \underline{\qquad}°C$$ −75.00

(Notice the − sign, which shows that the temperature has decreased.)
You now have all the information you need to get the specific
heat of the aluminum.

_____ cal   −169

_____ g Al   10.0

_____ °C = ΔT −75.00

We can use the shortcut way of getting the specific heat, as we
did in Section A. We simply noted the dimensions of specific
heat are

$$\frac{\underline{\qquad}}{(\underline{\quad})\,(\underline{\quad})}$$ cal

g,   °C

and put in our values in such a way as to match these dimensions. Thus we have

$$-169 \text{ cal}$$
$$10.0 \text{ g}, \quad -75.00°C$$
$$\frac{0.225 \text{ cal}}{\text{g} °C}$$

specific heat $= \dfrac{\underline{\hspace{2cm}}}{(\underline{\hspace{1.5cm}}) \ (\underline{\hspace{1.5cm}})}$

$$= \underline{\hspace{2cm}}$$

(Notice that the two − signs have cancelled. The specific heat is always a + number.)

Calculating the Heat Capacity of the Calorimeter, K_c

Now we are going to introduce a new term, *heat capacity*. This is the general term for the heat required to raise the temperature of a substance 1°C. If the heat capacity is expressed per gram of substance, it is called *specific heat*. If it is expressed per mole of substance, it is called the *molar hear capacity*. If the substance has only one possible weight, you just say *heat capacity*. Since each calorimeter has a weight that doesn't change (and also loses the same amount of heat to the surroundings), you can simply say *heat capacity of the calorimeter*. Its value is a constant, and the symbol K_c is used. It has as its dimensions

$$\frac{\text{cal}}{°C}$$

Therefore, if you want to know how much heat the calorimeter absorbs and gives off to its surroundings, you have to multiply the value of K_c by ΔT the change in temperature of the calorimeter. This amount of heat must be added to the heat gained or lost by the water contained in the calorimeter to give the total heat change of the calorimeter.

$$Q_{\text{calor.}} = Q_w + K_c \times \Delta T = Q_w + Q_c$$

The abbreviation Q_c is used to represent the heat gained by the calorimeter and lost to the surroundings.

Question 8 on the pretest told you to assume that the body of the calorimeter holding the water didn't absorb or give off any heat. Also, it implied that there was no heat lost to or gained from the surroundings. If you are working in the laboratory with a thermometer that allows you to read temperatures only to ±0.3°C, and you are using a styrofoam cup to hold the water, you can probably assume this. The styrofoam absorbs or loses heat very slowly, so it provides excellent insulation from the surroundings. Any differences caused by these changes would be too small to read on your thermometer. However, if you are doing high precision work using a thermometer that reads to ±0.02°C and you are using a calorimeter made of metal, you are going to have to worry about the heat gained or lost by the calorimeter itself. Then you'll have to use the equation that was shown above:

$$Q_{\text{calor.}} = Q_w + \underline{\qquad} \times \Delta T$$

$$= \underline{\qquad} + \underline{\qquad}$$

K_c

$Q_w,\ Q_c$

Then, when you write the equation used in calorimetry,

$$Q_{\text{calor.}} = -Q_{\text{process}}$$

you have to write

$$Q_w + Q_c = -Q_{\text{process}}$$

Or, expressing Q_c in terms of the calorimeter constant K_c, you have

$$Q_w + (\underline{\qquad}) = -Q_{\text{process}}$$

$K_c \times \Delta T$

This last equation gives you a way to determine the K_c for a calorimeter. What you do is perform some process in the calorimeter whose heat is known. Thus you know Q_{process}. You measure the ΔT for the water in the calorimeter, and from the weight of the water and the specific heat of water, you calculate Q_w. Then, if you subtract Q_w from both sides of the equation,

$$Q_w + Q_c = -Q_{\text{process}}$$

you have

$$\underline{\qquad} = -Q_{\text{process}} - Q_w.$$

Q_c

Since $Q_c = \underline{\qquad} \times K_c$, you can calculate K_c by $\underline{\qquad}$ both

ΔT, dividing

sides of this equation by $\underline{\qquad}$.

ΔT

$$\frac{Q_c}{\underline{\qquad}} = K_c$$

ΔT

It is quite correct to assume that the ΔT for the calorimeter is identical to the ΔT for the water that it contains. Therefore you get the values for ΔT by subtracting the initial temperature of the water in the calorimeter from its final temperature.

Here is an example that shows how this is done experimentally. The process that produces a known amount of heat is the cooling down of a piece of aluminium. You can calculate the heat that this

process produces if you know three things: the $\underline{\qquad}$ of

weight

aluminum, the specific heat of aluminum, and the $\underline{\qquad}$ in its temperature.

change

In this example, the Al weighs 10.0 g, and its specific heat is

known to be 0.225 $\underline{\qquad}$. Its initial temperature is 10.25°C, and it cools in the calorimeter until the temperature of the water in the calorimeter is 27.05°C. This last temperature is the final temperature of the aluminum, the water in the calorimeter, and the

$\dfrac{\text{cal}}{(\text{g Al}) (°\text{C})}$

calorimeter body. To get ΔT for the aluminium, you subtract its

initial

_____ temperature from its final temperature

initial

$$\Delta T = T_{final} - T\text{_____}$$

27.05, 102.25

$$= \text{_____}°C - \text{_____}°C$$

-75.20

$$= \text{_____}°C$$

From these three pieces of information you can calculate the heat $Q_{process}$ given off by the aluminum.

10.0 g Al

$$?\ Q_{process} = ?\ cal = \text{_____}$$

$\dfrac{0.225\ cal}{(g\ Al)\ (°C)}$

$$= 10.0\ g\ Al \times \text{_____}$$

$-75.20°C$

$$= 10.0\ g\ Al \times \frac{0.225\ cal}{(g\ Al)\ (°C)} \times \text{_____}$$

$-169\ cal$

$$= \text{_____}$$

The next thing that you must determine is Q_w, the heat absorbed by the water in the calorimeter. You need to know three things: the

specific heat

weight of the water, the _____ of water, and the ΔT of the water. In this example, the water weighs 25.0 g, and its

$\dfrac{1\ cal}{(g\ water)\ (°C)}$

specific heat is _____. The initial temperature is 20.50°C, and its final temperature is 27.05°C. The ΔT for water is

27.05, 20.50

$$\Delta T = \text{_____}°C - \text{_____}°C$$

6.55

$$= \text{_____}°C$$

With these three pieces of information you can calculate Q_w.

25.0 g water

$$?\ Q_w = ?\ cal = \text{_____}$$

$\dfrac{1\ cal}{(g\ water)\ (°C)}$

$$= 25.0\ g\ water \times \text{_____}$$

$6.55°C$

$$= 25.0\ g\ water \times \frac{1\ cal}{(g\ water)\ (°C)} \times \text{_____}$$

164 cal

$$= \text{_____}$$

You now have $Q_{process}$ and Q_w, and you can calculate Q_c from the equation

$$Q_c = -(Q_{process}) - Q_w$$

$-(-169),\ 164$

$$= \text{_____}cal - \text{\_\_\_\_\_}cal$$

$+5$

$$= \text{\_\_\_\_\_}cal$$

[If you have forgotten the sign convention, it is that $-(-)$ and $+(+)$ both give $+$. But $-(+)$ or $+(-)$ both give $-$.]

It is now possible to calculate K_c from the equation

$$K_c = \frac{Q_c}{\underline{\hspace{1.5em}}}$$

The value for ΔT is the same that it was for the water in the

calorimeter, which was _____. Putting in the values for Q_c and
ΔT gives you

$$K_c = \frac{\overline{\hspace{1.5em}}}{\underline{\hspace{1.5em}}}$$

$$= \underline{\hspace{1.5em}} \quad \text{cal/°C (to the correct number of significant figures)}$$

Notice that the sign of K_c is $+$. Just as specific heat is always
positive, any heat capacity is positive. Once you have determined
K_c for a calorimeter, you can use it any time you are using that
particular calorimeter.

Using the Heat Capacity of the Calorimeter in Calculations

Question 9 of the pretest is an example of how you can calculate
the heat of a process using a calorimeter whose heat capacity
is known.

What is the heat of fusion for water in kcal per mole if 25.0 g of
ice at 0.00°C that is put into a calorimeter containing 300 g of
water at 21.16°C melts the ice and lowers the temperature of the
water in the calorimeter to 13.40°C? The heat capacity of the
calorimeter, K_c, equals 0.80 cal/°C.

As you did in pretest question 8, you must determine the heat
change of the calorimeter. But in this case, it will be the sum of

Q_w and _____. In Question 8, we assumed that Q_c was equal to
zero.

$$Q_{\text{calor.}} = Q_w + \underline{\hspace{2em}}$$

You have all the information that you need to calculate Q_w. You

have the weight of the water, _____ g, you know the specific heat

of water, _____, and you can calculate ΔT of the water

by subtracting the _____ temperature of the water from the

_____ temperature of the water.

$$\Delta T = \underline{\hspace{2em}}°C - \underline{\hspace{2em}}°C$$

$$= \underline{\hspace{2em}}°C$$

Margin answers:

ΔT

6.55°C

$+5$

6.55

0.8

Q_c

Q_c

300

$\dfrac{1 \text{ cal}}{(\text{g water)(°C)}}$

initial

final

13.40,   21.16

-7.76

Thus

$$\dfrac{1 \text{ cal}}{\text{(g water) (°C)}}$$

$$?Q_w = ? \text{ cal} = 300 \text{ g water} \times \underline{\hspace{2cm}}$$

$-7.76°\text{C}$

$$= 300 \text{ g water} \quad \dfrac{1 \text{ cal}}{\text{(g water) (°C)}} \times \underline{\hspace{1.5cm}}$$

-2328

$$= \underline{\hspace{1.5cm}} \text{ cal}$$

To get Q_c, you use the heat capacity of the calorimeter

$$? Q_c = ? \text{ cal} = \dfrac{0.80 \text{ cal}}{°\text{C}}$$

To cancel the unwanted °C, you use the ΔT for the calorimeter, which is the same as the ΔT for the water that it contains,

-7.76

$$\underline{\hspace{1.5cm}} °\text{C}.$$

$-7.76°\text{C}$

$$= \dfrac{0.80 \text{ cal}}{°\text{C}} \times \underline{\hspace{1.5cm}}$$

-6.2 cal

$$= \underline{\hspace{1.5cm}}$$

You now have Q_w and Q_c, so you can get $Q_{\text{calor.}}$.

$+$

$$Q_{\text{calor.}} = Q_w \underline{\hspace{0.6cm}} Q_c$$

$-2328, \ -6.2$

$$= \underline{\hspace{1.5cm}} \text{ cal} + \underline{\hspace{1cm}} \text{ cal}$$

-2334

$$= \underline{\hspace{1.5cm}} \text{ cal}$$

Once you know the heat given off by the calorimeter, you can tell the heat picked up by the process.

$-Q_{\text{process}}$

$$Q_{\text{calor.}} \quad = \underline{\hspace{2cm}},$$

$+2334 \text{ cal}$

$$\underline{\hspace{2cm}} = Q_{\text{process}}$$

The process here is one of the two-step processes that we discussed at the end of Section B in this chapter. Here you first change the phase of the substance (ice to water), and then you raise its temperature. By Hess's law, the heat for the total process

sum

is the $\underline{\hspace{1cm}}$ of the heats of the individual steps.

$$Q_{\text{process}} = Q_{\text{fusion}} + Q_{\text{temp. change}}$$

You already know Q_{process}, and it is very easy to calculate

$Q_{\text{temp. change}}$

$Q_{\text{temp. change}}$ for water. So to get Q_{fusion}, you subtract $\underline{\hspace{2cm}}$ from Q_{process}.

$Q_{\text{temp. change}}$

$$Q_{\text{process}} - \underline{\hspace{2cm}} = Q_{\text{fusion}}$$

To calculate the heat involved in the temperature change, you need three pieces of information: the weight of the water that has

25.0

come from the ice, $\underline{\hspace{1cm}}$g, the specific heat of water, and the

_____ of the water, The ice, which has liquefied, is at _____°C, and its temperature rises to the final temperature of the calorimeter,

_____°C. So

$$\Delta T = \underline{\quad\quad}°C - \underline{\quad\quad}°C$$

$$= \underline{\quad\quad}°C$$

Therefore

$$? Q_{temp.\,change} = ? \text{ cal} = \underline{\quad\quad}$$

$$= 25.0 \text{ g water} \times \underline{\quad\quad\quad}$$

$$= 25.0 \text{ g water} \times \frac{1 \text{ cal}}{(\text{g water}) (°C)} \times \underline{\quad\quad}$$

$$= \underline{\quad\quad} \text{ cal}$$

Now you can calculate Q_{fusion}, the heat from the melting of ice at 0°.

$$Q_{process} \underline{\quad\quad\quad\quad} = Q_{fusion}$$

$$\underline{\quad\quad} \text{ cal} - \underline{\quad\quad} \text{ cal} =$$

$$\underline{\quad\quad\quad\quad} =$$

But the value you have calculated is the heat required to melt 25.0 g of ice at 0.00°C. The question asked for the value per mole. This is a very simple dimensional-analysis problem.

$$? \text{ cal} = \underline{\quad\quad}$$

You can use the _____ weight to get from moles to grams.

$$? \text{ cal} = 1 \text{ mole water} \times \underline{\quad\quad\quad\quad}$$

And the preceding calculation has given you a conversion factor that relates calories to grams of water, _____.

$$? \text{ cal} = 1 \text{ \sout{mole water}} \times \frac{18.0 \text{ g water}}{\text{\sout{mole water}}} \times \underline{\quad\quad\quad\quad}$$

$$= \underline{\quad\quad} \text{ cal} = \underline{\quad\quad} \text{ kcal}$$

(It is best to use kcal and get the correct number of significant figures).

This is about as far as we will go with heats that are involved in chemical and physical processes. There is a great deal more to the story, but this is a good start for you. One word of caution about calorimetry. We have had only pure water as the liquid in our calorimeter, and we know its specific heat. However, if you have some process that occurs in the calorimeter which involves a substance that dissolves in the water, the specific heat of the solution that is formed will be different from that of pure water. When this comes up, you will be told the specific heat of the solution.

Right margin answers:

ΔT, 0.00

13.40

13.40, 0.00

13.40

25.0 g

$\dfrac{1 \text{ cal}}{(\text{g water}) (°C)}$

13.40°C

335

$- Q_{temp.\,change}$

2334, 335

1999 cal

1 mole

molecular

$\dfrac{18.0 \text{ g water}}{\text{mole water}}$

$\dfrac{1999 \text{ cal}}{25.0 \text{ g water}}$

$\dfrac{1999 \text{ cal}}{25.0 \text{ g water}}$

1439, 1.4

PROBLEM SET

1. How many calories are needed to raise the temperature of 500 g of water from 22.15°C to 24.70°C?

2. How many calories are released when 30.0 ml of chloroform cool 18.0°C? The specific heat of chloroform is 0.232 cal/g°C, and its density is 1.50 g/ml.

3. The specific heat of benzene is 0.438 cal/g°C. Suppose you remove 350 calories from 20.0 g of benzene at 45.0°C. What does its temperature become?

4. What is the specific heat of Cu metal if 15.3 cal are needed to raise the temperature of 15.0 g of Cu metal from 22.0°C to 33.0°C?

5. How many kilocalories are needed to vaporize 85.0 g of water at 100°C to steam at 100°C? The molar heat of vaporization of water is 9.71 kcal/mole.

6. How many kcal are needed to change 5.00 g of water at 25.0°C to steam at 100.0°C? The molar heat of vaporization is 9.71 kcal/mole.

*7. How many calories are required to convert 10.0 g of solid ethyl alcohol at −180.3°C to its vapor at its boiling point of 78.3°C? The melting point of solid ethyl alcohol is −117.3. The specific heat of solid ethyl alcohol is 0.232 cal/g°C, and the specific heat of liquid ethyl alcohol is 0.550 cal/g °C. The heat of fusion is 52.1 cal/g, and the heat of vaporization is 204 cal/g.

8. The heat of reaction for the burning of ethene (C_2H_4),

$$C_2H_4(g) + 3O_2(g) = 2CO_2(g) + 2H_2O(l),$$

is −337 kcal/mole. How many kilocalories result from burning 450 g of ethene?

9. Use the table of Standard Heats of Formation on page 344 to determine the ΔH_f^0 for ethene from the equation and the data in question 8.

10. Determine the heat of reaction for

$$CuCl(s) + \tfrac{1}{2}Cl_2(g) = CuCl_2(s)$$

by using the following two reactions.

$$Cu(s) + Cl_2(g) - 49.2 \text{ kcal} = CuCl_2(s)$$

$$Cu(s) + \tfrac{1}{2}Cl_2(g) - 32.4 \text{ kcal} = CuCl(s)$$

11. Use the table of Standard Heats of Formation on page 344 to determine the heat of reaction for

$$3NO_2(g) + H_2O(l) = 2HNO_3(aq) + NO(g)$$

To get the ΔH_f^0 for $HNO_3(aq)$, you must use the values for

$$H^+(aq) + NO_3^-(aq) = HNO_3(aq).$$

12. What is the heat generated by a process that occurs in a calorimeter containing 50.0 g of water at 22.5°C if the temperature of the calorimeter increases to 31.0°C? The heat capacity of the calorimeter, K_c, is 1.5 cal/°C. Be careful with the sign of the answer.

13. Suppose that you added 35.0 g of water at 99.50°C to a calorimeter containing 150 g of water at 23.0°C and the temperature in the calorimeter rose to 35.60°C. What would be the heat capacity of the calorimeter?

14. What is the specific heat of a metal if, when a 50.3-g piece of the metal at 100.0°C is put into a calorimeter containing 150.0 g of water at 23.0°C, the temperature in the calorimeter increases to 24.8°C? The heat capacity of this calorimeter is 6.5 cal/°C.

15. Sucrose, cane sugar, is consumed by the body to produce energy in the form of heat. The reaction is

$$C_{12}H_{22}O_{11}(s) + 12O_2(g) - 1349 \text{ kcal} = 12CO_2(g) + 11H_2O(l)$$

How many kilocalories can be gotten from a teaspoon of sugar? There are 80 teaspoons per pound. To save you a little time, the molecular weight of sucrose is 342 g/mole and there are 454 g per pound.

16. Using the table of Standard Heats of Formation on page 344, calculate the ΔH_f^0 for sucrose from the information in question 15.

*17. The heat capacity K_c of a calorimeter was determined by using the following reaction as the process that produces a known amount of heat

$$HCl(aq) + NaOH(aq) - 13.7 \text{ kcal} = NaCl(aq) + H_2O(l)$$

You add 50.0 ml of a 2.00 M solution of HCl (if you have not done Chapter 12 yet, 2.00 M means 2.00 moles per liter of solution) at 22.0°C to an empty calorimeter. Then you add 50.0 ml of a 2.00 M solution of NaOH, also at 22.0°C, to the calorimeter. The calorimeter now contains a solution of NaCl rather than pure water. The density of this solution is 1.04 g/ml, and its specific heat is 0.940 cal/g°C. The temperature of the NaCl solution in the calorimeter rises to 35°C. What is K_c for the calorimeter?

PROBLEM SET ANSWERS

1. $? \text{ cal} = 500 \text{ g water} \times \dfrac{1 \text{ cal}}{(\text{g water})(°C)} \times 2.55°C$

    $= 1275 \text{ cal} = 1.28 \times 10^3 \text{ cal}$

    $[\Delta T = 24.70°C - 22.15°C = 2.55°C]$

2. $? \text{ cal} = 30.0 \text{ ml CHCl}_3 \times \dfrac{1.50 \text{ g CHCl}_3}{\text{ml CHCl}_3} \times \dfrac{0.232 \text{ cal}}{(\text{g CHCl}_3)(°C)} \times -18.0°C$

    $= -188 \text{ cal}$

3. $? \Delta T = ?°C = -350 \text{ cal} \times \dfrac{(\text{g benzene}) (°C)}{0.438 \text{ cal}} \times \dfrac{1}{20.0 \text{ g benzene}}$

   $= -40.0°C$

   $T_{final} = T_{initial} + \Delta T = 45°C - 40.0°C = 5.0°C$

4. Specific heat $= \dfrac{? \text{ cal}}{(\text{g Cu}) (°C)} = \dfrac{15.3 \text{ cal}}{(15.0 \text{ g Cu}) (11.0°C)} = \dfrac{0.0927 \text{ cal}}{(\text{g Cu}) (°C)}$

   $[\Delta T = 33.0°C - 22.0°C = 11.0°C]$

5. $? \text{ kcal} = 85.0 \text{ g water} \times \dfrac{\text{mole water}}{18.0 \text{ g water}} \times \dfrac{9.71 \text{ kcal}}{\text{mole water}}$

   $= 45.9 \text{ kcal}$

6. This is a two-step process. First water must go from 25.0°C to 100.0°C,

   so $\Delta T = 75.0°C$, 5.00 g water, $\dfrac{1 \text{ cal}}{(\text{g water}) (°C)}$

   $? \text{ cal} = 5.00 \text{ g water} \times \dfrac{1 \text{ cal}}{(\text{g water}) (°C)} \times 75.0°C = 375 \text{ cal}$

   Then the water must be vaporized.

   $? \text{ cal} = 5.00 \text{ g water} \times \dfrac{\text{mole water}}{18.0 \text{ g water}} \times \dfrac{9.71 \times 10^3 \text{ cal}}{\text{mole water}} = 2697 \text{ cal}$

   The total heat is the sum of the steps.

   $? \text{ cal} = 375 \text{ cal} + 2697 \text{ cal} = 3072 \text{ cal} = 3.07 \times 10^3 \text{ cal}$

7. This is a four-step process. Step 1: The solid must be warmed to its melting point. Step 2: The solid must be converted to a liquid. Step 3: The liquid must be warmed to its boiling point. Step 4: The liquid must be converted to a gas.

   Step 1. 10.0 g alc; $\dfrac{0.232 \text{ cal}}{(\text{g alc}) (°C)}$;

   $\Delta T = -180.3°C - (-117.3°C) = 63.0°C$

   $? \text{ cal} = 10.0 \text{ g alc} \times \dfrac{0.232 \text{ cal}}{(\text{g alc}) (°C)} \times 63.0°C = 146 \text{ cal}$

   Step 2. $? \text{ cal} = 10.0 \text{ g alc} \times \dfrac{52.1 \text{ cal}}{\text{g alc}} = 521 \text{ cal}$

   Step 3. 10.0 g alc; $\dfrac{0.550 \text{ cal}}{(\text{g alc}) (°C)}$;

   $\Delta T = 78.3°C - (-117.3°C) = 195.6°C$

$$? \text{ cal} = 10.0 \text{ g ale} \times \frac{0.550 \text{ cal}}{(\text{g ale}) (°\text{C})} \times 195.6°\text{C} = 1076 \text{ cal}$$

Step 4. $? \text{ cal} = 10.0 \text{ g ale} \times \dfrac{204 \text{ cal}}{\text{g ale}} = 2040 \text{ cal}$

The total heat is the sum of the heats of the individual steps.

$? \text{ cal} = 146 \text{ cal} + 521 \text{ cal} + 1076 \text{ cal} + 2040 \text{ cal}$

$\qquad = 3783 \text{ cal} = 3.78 \times 10^3 \text{ cal}$

8. $? \text{ kcal} = 450 \text{ g ethene} \times \dfrac{\text{mole ethene}}{28.0 \text{ g ethene}} \times \dfrac{337 \text{ kcal}}{\text{mole ethene}} = 5.42 \times 10^3 \text{ kcal}$

9. $C_2H_4(g) + 3O_2(g) \qquad -337 \text{ kcal} = 2CO_2(g) \qquad + 2H_2O(l)$

$(? \text{ kcal}) + 3(0.00 \text{ kcal}) -337 \text{ kcal} = 2(-94.1 \text{ kcal}) + 2(-68.3 \text{ kcal})$

Sum of heats of products: $2(-94.1 \text{ kcal}) + 2(-68.3 \text{ kcal}) = -324.8 \text{ kcal}$

Sum of heats of reactants:

$? \text{ kcal} + 0.00 \text{ kcal} + (-337 \text{ kcal}) = (? - 337) \text{ kcal}$

The sum of the heats on the reactant side of the equation equals the sum on the product side.

$? \text{ kcal} - 337 \text{ kcal} = -324.8 \text{ kcal}$

$? \text{ kcal} = -324.8 \text{ kcal} - (-337 \text{ kcal}) = +12 \text{ kcal}$

10.   You must reverse the second equation in order to have the sum of the equations be the equation of interest.

$Cu(s) + Cl_2(g) - 49.2 \text{ kcal} = CuCl_2(s)$

$CuCl(s) = -32.5 \text{ kcal} + \tfrac{1}{2}Cl_2(g) + Cu(s)$

Adding these gives

$Cu(s) + Cl_2(g) - 49.2 \text{ kcal} + CuCl(s)$

$= CuCl_2(s) - 32.4 \text{ kcal} + \tfrac{1}{2}Cl_2(g) + Cu(s)$

Cancelling the terms that occur on both sides of the equals sign, subtracting $\tfrac{1}{2}Cl_2(g)$, and adding 32.4 kcal to both sides yields

$CuCl(s) + \tfrac{1}{2}Cl_2(g) - 16.8 \text{ kcal} = CuCl_2(s)$

11.   First get ΔH_f^0 for $HNO_3(aq)$:

$H^+(aq) \quad + NO_3^-(aq) \quad = HNO_3(aq)$

$(0.0 \text{ kcal}) + (-49.4 \text{ kcal}) = (? \text{ kcal})$

Its value must be -49.4 kcal, since the sums of the heats on both sides of the equation are equal. Then you have all the standard heats of formation to put in the equation.

$$3NO_2(g) + H_2O(l) = 2HNO_3(aq) + NO(g)$$

$$3(8.1 \text{ kcal}) + (-68.3 \text{ kcal}) = 2(-49.4 \text{ kcal}) + (21.6 \text{ kcal})$$

The sum of the heats of formation of the products is

$$2(-49.4) + 21.6 = -77.2 \text{ kcal}$$

The sum of the heats of formation of the reactants is

$$-77.2 + (-68.3) = -44.0 \text{ kcal}$$

Subtracting the reactants from the products gives

$$-77.2 \text{ kcal} - (-44.0 \text{ kcal}) = -33.2 \text{ kcal}$$

12. 50.0 g water; $\dfrac{1 \text{ cal}}{(\text{g water}) (°C)}$; $\Delta T = 31.0°C - 22.5°C = 8.5°C$

The heat gained by the calorimeter is the sum of the heat gained by the water and the heat gained by the body of the calorimeter.

$$? \, Q_w = ? \text{ cal} = 50.0 \, \cancel{\text{g water}} \times \frac{1 \text{ cal}}{\cancel{(\text{g water})} \, \cancel{(°C)}} \times 8.5\cancel{°C} = 425 \text{ cal}$$

$$? \, Q_c = ? \text{ cal} = 8.5\cancel{°C} \times \frac{1.5 \text{ cal}}{\cancel{°C}} = 13 \text{ cal}$$

$$Q_{\text{calor.}} = Q_w + Q_c = 425 \text{ cal} + 13 \text{ cal} = 438 \text{ cal}$$

Since $Q_{\text{calor.}} = -Q_{\text{process}}$, then $Q_{\text{process}} = -438 \text{ cal}$.

13. The process that is occurring in the calorimeter is the cooling of 35.0 g of water from 99.50°C to 35.60°C, so $\Delta T = -63.90°C$.

$$? \text{ cal} = 35.0 \text{ g water} \times \frac{1 \text{ cal}}{(\text{g water}) \, (°C)} \times -63.90°C = -2236 \text{ cal}$$

You can calculate the heat Q_w picked up by the water in the calorimeter, when you know the weight of the water, 150 g; its temperature change,

$35.60°C - 23.00°C = 12.60°C$; and the specific heat of water.

$$? \, Q_w = ? \text{ cal} = 150 \, \cancel{\text{g water}} \times \frac{1 \text{ cal}}{\cancel{(\text{g water})} \, \cancel{(°C)}} \times 12.60\cancel{°C} = 1890 \text{ cal}$$

Since $Q_{\text{calor.}} = -Q_{\text{process}}$, we have $Q_{\text{calor.}} = +2236 \text{ cal}$, and the $Q_{\text{calor.}}$ is the sum of Q_w and Q_c.

$$Q_{\text{calor.}} = Q_w + Q_c$$

so $2236 \text{ cal} = 1890 \text{ cal} + Q_c$, or $346 \text{ cal} = Q_c$. The heat capacity is calories per °C change in the temperature of the calorimeter, so

$$K_c = \frac{346 \text{ cal}}{12.60°\text{C}} = \frac{27.5 \text{ cal}}{°\text{C}}$$

14. First you must determine the heat gained by the calorimeter. This is the sum of Q_w and Q_c. The ΔT is $24.8°\text{C} - 23.0°\text{C} = 1.8°\text{C}$.

$$? \, Q_w = ? \text{ cal} = 150.0 \text{ g water} \times \frac{1 \text{ cal}}{(\text{g water}) \, (°\text{C})} \times 1.8°\text{C} = 270 \text{ cal}$$

$$? \, Q_c = ? \text{ cal} = 1.8°\text{C} \times \frac{6.5 \text{ cal}}{°\text{C}} = 12 \text{ cal}$$

$$Q_{\text{calor.}} = Q_w + Q_c = 270 \text{ cal} + 12 \text{ cal} = 282 \text{ cal}$$

Then you know that $Q_{\text{calor.}} = -Q_{\text{process}}$, so $-282 = Q_{\text{process}}$.

To get the specific heat, you must know the weight of the substance, the heat of the process, and $\Delta T = 24.8°\text{C} - 100.0°\text{C} = -75.2°\text{C}$.

$$\text{Specific heat} = \frac{? \text{ cal}}{(\text{g metal}) \, (°\text{C})} = \frac{-282 \text{ cal}}{(50.3 \text{ g metal}) \, (-75.2°\text{C})}$$

$$= \frac{0.0746 \text{ cal}}{(\text{g metal}) \, (°\text{C})}$$

15. $? \text{ kcal} = 1 \text{ tsp} \times \dfrac{\text{lb sucrose}}{80 \text{ tsp}} \times \dfrac{454 \text{ g sucrose}}{\text{lb sucrose}} \times \dfrac{\text{mole sucrose}}{342 \text{ g sucrose}}$

$$\times \frac{1349 \text{ kcal}}{\text{mole sucrose}} = 22.4 \text{ kcal}$$

(Nutritionists call our kcal a Calorie.)

16. $C_{12}H_{22}O_{11}(s) + 12O_2(g) - 1349 \text{ kcal} = 12CO_2(g) + 11H_2O(l)$

$(? \text{ kcal}) + 12(0.00 \text{ kcal}) - 1349 \text{ kcal} = 12(-94.1 \text{ kcal}) + 11(-68.3 \text{ kcal})$

$[? + O + (-1349)] \text{ kcal} = [-1129 + (-751)] \text{ kcal}$

$? \text{ kcal} = (-1129 - 751 + 1349) \text{ kcal}$

$= -531 \text{ kcal}$

17. First you determine the heat produced by the process.

$? \, Q_{\text{process}} = ? \text{ cal}$

$$= 50.0 \text{ ml soln.} \times \frac{10^{-3} \text{ liter soln.}}{\text{ml soln.}} \times \frac{2.00 \text{ mole HCl}}{\text{liter soln.}}$$

$$\times \frac{-13.7 \text{ kcal}}{\text{mole HCl}} \times \frac{10^3 \text{ cal}}{\text{kcal}} = -1370 \text{ cal}$$

Since the number of moles of HCl present is the same as the number of moles of NaOH, you could use either solution in the calculation.

Next you must determine the heat absorbed by the contents of the calorimeter. In this case, it is not water but a solution of NaCl, and this solution has a different heat capacity from that of pure water. You also have to determine the weight of this solution from its density.

$$? \text{ weight soln.} = ? \text{ g soln.} = 100.0 \text{ ml soln.} \times \frac{1.04 \text{ g soln.}}{\text{ml soln.}}$$

$$= 104 \text{ g soln.}$$

$$[\Delta T = 35.0°C - 22.0°C = 13.0°C]$$

$$? Q_{soln.} = ? \text{ cal} = 104 \text{ g soln.} \times \frac{0.940 \text{ cal}}{(\text{g soln.})(°C)} \times 13.0°C = 1271 \text{ cal}$$

Since $-Q_{process} = Q_{calor.} = Q_{soln.} + Q_c$, you have

$$1370 \text{ cal} = 1271 \text{ cal} + Q_c \quad \text{or} \quad 99 \text{ cal} = Q_c$$

Since the heat capacity is per °C,

$$K_c = \frac{? \text{ cal}}{°C} = \frac{99 \text{ cal}}{13.0°C} = \frac{7.61 \text{ cal}}{°C}$$

Chapter 18
Faraday's Law Calculations

PRETEST

You may be able to do the material in this chapter if you know a few new conversion factors. They are given below.

$$\frac{1 \text{ faraday}}{\text{mole electrons}} \qquad \frac{1 \text{ faraday}}{96{,}500 \text{ coulombs}} \qquad \frac{1 \text{ coulomb}}{(\text{amp})(\text{sec})}$$

$$\frac{(\text{amp})(\text{ohm})}{\text{volt (v)}} \qquad \frac{(\text{amp})(\text{volt})}{\text{watt (w)}} \qquad \frac{10^3 \text{ watt}}{\text{kilowatt (kw)}}$$

1. How many grams of Cl_2 will be formed using a 6.0 amp current for 100 seconds? You are electrolyzing a solution of NaCl.

2. How many hours will be required to deposit 11.2 g of Cd metal from a solution of $CdSO_4$ using a 1.75 amp current? The atomic weight of Cd is 112.4 g/mole.

3. How many amps current will be required to produce 15.0 liters of O_2 gas at 27°C and collected over water at a total pressure of 766 torr if you have just 300 minutes to do it? The vapor pressure of water at this temperature is 26 mm Hg. The half-reaction considered is at the anode

   $2H_2O = O_2 + 4H^+ + 4 \text{ electrons}^-$.

4. What is the resistance in ohms if 20.0 volts are required to plate 0.010 mole of Au from an $AuCl_3$ solution in 2.00 hours?

5. What is the cost for the electricity needed to plate 1.00 lb of Cu metal from a $CuSO_4$ solution? The electricity is supplied at 110 volts and costs $0.03 per kilowatt hour. The atomic weight of Cu is 63.5 g/mole.

PRETEST ANSWERS

If you missed questions 1–3, you will find this material covered in Section A. Questions 4 and 5 are covered in Section B. All these problems appear as examples in the body of the chapter.

1. 0.22 g Cl_2
2. 3.05 hr
3. 12.7 amps
4. 50 ohms
5. $1.26

If you have done Chapter 13, Sections B and C, you may recall that there is a type of chemical reaction which involves the gain of electrons by one reactant and the loss of an equivalent number of electrons by another reactant. The loss of electrons is called *oxidation* (the oxidation number increases) and the gain of electrons is called *reduction* (the oxidation number decreases).

It is possible to conduct this type of reaction using an electrical current. If you pass a current through a system in which the particles are charged (that is, they are ions), the positively charged ion will migrate to the negatively charged electrode (cathode) and will gain electrons. It will thus be _____. The negatively charged | reduced

ion will migrate toward the _____ charged electrode (anode) | positively

and will lose _____ to the anode. At the anode, then, you will | electrons

have _____ of the ion. | oxidation

You can write balanced equations for the two separate reactions, called the *half-reactions*, that occur at the two electrodes; but, unlike our previous balanced equations, one of the substances in the balanced half-reaction will be electrons. Consider the reaction which occurs when you pass a current through molten AgCl. The molten salt consists of Ag^+ ions and Cl^- ions. The Ag^+ ion will

migrate to the _____ electrode (cathode) where it will gain one | negative

_____ to produce Ag metal. The balanced equation for this | electron

cathode reaction is

$$Ag^+ + 1 \text{ electron}^- = Ag.$$

(Notice that an electron has a -1 charge. The charges on both sides of the equation are balanced.)

The Cl^- ion will migrate to the positive electrode (_____) where | anode

it can lose an electron. However, the product of the reaction is not Cl atom but rather Cl_2 molecule. The balanced half-reaction is

$$2Cl^- = Cl_2 + 2 \text{ electrons}^-.$$

Since the Ag^+ gained an electron we say that it was _____ at | reduced

the cathode. Since the Cl^- lost an electron at the _____ we say | anode

that it was _____. It is now possible to set up two "bridge" | oxidized

conversion factors from these two half-reactions in exactly the same way that we have "bridges" from a normal balanced equation. The only difference will be that one of the substances in the "bridge" will be electrons. Thus, we can write

$$\frac{1 \text{ mole electrons}}{\text{mole Ag}},$$

or

$$\frac{\text{_____ electrons}}{\text{mole } Cl_2}.$$ | 2 moles

In working problems dealing with the amount of one product, you will find that you only have to set up the balanced half-reaction which contains the substances in which you are interested. Thus, if a problem asks how much Cr you can produce under certain conditions for a $CrCl_3$ solution, you need only write the half-reaction from the Cr^{+3} ion.

3 electrons⁻

$$Cr^{+3} + \underline{\hspace{3cm}} = Cr$$

The "bridge" conversion factor from this half-reaction will be

3 moles electrons

$$\frac{\underline{\hspace{2.5cm}}}{mole\ Cr} .$$

If the problem asks how much Cl_2 you can produce under certain conditions, you would write the balanced half-reaction

2 Cl⁻

$$\underline{\hspace{2cm}} = Cl_2 + 2\ electrons^-$$

To refresh your memory, the reduction of the Cr^{+3} occurs at the

cathode,   oxidation

$\underline{\hspace{2cm}}$ and the $\underline{\hspace{2cm}}$ of the Cl⁻ occurs at the anode.

A mole of electrons is called a faraday (F). You can therefore write the conversion factor for the balanced half-reaction

$$Cr^{+3} + 3\ electrons^- = Cr$$

as

3

$$\frac{\underline{\hspace{0.4cm}}F}{Mole\ Cr} ,$$

and for the balanced half-reaction

$$2Cl^- = Cl_2 + 2\ electrons^-$$

you can write the conversion factor,

F

mole Cl_2

$$\frac{2\ \underline{\hspace{1cm}}}{\underline{\hspace{1.5cm}}} .$$

You will need conversion factors to relate a faraday to more convenient electrical terms. The first such factor relates the faraday to coulombs; there are 96,500 coulombs per faraday

$$\boxed{\frac{96,500\ coulombs}{F}} .$$

A coulomb is defined as that amount of current you have when you have 1 ampere for 1 second. So your second conversion factor is

$$\boxed{\frac{1\ coulomb}{(amp)\ (sec)}} .$$

This is another conversion factor which contains more than two different units.

You now have all the information that you will need to solve simple electrolysis problems.

SECTION A   Problems with Amps, Time, and Amount of Substance

You will frequently be asked to determine the amount of a product that can be prepared (or the amount of a reactant that is consumed) by electrolysis of a certain substance. Electrolysis simply means using an electrical current to perform a chemical reaction. You may be given information concerning the current you are using and the time you allow for the reaction to proceed, and you will be asked for the amount of product.

Question 1 on the pretest is a very simple example of this type. How many grams of Cl_2 will be formed using a 6.0 amp current for 100 seconds? You are electrolyzing a solution of NaCl.

Your first step will be to write the balanced half-reaction containing the substance of interest. In this case you are interested in Cl_2 which must have been produced by the oxidation of ____ ions, so you can write

$$\text{\_\_\_\_\_} = Cl_2 + 2 \text{ electrons}^-.$$

From this balanced equation you can write the "bridge" conversion factor containing Cl_2 and moles of _____. (Remember that a mole of electrons is called a _____.) This "bridge" will be

$$\frac{\text{mole } Cl_2}{\text{\_ F}}.$$

The information given in the problem is that the current is 6.0 amps and the time is _____, so you start the problem in the usual way by writing

$$? \text{\_\_\_\_\_} = 6.0 \text{ amp} \times 100 \text{ sec}.$$

The two unwanted dimensions are _____ and _____, and both of these appear in the conversion factor for coulombs. You therefore multiply by this conversion factor.

$$? \text{ g } Cl_2 = 6.0 \text{ amp} \times 100 \text{ sec} \times \frac{\text{\_\_\_\_\_}}{\text{(amp) (sec)}}$$

The unwanted dimension now is _____, which you can convert to faradays.

$$? \text{ g } Cl_2 = 6.0 \text{ amp} \times 100 \text{ sec} \times \frac{1 \text{ coulomb}}{\text{(amp) (sec)}}$$

$$\times \frac{\text{\_\_\_\_\_}}{96,500 \text{ coulombs}}$$

Cl^-

$2 Cl^-$

electrons
faraday

2

100 sec

g Cl_2
amps,   seconds

1 coulomb

coulombs

1 F

Now you are at your "bridge," which connects electricity with the product of the reaction.

$$? \text{ g Cl}_2 = 6.0 \text{ amp} \times 100 \text{ sec} \times \frac{1 \text{ coulomb}}{\text{(amp) (sec)}}$$

$$\times \frac{1 \text{ F}}{96{,}500 \text{ coulombs}} \times \frac{\text{mole Cl}_2}{\underline{\quad}}$$

2 F

To get to the dimensions of your answer you simply multiply by

molecular

the _____ weight conversion factor, 71.0 g Cl_2/mole Cl_2.

The final answer will then be (without all the dimensions which have cancelled)

$$? \text{ g Cl}_2 = 6.0 \times 100 \times 1 \times \frac{1}{96{,}500} \times \frac{1}{2} \times 71.0 \text{ g Cl}_2$$

$$= 0.22 \text{ g Cl}_2.$$

Question 2 on the pretest is very similar, except that the time is asked for and both the amps and the amount of product are given. How many hours will it take to deposit 11.2 g of Cd metal from a solution of $CdSO_4$ using a 1.75 amp current? The atomic weight of Cd is 112.4 g/mole.

Once again you must write a balanced half-reaction which contains the substance of interest. In this question the substance

Cd

of interest is ___. The half-reaction will be

Cd++

_____ + 2 electrons⁻ = Cd.

(Incidentally, if you are having trouble determining what is the charge of the ion from the formula of a molecule, you will find Appendix I on nomenclature very helpful.) Now you can write your "bridge" conversion factor

mole Cd

$$\frac{\underline{\quad}}{2 \text{ F}}.$$

There are two pieces of information given in the problem,

11.2, 1.75 amp

_____ g Cd and _____. You will have to be careful which of the two you choose to start with. In this type of problem you start with the information given concerning the amount of substance, so you write

hr

$$? \underline{\quad} = 11.2 \text{ g Cd.}$$

Now you must multiply by conversion factors until you get to one of the dimensions in your "bridge." In this case you can do

atomic weight

that in one step. You simply multiply by the _____ of Cd in its inverted form.

mole Cd

$$? \text{ hr} = 11.2 \text{ g Cd} \times \frac{\underline{\quad}}{112.4 \text{ g Cd}}$$

You now have one of the units in your "bridge" so you next

multiply by _____. The remaining unwanted dimension is

_____, which you cancel by multiplying by the electrical

conversion factor, _____/faraday. Using the second

electrical conversion factor which relates coulombs to _____ and

_____, you will have

$$? \text{ hr} = 11.2 \text{ g Cd} \times \frac{\text{mole Cd}}{112.4 \text{ g Cd}} \times \frac{2 \text{ F}}{\text{mole Cd}}$$

$$\times \frac{96,500 \text{ coulombs}}{\text{F}} \times \frac{(\text{amp})(\text{sec})}{1 \text{ coulomb}} .$$

You can cancel the unwanted dimension amps by multiplying by

the _____ of the information given, 1/1.75 amp. Only _____

remain as an unwanted dimension, and the dimension of the

answer will be _____. You therefore do a very simple two-step

conversion. First you multiply by min/_____ and then by

hr/_____. Your final answer will then be

$$? \text{ hr} = 11.2 \times \frac{1}{112.4} \times 2 \times 96,500 \times 1 \times \frac{1}{1.75} \times \frac{1}{60} \times \frac{1 \text{ hr}}{60}$$

$$= 3.05 \text{ hr}$$

   Question 3 on the pretest is also a problem concerning amperes,
time, and amount of product. However, it is a little more
complicated since the amount of product to be formed is expressed
as the volume of a gas at known pressure and temperature. How
many amps current will it require to produce 15.0 liters of O_2 gas
at 27°C collected over water with a total pressure of 766 torr if you
have just 300 min to do it? The vapor pressure of water at this
temperature is 26 mm Hg. The half-reaction considered is at the
anode

   $$2H_2O \longrightarrow O_2 + 4H^+ + 4 \text{ electrons}^-.$$

   Your first step will be to write down your "_____" conversion

factor for the ____-reaction.

   $$\frac{\text{mole } O_2}{\underline{\quad\quad}}$$

The next step is to determine the number of _____ of O_2 from the

volume, pressure, and _____ of the gas. This is a standard
gas law calculation; if you are rusty, you can look back to Section
C in Chapter 10. The fact that the O_2 was collected over water,

---

(margin answers)

$\dfrac{2 \text{ F}}{\text{mole Cd}}$

faradays

96,500 coulombs

amps

seconds

reciprocal,   seconds

hours

60 sec,

60 min

bridge

half

4 F

moles

temperature

which has its own vapor pressure, means that to get the O_2 pressure you will have to subtract the partial pressure of water at the given temperature from the total pressure. You can find this covered in Section E, Chapter 10.

We can calculate the number of moles of O_2. First we will write down all the variables.

$$P = \left(766 \text{ torr} - 26 \text{ mm Hg} \times \frac{1 \text{ torr}}{\text{mm Hg}}\right)$$

760 torr

$$\times \frac{\text{atm}}{\underline{\hspace{1cm}}} = 0.974 \text{ atm}$$

300

$$T = \underline{\hspace{1cm}}°K$$

$$V = 15.0 \text{ liters}$$

$$? \text{ moles } O_2 = 15.0 \text{ liters } O_2 \times 0.974 \text{ atm}$$

(mole O_2) (°K)

0.0820 (liter) (atm), 300°K

$$\times \frac{\underline{\hspace{1cm}}}{\underline{\hspace{1cm}}} \times \frac{1}{\underline{\hspace{1cm}}}$$

$$= 0.594 \text{ mole } O_2$$

Now that you have the amount of product in a useful unit, the problem is very simple.

amps

$$? \underline{\hspace{1cm}} = 0.594 \text{ mole } O_2$$

The unwanted dimension appears in your "bridge" from the balanced half-reaction, so you can immediately use the "bridge."

$$? \text{ amps} = 0.594 \text{ mole } O_2 \times \underline{\hspace{1cm}}$$

$\dfrac{4 \text{ F}}{\text{mole } O_2}$

The unwanted dimension now is _____, which you can convert to coulombs.

faradays

$$? \text{ amp} = 0.594 \text{ mole } O_2 \times \frac{4 \text{ F}}{\text{mole } O_2} \times \underline{\hspace{1cm}}$$

$\dfrac{96,500 \text{ coulombs}}{\text{F}}$

Now the coulombs can be converted to (amp) (_____)

seconds

$$? \text{ amp} = 0.594 \text{ mole } O_2 \times \frac{4 \text{ F}}{\text{mole } O_2} \times \frac{96,500 \text{ coulombs}}{\text{F}}$$

$$\times \frac{(\text{amp}) (\text{sec})}{\text{coulombs}}$$

You now have the dimension that you want for your answer, but

seconds

you still have the unwanted dimension _____. However, there is one more piece of information given in the problem, the

minutes

specification that you have just 300 _____ to do it. If you convert

minutes

your unwanted seconds to _____, you can then cancel all the

unwanted dimensions. Therefore, you next multiply by _____, and

then you can multiply by the _____ of the information given, 300 min.

$$? \text{ amp} = 0.594 \;\cancel{\text{mole } O_2} \times \frac{4 \;\cancel{F}}{\cancel{\text{mole } O_2}} \times \frac{96,500 \;\cancel{\text{coulombs}}}{\cancel{F}}$$

$$\times \frac{(\text{amp}) \;\cancel{(\text{sec})}}{\cancel{\text{coulombs}}} \times \frac{\cancel{\text{min}}}{60 \;\cancel{\text{sec}}} \times \frac{1}{300 \;\cancel{\text{min}}}$$

$$= 12.7 \text{ amps}$$

$\dfrac{\text{min}}{60 \text{ sec}}$

reciprocal

Actually, these calculations involving Faraday's Law are extremely easy, even though they may have a great many conversion factors thrown into them.

SECTION B   Problems with Other Electrical Units

There are a few other electrical conversion factors which you may find in chemical problems. One of these relates the voltage to the resistance in ohms and the current in amps. The relationship is

Volts = amp × ohms.

As with any equality, you can express this as a conversion factor

equal to 1 by _____ both sides of the equality by (amp) (ohm). You then have the conversion factor

dividing

$$\boxed{\frac{1 \text{ volt}}{(\text{amp}) (\text{ohm})}} \cdot$$

There is still another electrical relationship which introduces another new unit, the watt. You may be familiar with watts since this is how the strength of our electric light bulbs is specified. The relationship is

Watts = amp × volts.

Written as a conversion factor, this is

$$\boxed{\frac{1 \text{ watt}}{(\text{volt}) (\text{amp})}} \cdot$$

If you look at your electric bill, you will see that you are paying for the current that you use in units of kilowatt-hours. A kilowatt is one of the few metric units used in the United States. From what you already know about metric nomenclature, a

kilowatt, kw, must be ____ watts, so a kilowatt-hour means that

you have used 1 _____ for 1 hour. Incidentally, the hyphen between the kilowatt and the hour is a short way of writing kilowatt × hour or (kilowatt) (hour). Here we will use the ( ) ( ) notation to avoid any confusion.

10^3

kilowatt

Question 4 on the pretest is an example of a problem which is concerned with the resistance in ohms. What is the resistance in ohms if 20.0 volts are required to plate 0.010 moles of Au from an $AuCl_3$ solution in 2.0 hours?

The first thing which you must do is to write the balanced

half-reaction | _____.

Au⁺³ | _____ + 3 electrons⁻ ⟶ Au

From this you can write the "bridge" conversion factor

$$\frac{\text{mole Au}}{\text{\_\_\_}}$$

3 F

There is also a conversion factor given in the problem. It is

$$\frac{0.010 \text{ mole Au}}{\text{\_\_\_\_\_}}$$

2.0 hr

Now you are ready to start the problem. The question asked is

ohms | "How many _____?" since the unit of resistance is ohms. You therefore write

? ohms = _____ .

20.0 | The information given in the problem is _____ volts, so you have

? ohms = 20.0 volts.

The unwanted dimension volts can be related to ohms by using the electrical conversion factor

$$\frac{\text{(amp) (ohm)}}{\text{\_\_\_\_\_}} .$$

1 volt

amps | The unwanted dimension now is _____, but this can be converted to your "bridge" in two steps. First you multiply by the

(amp) (sec) | conversion factor, coulombs/_____, and then you multiply by the conversion factor F/96,500 coulombs. You now have one of the units in your "bridge." Multiplying by your "bridge" will leave what is shown below.

$$? \text{ ohm} = 20.0 \text{ volts} \times \frac{\text{(amp) (ohm)}}{1 \text{ volt}} \times \frac{\text{coulombs}}{\text{(amp) (sec)}}$$
$$\times \frac{F}{96,500 \text{ coulombs}} \times \frac{\text{mole Au}}{3 F}$$

Now you can multiply by the conversion factor given in the

$$\frac{2.0 \text{ hr}}{0.010 \text{ mole Au}}$$

problem, _____, which will cancel the unwanted

dimension mole Au but still leave the two unwanted dimensions,

_____ and _____. Cancelling these is very simple: you multiply

by the two time conversion factors _____/min and _____/hr.
Thus, all unwanted dimensions have cancelled, so you will have

$$? \text{ ohms} = 20.0 \times 1 \times 1 \times \frac{1}{96,500} \times \frac{1}{3} \times \frac{2.0}{0.010}$$

$$\times\, 60 \times 60 \text{ ohms}$$

$$= 50 \text{ ohms.}$$

This type of problem is really difficult, but if you use only
conversion factors and keep cancelling unwanted dimensions, you
will eventually get the correct answer. If you happen to choose an
incorrect conversion factor (that is, one that you cannot use), you
will find that you must use it again, inverted, to get back on the
track. Nothing is lost except some time.

The next problem, question 5 on the pretest, is equally annoying.
What is the cost for the electricity needed to plate 1.00 lb of Cu
metal from a $CuSO_4$ solution? The electricity, which is supplied at
110 volts, costs $0.03 per kilowatt-hour.

As usual, your first step is to determine the balanced

_____.

$$Cu^{+2} + \underline{\hspace{3cm}} \longrightarrow Cu$$

You now have the "bridge" conversion factor mole Cu/____. The

question asked is "How many _____?" The information given for

the Cu is _____. Starting in the usual way, then, you write

? dollars = 1.00 lb Cu.

It is obvious that you must get to _____ Cu, so you multiply by

_____. Next you multiply by _____, the inverted form of
the atomic weight. Now you can use the "bridge" (electrical

variety), so you multiply by _____. Next you use the two
conversion factors, which get you to a useful electrical unit, amp:

first _____ and then _____.

The conversion factor given in the problem is _____, but
you must somehow get from (amp) (sec) to (kw) (hour). It is
simple to get from seconds to hours; to do this you need two

conversion factors, _____ and _____. Now you can use the
conversion factor given in the problem (which, incidentally, gives
you the dimensions of the answer, dollars), so you multiply by

_____.

Let's take a look and see where we are now. You, of course, should have all of this written down.

$$? \text{ dollars} = 1.00 \text{ lb Cu} \times \frac{454 \text{ g Cu}}{\text{lb Cu}} \times \frac{\text{mole Cu}}{63.5 \text{ g Cu}} \times \frac{2 \text{ F}}{\text{mole Cu}}$$

$$\times \frac{96{,}500 \text{ coulombs}}{\text{F}} \times \frac{(\text{amp}) (\text{sec})}{\text{coulombs}} \times \frac{\text{min}}{60 \text{ sec}}$$

$$\times \frac{\text{hr}}{60 \text{ min}} \times \frac{\$0.03}{(\text{kw}) (\text{hr})}$$

kilowatts

$\dfrac{\text{kw}}{10^3 \text{ watts}}$

amps

(amp) (volt)

volts

110 volts

dollars

There are two unwanted dimensions remaining. One is _____ in the denominator, which is easily converted to watts by

multiplying by _____, and way back in the solution there still

remains the unwanted dimension _____ in the numerator. The dimensions amps and watts can be cancelled simultaneously by

multiplying by the electrical conversion factor, 1 watt/_____.

Now the unwanted dimension is _____, but you are given the information in the problem that the current is supplied at

_____; so, if you multiply by this information, the only

dimension remaining is _____, which is the dimension of the answer.

$$? \text{ dollars} = 1.00 \times 454 \times \frac{1}{63.5} \times 2 \times 96{,}500 \times 1 \times \frac{1}{60}$$

$$\times \frac{1}{60} \times \$0.03 \times \frac{1}{10^3} \times 1 \times 110$$

$$= \$1.26$$

If you were able to follow your way through the ten conversion factors involved in this last problem, you've got it made. You have now at your finger tips an extremely powerful tool for solving some of the most complicated types of problems, be they in chemistry, in physics, or even shopping at the supermarket. Congratulations!

PROBLEM SET

1. How many grams of Ag metal can be plated from a solution of $AgNO_3$ by a 5.0 amp current flowing for 3.60×10^4 sec?

2. How many grams of Ni will be plated from an $NiCl_2$ solution in 6.00 min using a 10.0 amp current?

3. How many liters of Cl_2 gas at 3.0 atm pressure and $-23°C$ will be produced by the electrolysis of an NaCl solution if you use a 3.0 amp current for 40 min?

4. How many hours will it take to plate out 50.0 g Al from molten Al_2O_3 using a 50.0 amp current?

5. How many hours will it take to prepare 47.0 liters of H_2 gas measured over water at a total pressure of 735 torr and 35°C if you electrolyze H_2O with a 5.5 amp current? The vapor pressure of water at this temperature is 42 torr. The half-reaction at the cathode is

    $2H_2O + 2 \text{ electrons} \longrightarrow H_2 + 2OH^-.$

6. How many amps are needed to plate 100 g Cu from a $CuSO_4$ solution in 1.0 hr?

7. How many coulombs are required to plate 48.0 g Mg from molten $MgCl_2$?

8. How many seconds are needed to plate out 0.010 mole of Cd from a $CdSO_4$ solution using a 10.0 milliamp current?

9. How many dollars will the electricity cost to deposit 10.0 kg of Cr from a $CrCl_3$ solution if electricity costs $0.025 per (kw) (hr). The electricity is at 6.0 volts and 100 amps. (Be careful; there is too much information given.) Assume 100% efficiency for the reaction.

10. What voltage current must you have to plate 2.0 g of Au from an $AuCl_3$ solution in 120 min if the resistance is 5.0 ohms?

PROBLEM SET ANSWERS

You should have been able to do at least seven of the problems.

1. The half-reaction is $Ag^+ + 1 \text{ electron}^- \longrightarrow Ag$

    $$? \text{ g Ag} = 5.0 \text{ amp} \times 3.6 \times 10^4 \text{ sec} \times \frac{\text{coulombs}}{(\text{amp}) (\text{sec})} \times \frac{F}{96,500 \text{ coulombs}}$$

    $$\times \frac{\text{mole Ag}}{F} \times \frac{107.9 \text{ g Ag}}{\text{mole Ag}}$$

    $$= 2.0 \times 10^2 \text{ g Ag}$$

2. The half-reaction is $Ni^{+2} + 2$ electrons$^- \longrightarrow Ni$

$$? \text{ g Ni} = 10.0 \text{ amp} \times 6.00 \text{ min} \times \frac{60 \text{ sec}}{\text{min}} \times \frac{\text{coulombs}}{(\text{amp})(\text{sec})} \times \frac{F}{96,500 \text{ coulombs}}$$

$$\times \frac{\text{mole Ni}}{2\,F} \times \frac{58.7 \text{ g Ni}}{\text{mole Ni}}$$

$$= 1.09 \text{ g Ni}$$

3. The half-reaction is $2Cl^- \longrightarrow Cl_2 + 2$ electrons$^-$
   First you must calculate the number of moles of Cl_2 produced by the electrolysis.

$$? \text{ moles } Cl_2 = 3.0 \text{ amp} \times 40 \text{ min} \times \frac{60 \text{ sec}}{\text{min}} \times \frac{\text{coulombs}}{(\text{amp})(\text{sec})}$$

$$\times \frac{F}{96,500 \text{ coulombs}} \times \frac{\text{mole } Cl_2}{2\,F}$$

$$= 3.7 \times 10^{-2} \text{ mole } Cl_2$$

Then, using the given temperature and pressure, you can calculate the volume using the gas laws.

$$? \text{ liters } Cl_2 = 3.7 \times 10^{-2} \text{ mole} \times 250°K \times \frac{0.0820 \text{ (liter) (atm)}}{(\text{mole})(°K)} \times \frac{1}{3.0 \text{ atm}}$$

$$= 0.25 \text{ liter } Cl_2$$

4. The half-reaction is $Al^{+3} + 3$ electrons$^- \longrightarrow Al$

$$? \text{ hr} = 50.0 \text{ g Al} \times \frac{\text{mole Al}}{27.0 \text{ g Al}} \times \frac{3\,F}{\text{mole Al}} \times \frac{96,500 \text{ coulombs}}{F} \times \frac{(\text{amp})(\text{sec})}{\text{coulombs}}$$

$$\times \frac{\text{min}}{60 \text{ sec}} \times \frac{\text{hr}}{60 \text{ min}} \times \frac{1}{50.0 \text{ amp}}$$

$$= 2.98 \text{ hr}$$

5. First calculate the number of moles of H_2 from the Gas Law since three variables, P, V, and T are given.

$$? \text{ moles } H_2 = (4.70 \text{ liters}) (735 \text{ torr} - 42 \text{ torr}) \times \frac{\text{atm}}{760 \text{ torr}}$$

$$\times \frac{(\text{mole}) (°K)}{0.0820 \text{ (liter) (atm)}} \times \frac{1}{308°K}$$

$$= 1.70 \text{ moles } H_2$$

Then calculate the time required for 1.70 moles H_2.

$$? \text{ hr} = 1.70 \text{ moles } H_2 \times \frac{2\,F}{\text{mole } H_2} \times \frac{96,500 \text{ coulombs}}{F} \times \frac{(\text{amp})(\text{sec})}{\text{coulombs}}$$

$$\times \frac{\text{min}}{60 \text{ sec}} \times \frac{\text{hr}}{60 \text{ min}} \times \frac{1}{5.5 \text{ amp}}$$

$$= 17 \text{ hr}$$

6. The half-reaction is $Cu^{+2} + 2$ electrons$^- \longrightarrow Cu$

$$? \text{ amps} = 100 \text{ g Cu} \times \frac{\text{mole Cu}}{63.5 \text{ g Cu}} \times \frac{2 \text{ F}}{\text{mole Cu}} \times \frac{96{,}500 \text{ coulombs}}{\text{F}}$$

$$\times \frac{(\text{amp})(\text{sec})}{\text{coulombs}} \times \frac{\text{min}}{60 \text{ sec}} \times \frac{\text{hr}}{60 \text{ min}} \times \frac{1}{1.0 \text{ hr}}$$

$$= 84 \text{ amps}$$

7. The half-reaction is $Mg^{+2} + 2$ electrons$^- \longrightarrow Mg$

$$? \text{ coulombs} = 48.0 \text{ g Mg} \times \frac{\text{mole Mg}}{24.3 \text{ g Mg}} \times \frac{2 \text{ F}}{\text{mole Mg}} \times \frac{96{,}500 \text{ coulombs}}{\text{F}}$$

$$= 3.81 \times 10^5 \text{ coulombs}$$

8. The half-reaction is $Cd^{+2} + 2$ electrons$^- \longrightarrow Cd$

$$? \text{ sec} = 0.010 \text{ mole Cd} \times \frac{2 \text{ F}}{\text{mole Cd}} \times \frac{96{,}500 \text{ coulombs}}{\text{F}} \times \frac{(\text{amp})(\text{sec})}{\text{coulombs}}$$

$$\times \frac{\text{milliamp}}{10^{-3} \text{ amp}} \times \frac{1}{10.0 \text{ milliamp}} \quad (\text{Remember that "milli" means } 10^{-3}.)$$

$$= 1.9 \times 10^5 \text{ sec}$$

9. The half-reaction is $Cr^{+3} + 3$ electrons$^- \longrightarrow Cr$

$$? \text{ dollars} = 10.0 \text{ kg Cr} \times \frac{10^3 \text{ g Cr}}{\text{kg Cr}} \times \frac{\text{mole Cr}}{52.0 \text{ g Cr}} \times \frac{3 \text{ F}}{\text{mole Cr}} \times \frac{96{,}500 \text{ coulombs}}{\text{F}}$$

$$\times \frac{(\text{amp})(\text{sec})}{\text{coulombs}} \times \frac{\text{min}}{60 \text{ sec}} \times \frac{\text{hr}}{60 \text{ min}} \times \frac{\$0.025}{(\text{kw})(\text{hr})}$$

$$\times \frac{\text{kw}}{10^3 \text{ watt}} \times \frac{1 \text{ watt}}{(\text{amp})(\text{volt})} \times 6.0 \text{ volts}$$

$$= \$2.32$$

This was a difficult problem. The information given that you did not need was the number of amps.

10. The half-reaction is $Au^{+3} + 3$ electrons$^- \longrightarrow Au$

$$? \text{ volts} = 5.0 \text{ ohms} \times \frac{1 \text{ volt}}{(\text{amp})(\text{ohm})} \times \frac{(\text{amp})(\text{sec})}{\text{coulombs}} \times \frac{\text{min}}{60 \text{ sec}} \times \frac{2.0 \text{ g Au}}{120 \text{ min}}$$

$$\times \frac{\text{mole Au}}{197.0 \text{ g Au}} \times \frac{3 \text{ F}}{\text{mole Au}} \times \frac{96{,}500 \text{ coulombs}}{\text{F}}$$

$$= 2.0 \text{ volts}$$

This was a difficult problem, too. The trick is knowing where to start. Sometimes it will help if rather than starting with the information given, you start with some conversion factor which has the dimensions of the answer in its numerator. Here, you could have started with [1 volt/(amp) (ohm)].

Appendix I
Possible Oxidation Numbers
of Some Common Elements

Oxidation Numbers / Periodic Chart

*Indicates the "ic" oxidation number of metal ions
( )Indicates an uncommon oxidation state

	IA	IIA	IIIB	IVB	VB	VIB	VIIB	VIII	VIII	VIII	IB	IIB	IIIA	IVA	VA	VIA	VIIA	O
	H +1 (−1)																	
	Li +1	**Be** +2											**B** +3 (−5)	**C** +4 (−4) +2 (0) (−2)	**N** +5 (+1) +4 −1 +3 −3 (+2)	**O** −2 (−1) (−½)	**F** −1	
	Na +1	**Mg** +2											**Al** +3	**Si** +4 (+2) −4	**P** +5 (+1) +4 (−1) +3 (−2) (+2) −3	**S** +6 (+2) (+5)(+1) +4 −2 (+3)	**Cl** +7 −1 +5 +3 +1	
	K +1	**Ca** +2	**Sc** +3	**Ti** +4 +3 +2	**V** +5 +4 +3 +2	**Cr** +6 +3* +2	**Mn** +7 +4 +3* +2	**Fe** +3* +2	**Co** +3* +2	**Ni** +2	**Cu** +2* +1	**Zn** +2	**Ga** +3	**Ge** +4 +2 (−4)	**As** +5 +3 −3	**Se** +6 +4 (+2) −2	**Br** +5 +1 −1	
	Rb +1	**Sr** +2									**Ag** +1			**Sn** +4* +2 (−4)	**Sb** +5 +3 −3	**Te** +6 +4	**I** +7 +5 +1 −1	
	Cs +1	**Ba** +2								**Pt** +4* +2	**Au** +3* +1	**Hg** +2* +1		**Pb** +4* +2	**Bi** (+5)* +3 (−3)			

Some common anions (arranged in order of group number)

Group III A

BO_3^{-3}	Borate

Group IV A

CO_3^{-2}	Carbonate
SiO_4^{-4}	Silicate

Organic anions

CHO_2^{-}	Formate
$C_2H_3O_2^{-}$	Acetate
$C_2O_4^{-2}$	Oxalate
$C_8H_4O_4^{-2}$	Phthalate

Group V A

NO_3^{-}	Nitrate
NO_2^{-}	Nitrite
PO_4^{-3}	Phosphate
PO_3^{-3}	Phosphite
AsO_4^{-3}	Arsenate
AsO_3^{-3}	Arsenite

Group VI A

SO_4^{-2}	Sulfate
SO_3^{-2}	Sulfite
$S_2O_3^{-2}$	Thiosulfate
$S_2O_8^{-2}$	Persulfate

Group VII A

ClO_4^{-}	Perchlorate
ClO_3^{-}	Chlorate
ClO_2^{-}	Chlorite
ClO^{-}	Hypochlorite
BrO_3^{-}	Bromate
BrO^{-}	Hypobromite
IO_4^{-}	Periodate
IO_3^{-}	Iodate
IO^{-}	Hypoiodite

Group VI B

CrO_4^{-2}	Chromate
$Cr_2O_7^{-2}$	Dichromate

Group VII B

MnO_4^{-}	Permanganate

Miscellaneous

CN^{-}	Cyanide
SCN^{-}	Thiocyanate
O_2^{-2}	Peroxide
OH^{-}	Hydroxide

Appendix II
Simple Inorganic Nomenclature

Nomenclature, the naming of molecules, is the language of the chemist. When the chemist names a compound, he or she must describe it exactly and not leave any doubt as to what the formula is. One name describes just one molecule. It is important that all chemists agree on how to construct that one name. Unfortunately, there has not yet been a complete agreement on what system to use, so that some compounds have several possible names. Recently the International Union of Pure and Applied Chemistry set up a commission to work out a system that all chemists would adopt. It is the best system worked out so far, but there are still some exceptions. You will no doubt still find older systems used in your texts and on the labels of chemicals in your labs. This very brief discussion of nomenclature will try to give you enough information so that you will be able to name most of the compounds you see as formulas and also know what the formula is for most of the names you read, both in the new IUPAC system and also in the older systems.

Names and Symbols of the Elements

You must know the names of the atoms before you can start naming compounds whose formulas you know; or, vice versa, before you can write the formula for a compound whose name you know, you must know the symbols of the elements whose names are given. There are about 104 known elements at the present time, and each has a name and a symbol. Inside the front cover of this book you will find a list of 62 of the more common elements and their symbols. Most of the symbols are two letters (only the elements which have been known for a very long time have a single letter as their symbol). The two letters come from the name and, if possible, are the first two letters of the name. Thus the symbol for the element calcium is Ca; the symbol for the element beryllium is

Be, Co

_____; the symbol for the element cobalt is _____; the symbol for the

Os

element osmium is _____.

Sometimes two or more elements have the same first two letters in their names. For example, magnesium and manganese. In order to distinguish between these two, the symbol is made up of the first and third letters. Thus the symbol for magnesium is Mg, and

and the symbol for manganese is ____. The symbols for the two

Mn

elements chlorine and chromium would then be ____ and ____.

Cl, Cr

As we said before, some of the elements which have been known for a long time have only a single letter for their symbol. The

symbol for oxygen is O, for nitrogen is __, for fluorine is __, for

N, F

hydrogen is H, for iodine is __, for carbon is C, for sulfur is S, for phosphorus is P, etc.

I

Another group of elements are those which have symbols derived from their Latin name. (Latin was the language of science in the old days and even until recently medical doctors wrote their prescriptions in Latin.) Fortunately for us, there are only ten of these remaining. These are Fe, the symbol for iron (*ferrum*), K, the symbol for potassium (*kalium*), Pb, the symbol for lead (*plumbum*), Sn, the symbol for tin (*stannum*), Au, the symbol for gold (*aurum*), Ag, the symbol for silver (*argentum*), Cu, the symbol for copper (*cuprum*), Sb, the symbol for antimony (*stibium*), W, the symbol for tungsten (*wolfram*), and Hg, the symbol for mercury (*hydragyrum*).

You will have to memorize these names and symbols. One of the best ways to do this is to prepare a set of cards for yourself. On one side have the name of an element; on the other have its symbol. Then you can randomly go through the stack of cards, read what is on one side and say or write what is on the other, checking the reverse to see whether you were correct. The 62 elements at the beginning of this text should be a sufficient number to learn, and they will cover 99% of all the formulas you will come in contact with.

Oxidation Number

The concept of oxidation number is covered thoroughly in Section B, Chapter 13. If you have not already worked through this section, now would be a good time to do it. Just to check yourself, see if you are able to do the following.

The oxidation number of Cl^- is ____. The oxidation number of

-1

Na in NaCl is ____. The oxidation number of Fe in $FeCl_2$ is ____,

$+1$, $+2$

but the oxidation number of Fe in $FeCl_3$ is ____.

$+3$

The oxidation number of O in a compound is almost always ____.

-2

Therefore, the oxidation number of Mn in MnO_2 is ____, but when

$+4$

Mn is present as in the ion Mn^{+2}, its oxidation number is ____.

$+2$

The oxidation number of S in SO_4^{-2} is ____. (The sum of all the

$+6$

oxidation numbers must be −2 since the ion has a −2 charge. The

4

oxidation number for __ O's is −8.

Therefore, the S must be +6.) The oxidation number for S in

+4

H_2SO_3 is only ____.

If you are able to do all of the above, you are sufficiently skilled in determining oxidation numbers.

Appendix I is a periodic table of the elements. For the more common elements it shows the possible oxidation numbers that the elements may have in compounds. There is a zig-zag line running from the top of the periodic table at the element boron to the bottom of the table almost all the way to the right. This line forms a rough border between the elements that are considered metals (to the left) and the elements that are considered nonmetals (to the right). The metallic elements, when they form a compound, tend to give up electrons and become positive ions. The nonmetallic elements, when they form compounds, tend to pick up electrons

negative

and become _____ ions. As you can see, about 80% of all the elements are metallic.

The oxidation numbers for these metals are all positive. This is

give up

because in compounds they tend to _____ electrons and

positive

become _____ ions. Some of the metals only have one possible oxidation number. This means that they can only form one type of ion. Thus the element Na has only a +1 oxidation number.

Na⁺

This means it can only form the _____ ion. The element Ca can only have a +2 oxidation number. Therefore, it can only form the

Ca⁺²

_____ ion.

However, some of the metals have several possible oxidation numbers. Fe, for example, has both a +2 and a +3 oxidation

Fe⁺², Fe⁺³

number. It can therefore form the two ions _____ and _____. The element Au can have an Au⁺¹ and an Au⁺³ ion, so it has the two

+1, +3

possible oxidation numbers ____ and ____.

The periodic table can be some help in figuring out what the possible oxidation numbers for an element are. The up and down columns in the table are referred to as "groups," and each has a number. The long groups are called A Groups and the shorter columns in the center of the periodic table are the B Groups.

Look at the elements in Group IA. With the exception of H (which could also be put in Group VIIA), they can only have

+1

one possible oxidation number, ____. So, metals in Group IA can only have an oxidation number of +1. Look at the metals in

oxidation

Group IIA. The elements can only have one _____ number,

+2, group

which is ____. The oxidation number is the same as the _____ number. If you consider the metals in Group IIIA, the only possible

+3

oxidation number is ____. You did not consider B as a metal

because it is not to the _____ of the zig-zag line; nonetheless, the only positive oxidation number it can show is still $+3$.

left

Once you get to a group number higher than IIIA, you cannot use this simple rule. Some of the B groups do follow the generalization:

Group IIIB can only show an oxidation number of _____, and

$+3$

Group IIB can only show an oxidation number of _____. (Hg appears to be an exception since a $+1$ oxidation number is listed in the periodic table. This is because Hg will form a composite ion, Hg_2^{+2}, as well as its normal ion Hg^{+2}. This composite ion has a calculated $+1$ oxidation number.) All the rest of the Group B elements, except Ag, can show several possible oxidation numbers and, consequently, several ions of different charge. When you are very skilled in all of the reasons behind the location of the elements in the periodic table, you may be able to figure out just what positive charges are possible for an element; for the moment, however, you will simply have to rely on memory to know what charges you can have, and you can use the periodic table in Appendix I as your guide.

$+2$

What about the nonmetallic elements? If you look at the periodic table you will see that they all have negative oxidation numbers, and with the exception of O, C, and Si, only one possible negative value. You can easily determine what this negative oxidation number will be if you know the group number of the element. See if you can figure out how to do this.

Cl is in Group VIIA, and its negative oxidation number is -1. S is in Group VIA, and its negative oxidation number is -2. P is in Group VA, and its negative oxidation number is -3. You can determine the negative oxidation number by subtracting the group

number from ___. Consequently, the negative oxidation number for

8

B in Group IIIA will be _____, and N in Group VA will have a

-5

negative oxidation number of _____.

-3

You get the negative oxidation number for the nonmetallic

elements when they gain electrons to form _____ ions. However, with the exception of O and F, all these nonmetals also show several positive oxidation numbers. These do not result from the gaining of electrons in the sense that you form ions; rather, they are the result of sharing electrons with other nonmetals. At the very beginning of this section you worked out the oxidation number

negative

for S in SO_4^{-2}. It came out to be _____. This can be looked on as meaning that the S is sharing 6 electrons with the four O atoms. In the case of H_2SO_3 the oxidation number of the S was $+4$. This

$+6$

can be seen as the S _____ 4 electrons with the three O atoms. In the compound $Na_2S_2O_3$ the oxidation number of one S atom is

sharing

_____, so you can say that the S is sharing just _____ electrons.

$+2$, two

(This is an artifact of the system. Actually, the two S atoms are not identical in the compound.)

The highest possible value for the positive oxidation number of Cl in Group VIIA is +7; the highest possible positive oxidation number for S in Group VIA is 6; and, as you would expect, the

+5   highest positive oxidation number for N in Group VA is ____. For

+4   C in Group IVB it is ____. You can now figure out what the highest positive oxidation number will be.

The other possible positive oxidation numbers go down in steps of +2. Therefore, S in Group VIA has as its highest possible

+6   positive oxidation number ____; below that it can be +4, or it can

highest   be +2. Cl in Group VIIA has +7 as its _____ possible positive

+3   oxidation number, while the lower values it can show are +5, ____

+5   and +1. P in Group VA can show the values of ____ (the highest)

+3   and the lower value of ____. It does not show a +1 value in any known compounds. Silicon in Group IVA can show the two values,

+4, +2   ____ and ____.

A few of the nonmetals are exceptional. N in Group VA can show a +4 and a +2 oxidation number in just two compounds. The oxidation number of C is very confused in organic molecules and can have almost any value. O normally has a −2 oxidation number and shows no positive values (because of the way the system of counting numbers was set up). However, it forms two composite groups, O_2^{-2} (peroxide) and O_2^{-1} (superoxide), which occur only very rarely and then only with very metallic elements or hydrogen. F shows no positive oxidation numbers at all.

All, as you can see, is not a bed of roses. Once again, you will have to rely on your memory to a certain extent. Nonetheless, the periodic table can still be a good guide for the majority of oxidation states.

Naming Binary Compounds

When a molecule consists of only two different elements we say

binary   that we have a "binary" compound. Thus NaCl would be a _____

two   compound since it contains only the ____ elements, Na and Cl. $MgCl_2$ is also a binary compound since it contains just two

Cl, not   elements, Mg and ____. However, H_2SO_4 is ____ a binary

elements   compound because it contains three different _____, H, S, and O.

The binary compounds are the simplest to name. They will be made up of two elements, one of which is metallic and therefore

positive   electropositive since the metals have _____ oxidation numbers, and one which is electronegative, having a _____ oxidation number. The way you name this type of compound is first to name the electropositive element (the metal) and then follow this with the name of the nonmetal (the more electronegative), but you substitute "ide" for the last syllable. Thus NaCl would be called sodium chloride, $MgCl_2$ would be

_____ chloride, Na_2S would be sodium _____, and

CaC_2 would be called _____.

negative

magnesium, sulfide

calcium carbide

If the nonmetal name has a vowel in the next to last syllable you drop that too. Thus MgO is called magnesium oxide (not oxyide), BN is called boron nitride (not nitroide), and NaH is sodium

_____ (not hydroide). The name for AlP will be _____ phosphide (not phosphoride).

hydride, aluminum

In general, the metallic (electropositive) element will be the element which is further to the left and lower in the periodic table.

The electronegative element will be further to the _____ and higher in the periodic table. If both elements are more or less on the right in the periodic table, there is an order of increasing electronegative character which was agreed on for the purposes of nomenclature. It is the following.

right

B, Si, C, Sb, As, P, N, H, Te, Se, S, I, Br, Cl, O, F.

Try your hand at naming some compounds.

CaF_2 _____

AlH_3 _____

MgO _____

H_2S _____

Li_3N _____

Cs_2O _____

$ZnBr_2$ _____

SiC _____

calcium fluoride

aluminum hydride

magnesium oxide

hydrogen sulfide

lithium nitride

cesium oxide

zinc bromide

silicon carbide

Incidentally, there are a few ions which are made up of several atoms so commonly found as a group that they are considered, in so far as nomenclature goes, as a single element. The only positive one is the ammonium ion, NH_4^+, which is considered to be like an element in Group IA. Thus the compound NH_4Cl is named

ammonium _____, just as if it were binary. The most common negative one that you will see is the hydroxyl ion, OH^-, and, using it

in a binary compound, you drop the "-yl" and add _____. Thus

NaOH is called sodium _____, and the compound $Ca(OH)_2$

is called calcium _____. Another negative ion is CN^-, which

chloride

ide

hydroxide

hydroxide

potassium cyanide	is the cyano group. KCN would be named _____;
ammonium cyanide	the compound NH_4CN would be named _____;
ammonium hydroxide	and the name for the compound NH_4OH is _____.

Much less common negative ions are O_2^{-2} (the peroxy group), N_3^- (the azo group), and NH_2^- (the amido group). These form peroxides, azides, and amides respectively.

All of the binary compounds that we have considered so far have had the electropositive element chosen from the group which only had one possible oxidation number. However, if the metallic element has several possible oxidation numbers—that is, can form several different positive ions—you must name the compounds in such a manner that you know which of the several possible oxidation states is present in the molecule. The method adopted by the IUPAC commission is to place a Roman numeral in parentheses immediately after the electropositive element, indicating the oxidation number. This is called the "stock notation." Thus, $FeCl_3$

+3	is called iron(III) chloride. The oxidation number of Fe is ____, and
II	$FeCl_2$ is called iron(__) chloride because the oxidation number of
+2	Fe is ____ in this compound. You would name the compound CoF_2 cobalt(II) fluoride and the compound CoF_3
cobalt(III) fluoride, II	_____. The compound SnO is called tin(__)
+2	oxide since the oxidation number of the Sn is ____. If you missed this it was because you forgot that the oxidation number of O is
−2	always ____ when it occurs alone. The sum of the oxidation numbers of a noncharged group (a molecule) will always add up
zero	to _____, so the name of SnO_2 by the IUPAC system will be
tin(IV) oxide, −4	_____. The two O's have a total oxidation number of ____,
+4	and so the Sn must have an oxidation number of ____. How about
−1	the compound $Cu(OH)_2$? The hydroxyl group is an ion with a ____
−1	charge. Therefore, its oxidation number, as a group, must be ____. Since the molecule contains two of these hydroxide ions, the
+2	oxidation number of the Cu must be ____, and the IUPAC name
II	for the compound is copper(__) hydroxide. Here are a few compounds for practice.
manganese(II) chloride	$MnCl_2$ _____
titanium(IV) oxide	TiO_2 _____
iron(III) oxide	Fe_2O_3 _____

The last one, Fe_2O_3, is a little more difficult. You figure out the oxidation number of the Fe by saying that the three O's will have

$3 \times (-2) = -6$ for a total oxidation number. This must equal the positive oxidation number for two Fe's. So each Fe must have an

oxidation number of _____. Using this reasoning, name the next two.

      Au_2S _____

      Au_2S_3 _____

The sulfide ion has a -2 charge since it is a Group VIA element. Did you count correctly? Try one more like this.

      P_2O_5 _____

The oxidation number for the two P's must equal $5 \times 2 = 10$. Each is therefore $+5$.

   Although the Stock notation (the Roman numerals) can be used for this last compound, P_2O_5, it is generally acceptable to use a different method when both of the elements making up the binary compound are nonmetals. Remember that the nonmetals are found on the upper right of the periodic table. For molecules of this type the preferred method is to count the number of atoms present in *Greek*. Actually, the Greek numerals up to eight are not so hard.

mono	one	*penta*	five
di	two	*hexa*	six
tri	three	*hepta*	seven
tetra	four	*octa*	eight

Thus, the compound SO_2 would be named sulfur dioxide (the

mono is usually omitted). SO_3 would be called sulfur _____.

CO_2, which you know already, is called _____. One of the times when you write "mono" is for CO, which is called

carbon _____. The compound with which we started was P_2O_5. You could call this diphosphorus pentoxide, but it is customary to omit the di and simply call it phosphorus pentoxide. The compound Cl_2O_7 could be named by this method as

___chlorine _____oxide. Once again it is customary to omit the

initial ___ and simply call the compound _____.

   The oxides of nitrogen are so complex and numerous that it is necessary to put in all the prefixes except mono. Thus N_2O_4 is

called dinitrogen tetroxide, N_2O_5 is called _____

N_2O is called _____, and N_2O_3 is called dinitrogen trioxide.

   Try to name some compounds using this method.

      CCl_4 _____

      SiF_6 _____

      As_2O_3 _____

Answers (right column):

$+3$

gold(I) sulfide

gold(III) sulfide

phosphorus(V) oxide

trioxide

carbon dioxide

monoxide

di, hept

di, chlorine heptoxide

dinitrogen pentoxide

dinitrogen oxide

carbon tetrachloride

silicon hexafluoride

arsenic trioxide

silicon dioxide	SiO_2 _____
antimony trisulfide	Sb_2S_3 _____

There are a few binary compounds containing a metal with several possible oxidation numbers that are most often named by the "counting" method. Manganese(IV) oxide is commonly called

dioxide manganese _____; its formula is MnO_2. TiO_2 is usually called

dioxide, titanium tetrachloride titanium _____, and $TiCl_4$ is called _____.
Actually, the IUPAC commission gives this "counting" method as its first choice for naming all compounds where the electropositive element has several possible oxidation numbers, and it states that the Stock notation (Roman numerals) can only be used for compounds of metals with nonmetals. However, old habits die hard, and chemists are unaccustomed to seeing the name diiron

III trioxide for Fe_2O_3 rather than the Stock notation, iron(___) oxide.
There is one more way of naming compounds which have an electropositive element with several possible oxidation numbers. It is no longer recommended and slowly but surely it is disappearing, but you will still find it in older texts and on reagent bottles in the lab. What you do is name the metallic element (sometimes using its Latin name) and substitute "ic" for the last syllable if the element is in its higher oxidation state or substitute "ous" if it is in its lower oxidation state. The compound $HgCl_2$ is therefore called mercuric chloride (the mercury is in the $+2$ or higher oxidation state), and the compound Hg_2Cl_2 is called mercurous chloride (the mercury is

+1 in the ___ oxidation state which is its lower state).
Using this method of nomenclature, the compound $FeCl_3$ is called ferric chloride (the Latin name for iron is *ferrum*), and the

ferrous compound $FeCl_2$ would be named _____ chloride. The Fe in

+3 $FeCl_3$ has an oxidation number of ___, and in $FeCl_2$ it has an oxidation number of $+2$. The higher oxidation number is the "ic"

ous and the lower is the "___."
The Latin name for copper is *cuprum*, so the name of the compound CuO would be cupric oxide, and for the compound

cuprous Cu_2O it would be _____ oxide. The Latin name for tin is *stannum*. By this method of naming compounds you would say

stannic chloride that $SnCl_4$ is _____ and $SnCl_2$ is stannous chloride.
The Latin name for gold is *aurum*, so the names for the two oxides,

aurous oxide, auric oxide Au_2O and Au_2O_3, will be _____ and _____, respectively.
The big problem with this method is knowing which is the "ic" oxidation number and which is the "ous" oxidation number. The periodic table in Appendix I has an asterisk following the "ic" oxidation number. Another problem occurs when an element has

more than two positive oxidation numbers which represent positve ions that it can form. Ti, for example, can form a $+4$, $+3$, or $+2$ ion. Which is the "ic"? You cannot use this method at all in this case.

This notation system is so poor that we will not spend any more time with it.

There is a special group of binary compounds which when dissolved in water will release H^+ ions. Whenever you add H^+ ions to water, you say that you have an acid. Consequently, these compounds are named as acids when they are present in water. They will always contain H as one of their two elements and then some nonmetal. There are not very many of them. The way you name them is to write "hydro" for the hydrogen and then substitute "ic" for the last syllable of the nonmetal's name. You follow this with "acid." Thus, HCl dissolved in water is called hydrochloric

acid, and HF would be named _____. You will call   hydrofluoric acid

H_2Se _____, and for the compound HI dissolved in   hydroselenic acid

water you say hydriodic _____. (Notice that the "o" was dropped   acid
before the vowel "i" in iodic.) The important thing to remember is that when you have an acid whose name starts with "hydr" it is a

binary compound. It contains just ___ and one other nonmetallic   H
element.

There are other compounds containing H and one other element which have trivial names—that is, names which have no systematic

backing. The one that you surely know is H_2O. It is called _____.   water

The compound NH_3 also has a trivial name. It is called _____.   ammonia
There are some others which may appear but hardly need elaboration here.

Let's review now what we have done so far. We have been

dealing with binary compounds. These contain just _____ different   two
elements. You name the more electropositive element first and then

the electronegative element, substituting "_____" for the last   ide
syllable of its name. You can tell which is the electropositive

(metallic) element since it is located further to the _____ and   left
lower in the periodic table.

If the electropositive element has more than one possible positive oxidation state, you must indicate which occurs in the compound you are naming. There are three ways that this is done. The first

method is to write a _____ numeral in parentheses after the   Roman
name of the metal, the number being equal to the oxidation

_____. This system is called the _____ system, and it should   number, Stock

only be used if your binary compound is made up of a _____ and   metal
a nonmetal.

If your binary compound is made up of two nonmetals (the

right — elements which appear in the upper _____ in the periodic table),

count — you then _____ the number of each type of atom in Greek. Generally, you omit the "mono's" and frequently you omit "di" when it is the number of the first atom.

The old-fashioned method of naming a compound in which the electropositive element has several oxidation states is to name the

Latin, "ic" — metal (usually in _____) and substitute _____ for the last
"ous" — syllable in its name if it is in the higher oxidation state and _____ if it is in the lower oxidation state.

In all cases the electronegative (nonmetallic) element will have

"ide" — _____ as its last syllable. Therefore, any compound whose name

binary, two — ends in "ide" is a _____ compound and contains only _____ different elements. Aside from a few trivial names like water and ammonia, the only exception to this statement is the case where your binary compound contains H and is dissolved in water to give an acid solution. These compounds are named as acids. You write

H — "hydro" for the __ and then substitute "ic" for the last syllable of

acid — the nonmetal. You follow this with "_____" to show that it is an acid.

Now that you can name a compound whose formula is given, how about trying to reverse the process and find the formula for a compound whose name is given.

NaCl — Consider the compound sodium chloride, whose formula is _____.

+1 — The oxidation number of Na is always ____ since it is in Group IA in the periodic table. The negative oxidation number of Cl is

−1 — always ____ since it is in Group VIIA in the periodic table. The

zero — sum of the oxidation numbers of a molecule always equals _____.
Consider the compound silver oxide. Since the last syllable of

binary — the name is "ide" you know that it must be a _____ compound. The only possible oxidation number for silver, Ag, is +1. The

−2 — oxidation number of O when it is named oxide is ____. In order for

zero — the sum of the oxidation numbers for the molecule to be _____

Ag$_2$O — the formula must contain two Ag's. The formula therefore is_____.
How about aluminum sulfide? Aluminum is in Group IIIA in the

+3 — periodic table. Its only possible positive oxidation number is ____. Sulfur is in Group VIA in the periodic table. Its only negative

−2 — oxidation number will be ____. In order to have a formula which

2 — has a total oxidation number of zero, you will have to have __ Al's and __ S's. This way you have $2 \times (+3)$ and $3 \times (-2)$, so the | 3

formula would be _____. | Al_2S_3

See if you can work out the formula for calcium phosphide.

Since the last syllable is _____ you know you have a binary | "ide"

compound. Calcium only can have an oxidation number of ____. | $+2$

The only negative oxidation number possible for P is ____ since it | -3

is in Group ____ in the periodic table. Therefore the formula for the | VA

molecule must be _____. | Ca_3P_2

We might look at a few of the special ions that are considered as

a single element. Ammonium chloride has the formula _____ | NH_4Cl

since the NH_4^+ ion has an oxidation number of $+1$. Ammonium

sulfide would have the formula _____ since you must take two | $(NH_4)_2S$

of the NH_4^+ ions to balance the -2 oxidation number of S. Notice

how the NH_4^+ is placed in parentheses and the subscript written

after it. Recall the hydroxide ion OH^-. The formula for potassium

hydroxide would be _____ (the symbol for potassium is K). The | KOH

formula for magnesium hydroxide will be $Mg(OH)_2$ since the only

possible oxidation number for Mg is ____. It is in Group IIA in the | $+2$

periodic table. You will need two OH^- ions which have an

oxidation number of ____ each to balance the $+2$ of the Mg. | -1

Notice once again that you put the group in parentheses.

The CN^- ion is called _____. Since it has a -1 charge, it | cyanide

will have an oxidation number of ____. The formula of zinc | -1

cyanide will be _____ since zinc can only have an oxidation | $Zn(CN)_2$

number of $+2$.

The peroxide group is O_2^{-2}. As a group it will have an oxidation

number of ____ since its charge is -2. Then the formula of | -2

hydrogen peroxide must be _____ since you will need two H's to | H_2O_2

balance the -2 oxidation number of the peroxide ion. Calcium is

in Group IIA and can only have a ____ oxidation number. The | $+2$

formula for calcium peroxide will be _____ since only one Ca is | CaO_2

needed to balance the -2 oxidation number of the _____ ion. | O_2^{-2}

Actually, it is harder getting the formula for compounds

in which the more electropositive element has only one possible

oxidation number than it is for compounds where several oxidation

states exist. In the latter case you will be told in the name what the

oxidation state is or the number of atoms which occur in the

formula. For example, copper(I) chloride tells you that the

oxidation number for the Cu is ____. The formula would be | $+1$

CuCl, −1	_____ since the Cl has only a possible oxidation number of _____.
FeBr$_3$	Iron(III) bromide would have the formula _____ since you need
−1	three Br's, each with an oxidation number of ____ to balance the +3 oxidation number of the Fe.
+4	Tin(IV) oxide will have the formula SnO$_2$ since the Sn has ____ as its oxidation number and you will need two O's, each with a
−2	____ oxidation number, to balance the oxidation number of Sn. In order to write a formula from chromium(III) oxide you will have to realize that you cannot take one and one half oxygen atoms to get the needed −3 oxidation number. There is no way of having part of an atom. The only way that you can write a formula is to
two	have ____ Cr's and then you can balance the resulting +6
three	oxidation number with _____ O's. The formula for the compound
Cr$_2$O$_3$	will be _____. See if you can work out a formula for manganese(II) arsenide. Arsenic is in Group VA. Its only negative
−3, 8	oxidation number will be ____ (you subtract 5 from __). From the
+2	name you can tell that the oxidation number of the Mn is ____. Since it is only possible to write a formula with whole numbers of
three	atoms you take _____ Mn's, which will have a total oxidation
+6, two	number of ____, and ____ As's, which will have a total oxidation
−6	number of ____. Therefore, the formula will be Mn$_3$As$_2$.

You may find it helpful to write the molecule as ions as a start and then take enough of each to balance the total positive charge with the total negative charge.

First step: Mn^{+2}As^{-3}

Second step: Mn $\overset{+2}{\underset{3}{\text{Mn}}}$ $\overset{-3}{\underset{2}{\text{As}}}$

+6   −6

It is very simple if the name of the compound is given by the counting method. The formula for manganese dioxide will be

MnO$_2$, TiCl$_4$	_____. The formula for titanium tetrachloride will be _____.
UCl$_5$	The formula for uranium pentachloride will be _____, and the formula for octuranium trioxide will be U$_8$O$_3$. (Recall that *octa*
eight	is Greek for _____, and the final "a" is dropped before a vowel.)
S$_4$N$_4$	The formula for tetrasulfur tetranitride will be _____, and the
Si$_2$I$_6$	formula for disilicon hexaiodide will be _____. The name literally

tells you how many of each atoms are present (*mono* usually being understood).

The only problem you may have with these names is that occasionally the Greek number, *di*, for the first element may be omitted. Thus you say chlorine heptoxide for Cl_2O_7, and the

formula for phosphorus pentoxide is _____, not PO_5, which P_2O_5

would mean that the oxidation number for P is _____. It is   +10
not possible to have a positive oxidation number greater than

+7. The formula for arsenic trioxide will be _____ since the di- As_2O_3
in front of the arsenic is omitted.

If your compound is named by the old-fashioned "ic" and "ous" system, you will simply have to know what the oxidation number corresponding to these is for the metallic element. The only way is either to have a reference like a textbook or the periodic table in Appendix I. After a while you will simply remember some of the common ones like ferric (iron(III)), manganous (Mn(II)), stannic (Sn(IV)), and auric (Au(III)). You can see the advantage of the other two systems.

If you are asked for the formula of ferric chloride you will have

to know that the _____ oxidation state of Fe is +3. Then you   higher

can write the formula as _____ by the same reasoning you would $FeCl_3$
use if you were given the name iron(III) chloride for the same
compound. The formula of stannous oxide can be worked out if

you know that the _____ oxidation state of Sn is +2. The   lower

formula would be _____, and if you know that the oxidation   SnO
number corresponding to manganic is +3, you could write the

formula of manganic chloride as _____. $MnCl_3$

Naming Oxyacids

There is a group of acids which contain, besides H and a nonmetallic element, also O. These are called oxyacids or in the IUPAC commission's terminology, oxo acids. They are named by stating the name of the nonmetal and substituting "ic" for its last syllable if it is at its highest oxidation state, and "ous" for its last syllable if it is at the next lower oxidation state. This is then followed by the word "acid." Thus the name for HNO_3 is nitric acid, and the name for HNO_2 is nitrous acid. The oxidation number

for the N in HNO_3 is ____, and the oxidation number for the N in   +5

HNO_2 is ____.   +3

You can figure out that +5 is the highest oxidation number for N if you have a periodic table. N is in Group VA, and the highest

oxidation number is _____ to the group number.   equal

sulfuric acid	The name for H_2SO_4 is _____. (In the case of S and C you keep the last syllable.) You can tell that this is an "ic" acid
+6	since the oxidation number of S in H_2SO_4 is ____. (If you have forgotten how to calculate the oxidation number, check back to
sulfurous acid	Chapter 13, Section B.) The name of H_2SO_3 is_____.
+4	The S in H_2SO_3 is at a ____ oxidation number which is the next
lowest	_____ oxidation number to the +6.
	See if you can work out the name for H_2CO_3. The oxidation
+4	number for the C is ____. If you look in the periodic table, you see
IVA	that C is in Group _____. The highest oxidation number possible
+4, highest	in Group IVA is ____. Therefore, the C in H_2CO_3 is at its _____
carbonic acid	oxidation state, and the name of the acid is _____.
	The name for H_3PO_3 can be worked out if you calculate the
P, +3	oxidation number of the __, which works out to be ____. If you locate P in the periodic table you see that it is in Group VA and
+5	the highest oxidation number possible will be ____. But the
lower	oxidation number of the P in H_3PO_3 is two _____ than this.
phosphorous	Consequently, the name of H_3PO_3 must be _____ acid. (Incidentally, so that you don't confuse phosphorous the acid with phosphorus the element, when you are speaking you shift the accent; the acid is pronounced phos-PHOR-ous and the element is pronounced PHOS-phorus.)
	Some of the oxyacids that you are going to learn to name do not exist in the free state. However, as you will see in the following section, their salts (combinations with metallic ions) do exist. It will be very helpful if you learn how to name these imaginary acids.
	So far you have learned how to name an oxyacid when the nonmetallic element is in its highest oxidation state. You substitute
"ic"	_____ for the last syllable of the nonmetal's name. When it is in the next lower oxidation state (the oxidation number will be
two, "ous"	_____ less than for the highest), you substitute _____ for the last syllable. What will happen if there is an oxyacid where the oxidation number of the nonmetal is still lower than the "ous" acid? Chemists have decided to use the Greek word for beneath, which is *hypo* (hypodermic is "beneath the skin") as the start of the name. The acids are beneath the "ous."
	Consider the three oxyacids H_3PO_4 (oxidation number of P is
+5	____), H_3PO_3 (oxidation number of P is +3), and H_3PO_2
+1	(oxidation number of P is ____). The highest oxidation state
phosphoric	possible is +5, and so H_3PO_4 is called _____ acid. The

oxidation state two lower is $+3$, so H_3PO_3 is called _____ acid. The third acid, H_3PO_2, has an oxidation number two lower than the "ous" acid, and so it is named hypophosphorous acid.

| phosphorous |

Try one more. When you have the acid $H_2N_2O_2$ and calculate the oxidation number of N, you find it is ____. (Two N's are $+2$ so each N must be $+1$.) N is Group VA in the periodic table, so its highest possible oxidation number is $+5$. The next lower must be

| $+1$ |

____. Recall, the oxidation numbers go down in steps of two. But in the acid $H_2N_2O_2$ the oxidation number for N is only $+1$.

| $+3$ |

Therefore its name must be _____ acid.

| hyponitrous |

Actually, these hypo-ous oxidation states are very rare, except in the nonmetals in Group VIIA, Cl, Br, and I. Here a further problem arises, for there are four different positive oxidation states, $+7$, $+5$, $+3$, and $+1$. Our "ic," "ous," and "hypo-ous" system only allows for three different oxidation states. Since Br never shows a $+7$ oxidation number, chemists have decided to consider the $+5$ oxidation number the "ic" (highest), the $+3$

(two lower) the "____," and the $+1$ the "hypo-ous." The $+7$ was then named "greater than 'ic'" (in Greek, of course). This is "hyper," but to avoid confusion it was shortened to "per." Thus

| ous |

the acid $HClO_4$, where the oxidation number of Cl is ____, is called perchloric acid. The acid HIO_4, where the I has an oxidation

| $+7$ |

number of ____, will be named _____ acid. As a matter of fact, the "per-ic" name is reserved for oxyacids where the element has an oxidation number of $+7$. Thus the name of the acid

| $+7$, periodic |

$HMnO_4$ will be _____ acid since the oxidation number

| permanganic |

of the Mn is ____. If you check your periodic table, you will see that Mn is in Group VIIB. It will show some of the oxidation states similar to Group VIIA. The B Groups in the periodic table frequently have elements which are similar to the elements in the A Group of the same number.

| $+7$ |

Cr is in Group VIB, so you might expect it to have oxyacids similar to those found for elements in Group ____. The highest

| VIA |

oxidation number will be ____, and, consequently, the name for

| $+6$ |

H_2CrO_4 will be _____ acid.

| chromic |

There is another common oxyacid of Cr, $H_2Cr_2O_7$. The oxidation number for one Cr is still ____. Therefore, you must name this also chromic acid. However, to distinguish it from H_2CrO_4 it is called dichromic acid. The "di" tells you that there are two Cr's.

| $+6$ |

Let's put all of this into a table so that you will have a ready reference to the names for the oxidation numbers of elements in various groups.

Oxidation Number

Name	Group	III A	IV A	V A	VI A and B	VII A and B
per_____ic						+7
_____ic		+3	+4	+5	+6	+5
_____ous			+2	+3	+4	+3
hypo_____ous				+1	+2	+1

You should now be able to name any oxyacid if you are given its formula, and you can look at a periodic table to see which group the nonmetal is in. How about trying a few to see just how well you do?

In order to find the name for H_4SiO_4 you start by locating ____ in the periodic table. It is in Group IVA. You know that the highest oxidation number must therefore be ____. Next you calculate the oxidation number of Si in H_4SiO_4. It is ____. Therefore, the name must be _____ acid since you are at the highest oxidation number.

In order to find the name of H_3AsO_3 you start by locating As in the _____ table. It is in Group VA. The highest oxidation state will be ____. You then calculate the _____ number of the ____ in H_3AsO_3. It is +3. This is two below the value for the highest possible number, so the name will be _____ acid.

If you want the name for H_3BO_3 you first find that B is in Group _____ in the periodic table. The highest oxidation number will be ____. The oxidation number for B in H_3BO_3 is ____, and so the name must be _____ acid since you are at the highest oxidation number.

If you want the name for $HClO_2$ you first locate ____ in Group _____ in the periodic table. From this you know that the highest possible oxidation number will be ____. The oxidation number for the Cl in $HClO_2$ is only ____. Recall that for Group VII the ____ oxidation number was taken to be the "ic." The oxidation state of Cl in $HClO_2$ is ____ below the "ic" value, and so the name for $HClO_2$ must be _____ acid.

What about the name for HBrO? Br is in Group _____ also. The oxidation number for Br in HBrO is only ____. This is ____

Si

+4

+4

silicic

periodic

+5, oxidation

As

arsenious

IIIA

+3, +3

boric

Cl

VIIA

+7

+3, +5

two

chlorous

VIIA

+1, four

lower than the "ic" oxidation value for Group VII, and so the name

must be _____ acid.

hypobromous

Getting the name for an oxyacid whose formula is given is relatively simple. However, reversing the procedure is much more of a problem, and, although it is possible to logically figure out the number of H's and O's in the formula, you would need a considerable background in the electronic structure of the atoms and the chemistry of the formation of the acid molecules. At the present time it will be far, far easier if you simply memorize the more common formulas. The table below includes most of the acids that you will find in first-year chemistry. They are in order of the groups in which the nonmetallic element occurs.

Common Oxyacids

Boric acid	H_3BO_3	Perchloric acid	$HClO_4$
Carbonic acid	H_2CO_3	Chloric acid	$HClO_3$
Silicic acid	H_4SiO_4	Chlorous acid	$HClO_2$
Nitric acid	HNO_3	Hypochlorous acid	$HClO$
Nitrous acid	HNO_2	Bromic acid	$HBrO_3$
Phosphoric acid	H_3PO_4	Bromous acid	$HBrO_2$
Phosphorous acid	H_3PO_3	Hypobromous acid	$HBrO$
Arsenic acid	H_3AsO_4	Periodic acid	HIO_4
Arsenious acid	H_3AsO_3	Iodic acid	HIO_3
Sulfuric acid	H_2SO_4	Hypoiodous acid	HIO
Sulfurous acid	H_2SO_3	Permanganic acid	$HMnO_4$
Thiosulfuric acid	$H_2S_2O_3$	Acetic acid	$H(C_2H_3O_2)$
Chromic acid	H_2CrO_4	Oxalic acid	$H_2(C_2O_4)$
Dichromic acid	$H_2Cr_2O_7$		

Naming Salts of Oxyacids

Any or all of the H's in an oxyacid can be lost as H^+ ions leaving behind a negatively charged ion (an anion). These anions can be named on the basis of the name of the acid from which they come. If an acid is named -ic, you replace the "ic" with "ate." The

NO_3^- anion comes from HNO_3, and the name of the acid is _____ acid (the N is at $+5$ oxidation number). The name of the anion will be nitrate, the formula for carbonic acid is H_2CO_3, and

nitric

therefore the name of the anion CO_3^{-2} is _____. The anion

carbonate

from H_3PO_4, phosphoric acid, will be _____, and it is named phosphate. (Did you get the correct charge on the anion? The

PO_4^{-3}

H_3PO_4 has lost 3 H^+'s, so the anion must be left with a _____ charge.) How about the name for the anion MnO_4^-? This comes

-3

from the imaginary acid $HMnO_4$, which has Mn in the _____

$+7$

permanganic	oxidation state. The acid is called _____ acid, so the anion will be named permanganate. You have simply substituted
ate	"_____" for "ic."
	If the acid from which the anion is derived is an "ous" acid, you replace the "ous" with "ite" when you name the anion. Thus the name for NO_2^-, which comes from nitrous acid, HNO_2, will be
nitrite, +4	_____. Sulfurous acid, which has S in the ____ oxidation
SO_3^{-2}	state, will have an anion, _____, which is named sulfite. Can you figure out the name for the ClO^- anion? It is derived from HClO,
hypochlorous	which is named _____ acid. If you substitute "ite" for
hypochlorite	the "ous," the name for the anion becomes _____.

You do not have to backtrack to the acid to determine the oxidation number of the nonmetal. Recall that when you determine oxidation numbers the sum for all the atoms present must equal the charge on the entire group. If you have a molecule the charge is |
zero	_____. If you have an ion the charge appears as a superscript above and to the right of the ion. Thus the oxidation number for
+7	Cl in ClO_4^- will be ____ since the four O's have a total oxidation
−8	number of ____; and to end up with the charge −1 the one Cl must have an oxidation number of +7. If you remember that the acid in the +7 oxidation state is called per-ic, you will then know that the anion must be named per-ate. ClO_4^- is therefore
perchlorate	named _____.

Try one. You want the name for the anion PO_3^{-3}. The oxidation |
| +3 | number for the P will be ____. This is two lower than the highest |
| +5 | possible oxidation number, which is ____ (P is in Group VA), and |
| phosphite | so the name of the anion must be _____.

If the acid does not give up all its H's as H^+ ions, you must show that by noting the number of H's left behind. Thus the $H_2PO_4^-$ anion is called the dihydrogen phosphate ion, and the anion $HAsO_4^{-2}$ is named monohydrogen arsenate. If the acid only has two H's, then either all are given up or just one is left behind. There is no other possibility. Then you don't bother writing the "mono." HCO_3^- is called simply hydrogen carbonate, and HSO_4^- is called |
| hydrogen | simply _____ sulfate.

Now do a few of these to see how simple the procedure is. |
borate	BO_3^{-3} _____
periodate	IO_4^- _____
nitrite	NO_2^- _____
hypobromite	BrO^- _____

HCO_3^- _____ hydrogen carbonate

$H_2PO_3^-$ _____ dihydrogen phosphite

CrO_4^{-2} _____ chromate

$Cr_2O_7^{-2}$ _____ dichromate

If you had some problems, try this next group. They are in such an order that you will find it simpler.

CO_3^{-2} _____ Carbonate

HCO_3^- _____ Hydrogen carbonate

SO_4^{-2} _____ Sulfate

HSO_4^- _____ Hydrogen sulfate

SO_3^{-2} _____ Sulfite

HSO_3^- _____ Hydrogen sulfite

NO_3^- _____ Nitrate

NO_2^- _____ Nitrite

PO_4^{-3} _____ Phosphate

HPO_4^{-2} _____ Monohydrogen phosphate

$H_2PO_4^-$ _____ Dihydrogen phosphate

PO_3^{-3} _____ Phosphite

HPO_3^{-2} _____ Monohydrogen phosphite

$H_2PO_3^-$ _____ Dihydrogen phosphite

ClO_4^- _____ Perchlorate

MnO_4^- _____ Permanganate

ClO_3^- _____ Chlorate

BrO_2^- _____ Bromite

IO _____ Hypoiodite

You may have noticed the two organic acids which were included in the table of common oxyacids in the preceding section. They were acetic acid, $HC_2H_3O_2$, and oxalic acid, $H_2C_2O_4$. These form the anions acetate, $C_2H_3O_2^-$, and oxalate, $C_2O_4^{-2}$. There are no lower oxidation states for them.

Now we are interested in naming the salt which results when the negatively charged anion combines with a positively charged ion (cation) of a metallic element. Recall that the oxidation number of

an ion is equal to its _____.   charge

The way salts are named is to simply name the metallic element and then name the anion. If the metallic element has only one possible oxidation number (i.e., will produce only one possible cation), that is all that is needed. If, however, the metallic element has several possible cations of different charges (has several possible oxidation numbers), then you will have to use either the Stock notation or the counting method to distinguish between them. Some examples will make this very clear.

The compound $NaNO_3$ is called sodium nitrate. The metallic

Na, +1

element is _____; it is in Group IA, and it can only have a _____

nitrate

oxidation number. The NO_3^- anion is called _____.

carbonate

The compound $CaCO_3$ is called calcium _____. The cation, Ca^{+2}, is in Group IIA in the periodic table and can only

+2

have a _____ charge (oxidation number). The CO_3^{-2} is called carbonate.

aluminum phosphate

The compound $AlPO_4$ is called _____. The

Al

cation _____ is in Group IIIA in the periodic table and can show

+3, phosphate

only a _____ charge. The name of the anion is _____.

potassium dihydrogen phosphite

The compound KH_2PO_3 is _____ since K can only have a +1 charge. The name of the anion is

dihydrogen phosphite

_____.

calcium nitrate

The name of the compound $Ca(NO_3)_2$ is _____. Since Ca is in Group IIA in the periodic table it can only have a

+2

cation with a _____ charge. Notice that there are two NO_3^- anions in the formula, which is shown by placing parentheses around the

2

NO_3^- and writing a subscript __ after the parentheses. Two of the

−1

anions, each with a charge of _____, were needed to balance the

+2

charge of the Ca^{+2} cation, which is _____.

sodium sulfate

Consider the compound Na_2SO_4. This is called _____. The reason that two Na^+ cations are required in the formula is that

+2, −2

you will need a _____ charge to balance the _____ charge of the SO_4^{-2} anion.

The formula for aluminum sulfate is $Al_2(SO_4)_3$. Since the charge

+3

on the Al^{+3} cation is _____ and the charge on the SO_4^{-2} anion is

−2

_____, you must take two Al^{+3} cations, which will have a total

+6

charge of _____, to balance three SO_4^{-2} anions, which will have a

−6

total charge of _____. The charge of the whole molecule must equal

zero

_____.

So far we have examined salts whose cation has only one possible oxidation number. This is the simplest case. Consider the two salts Cu_2CO_3 and $CuCO_3$. They cannot both be called copper carbonate, so we must somehow note that the Cu is at a different oxidation number in the compounds. The Stock notation will do this very well. You simply must determine the oxidation number (charge) of the Cu cation in the compound. You must therefore remember that the CO_3^{-2} anion has a ____ charge. Then if you look | −2

at the first compound, Cu_2CO_3, you see that it takes ____ Cu's to | two

balance the ____ charge of the CO_3^{-2} anion. Therefore, one Cu | −2

must have a charge of ____. Then you can name the compound | +1

copper(__) carbonate. | I

In the second compound, $CuCO_3$, only one Cu is needed to

balance the ____ charge of the CO_3^{-2} anion. Therefore, the charge | −2

on the Cu must be ____. This compound can then be named | +2

_____ carbonate. | copper(II)

Let's try another example like this. There are two compounds $SnSO_4$ and $Sn(SO_4)_2$. The first thing you must remember is that

the charge on the SO_4^{-2} anion is ____. Then you see that in the | −2

first compound, $SnSO_4$, only ____ sulfate anion is needed to | one

balance the charge of one Sn. The charge on the Sn must be ____, | +2

so the name will be _____ sulfate. | tin(II)

In the second compound, $Sn(SO_4)_2$, ____ sulfate anions balance | two

the charge of one Sn. The charge on the Sn must be ____. The | +4

name of the compound is therefore _____. | tin(IV) sulfate

How about $Fe_3(PO_4)_2$ and $FePO_4$? You will have to remember

that the phosphate anion has a ____ charge. Then, examining the | −3

first formula, $Fe_3(PO_4)_2$, you see that you have ____ PO_4^{-3} anions. | two

This is a total negative charge of ____, which exactly balances | −6

_____ Fe cations. Therefore, each Fe cation must have a charge | three

of ____. The name of the compound will be _____. | +2, iron(II) phosphate

In the second compound, $FePO_4$, only one Fe is needed to

balance the ____ charge of one PO_4^{-3} anion. The charge on this | −3

one Fe then must be ____ and the name of the compound is | +3

_____. | iron(III) phosphate

This isn't really too difficult if you remember the charges on the various anions and their correct names, merely a matter of practice. Chemical nomenclature is like any language; you simply learn the vocabulary.

Now we will try to reverse this whole process. You will write the formula for a compound whose name is known. Thus the formula for potassium nitrate will be KNO_3. To do this you have to know

+1

that the only possible charge of potassium is _____ and that the

NO_3^-

formula of the nitrate anion is _____. What about the formula for sodium sulfate? First you must know that the only possible charge

+1

on the sodium cation is _____ and the formula for the sulfate anion

SO_4^{-2}, -2, two

is _____, which has a _____ charge. You will then need _____ sodium cations to balance this -2 charge. The formula must be

Na_2SO_4

_____.

To get the formula for calcium chlorate you must first know that

+2

the only possible charge on calcium is _____. Then you must remember the formula and charge for the chlorate anion. It is ClO_3^-.

two, +2

You will need _____ ClO_3^- anions to balance the _____ charge of the

$Ca(ClO_3)_2$

Ca^{+2} cation, and the formula will be _____.

Here is another example to really tax your memory. What is the formula of ammonium dichromate? First you must recall the

NH_4^+

formula and charge of the ammonium ion. It is _____. Next you must know the formula and charge for the dichromate anion. It is

$Cr_2O_7^{-2}$

_____. Now setting up the formula is simple. You will need

two, -2

_____ NH_4^+ cations to balance the _____ charge on $Cr_2O_7^{-2}$, and the

$(NH_4)_2Cr_2O_7$

formula must be _____.

Let's see how you handle a compound like aluminum sulfate. The

+3

aluminum cation always has a charge of _____. The formula for the

SO_4^{-2}

sulfate anion is _____, and it has a -2 charge. In order to get a formula which has as many positive charges as negative charges

two

you will have to have _____ Al^{+3} cations and three SO_4^{-2} anions,

$Al_2(SO_4)_3$

and the formula will be _____.

You may prefer to set the formulas up so that this is easily seen. In this example you would write the ions first. Then you would see that you must take two Al^{+3} to balance three $(SO_4)^{-2}$.

First step: Al^{+3} $(SO_4)^{-2}$

Second step: $Al \underset{2}{\overset{+3}{\diagdown}} (SO_4) \underset{3}{\overset{-2}{\diagup}}$

$+6$ -6

If you have a compound containing a cation with several possible charges, life is a little easier since the charge on the cation will be given in the name. Thus the compound cobalt(II) phosphate

+2

has cobalt cations with a _____ charge. You must only remember

that the formula and charge for the phosphate anion is _____.
Now you can balance the positive charges and negative charges by

taking _____ Co^{+2} cations and two PO_4^{-3} anions. The formula

will be _____.

PO_4^{-3}

three

$Co_3(PO_4)_2$

    Although it would be possible to name these salts by the
counting method, that is, *mono, di, tri*, etc., this is rarely done.
Since salts are compounds which are made up of a positively
charged metallic ion and a negatively charged anion, the Stock
notation is almost always used. It sounds a little strange to say
"tricobalt diphosphate" rather than cobalt(II) phosphate.
Occasionally a counting name will turn up, like lead tetra-acetate

for $Pb(C_2H_3O_2)_4$. This could also be named lead(___) acetate by
the Stock notation system.

IV

    You have now been exposed to sufficient nomenclature to get
you well underway in first-year chemistry. Your job now is simply
to practice sufficiently to learn the vocabulary.

Appendix III
Four-Place Logarithms

Four-Place Logarithms

	0	1	2	3	4	5	6	7	8	9
1.0	.0000	.0043	.0086	.0128	.0170	.0212	.0253	.0294	.0334	.0374
1.1	.0414	.0453	.0492	.0531	.0569	.0607	.0645	.0682	.0719	.0755
1.2	.0792	.0828	.0864	.0899	.0934	.0969	.1004	.1038	.1072	.1106
1.3	.1139	.1173	.1206	.1239	.1271	.1303	.1335	.1367	.1399	.1430
1.4	.1461	.1492	.1523	.1553	.1584	.1614	.1644	.1673	.1703	.1732
1.5	.1761	.1790	.1818	.1847	.1875	.1903	.1931	.1959	.1987	.2014
1.6	.2041	.2068	.2095	.2122	.2148	.2175	.2201	.2227	.2253	.2279
1.7	.2304	.2330	.2355	.2380	.2405	.2430	.2455	.2480	.2504	.2529
1.8	.2553	.2577	.2601	.2625	.2648	.2672	.2695	.2718	.2742	.2765
1.9	.2788	.2810	.2833	.2856	.2878	.2900	.2923	.2945	.2967	.2989
2.0	.3010	.3032	.3054	.3075	.3096	.3118	.3159	.3160	.3181	.3201
2.1	.3222	.3243	.3263	.3284	.3304	.3324	.3345	.3365	.3385	.3404
2.2	.3424	.3444	.3464	.3483	.3502	.3522	.3541	.3560	.3579	.3598
2.3	.3617	.3636	.3655	.3674	.3692	.3711	.3729	.3747	.3766	.3784
2.4	.3802	.3820	.3838	.3856	.3874	.3892	.3909	.3927	.3945	.3962
2.5	.3979	.3997	.4014	.4031	.4048	.4065	.4082	.4099	.4116	.4133
2.6	.4150	.4166	.4183	.4200	.4216	.4232	.4249	.4265	.4281	.4298
2.7	.4314	.4330	.4346	.4362	.4378	.4393	.4409	.4425	.4440	.4456
2.8	.4472	.4487	.4502	.4518	.4533	.4548	.4564	.4579	.4594	.4609
2.9	.4624	.4639	.4654	.4669	.4683	.4698	.4713	.4728	.4742	.4757
3.0	.4771	.4786	.4800	.4814	.4829	.4843	.4857	.4871	.4886	.4900
3.1	.4914	.4928	.4942	.4955	.4969	.4983	.4997	.5011	.5024	.5038
3.2	.5051	.5065	.5079	.5092	.5105	.5119	.5132	.5145	.5159	.5172
3.3	.5185	.5198	.5211	.5224	.5237	.5250	.5263	.5276	.5289	.5302
3.4	.5315	.5328	.5340	.5353	.5366	.5378	.5391	.5403	.5416	.5428
3.5	.5441	.5453	.5465	.5478	.5490	.5502	.5514	.5527	.5539	.5551
3.6	.5563	.5575	.5587	.5599	.5611	.5623	.5635	.5647	.5658	.5670
3.7	.5682	.5694	.5705	.5717	.5729	.5740	.5752	.5763	.5775	.5786
3.8	.5798	.5809	.5821	.5832	.5843	.5855	.5866	.5877	.5888	.5899
3.9	.5911	.5922	.5933	.5944	.5955	.5966	.5977	.5988	.5999	.6010
4.0	.6021	.6031	.6042	.6053	.6064	.6075	.6085	.6096	.6107	.6117
4.1	.6128	.6138	.6149	.6160	.6170	.6180	.6191	.6201	.6212	.6222
4.2	.6232	.6243	.6253	.6263	.6274	.6284	.6294	.6304	.6314	.6325
4.3	.6335	.6345	.6355	.6365	.6375	.6385	.6395	.6405	.6415	.6425
4.4	.6435	.6444	.6454	.6464	.6474	.6484	.6493	.6503	.6513	.6522
4.5	.6532	.6542	.6551	.6561	.6571	.6580	.6590	.6599	.6609	.6618
4.6	.6628	.6637	.6646	.6656	.6665	.6675	.6684	.6693	.6702	.6712
4.7	.6721	.6730	.6739	.6749	.6758	.6767	.6776	.6785	.6794	.6803
4.8	.6812	.6821	.6830	.6839	.6848	.6857	.6866	.6875	.6884	.6893
4.9	.6902	.6911	.6920	.6938	.6937	.6946	.6955	.6964	.6972	.6981
5.0	.6990	.6998	.7007	.7016	.7024	.7033	.7042	.7050	.7059	.7067
5.1	.7076	.7084	.7093	.7101	.7110	.7118	.7126	.7135	.7143	.7152
5.2	.7160	.7168	.7177	.7185	.7193	.7202	.7210	.7218	.7226	.7235
5.3	.7243	.7251	.7259	.7267	.7275	.7284	.7292	.7300	.7308	.7316
5.4	.7324	.7332	.7340	.7348	.7356	.7364	.7372	.7380	.7388	.7396
5.5	.7404	.7412	.7419	.7427	.7435	.7443	.7451	.7459	.7466	.7474
5.6	.7482	.7490	.7497	.7505	.7513	.7520	.7528	.7536	.7543	.7551
5.7	.7559	.7566	.7574	.7582	.7589	.7597	.7604	.7612	.7619	.7627
5.8	.7634	.7642	.7649	.7657	.7664	.7672	.7679	.7686	.7694	.7701
5.9	.7709	.7716	.7723	.7731	.7738	.7745	.7752	.7760	.7767	.7774

	0	1	2	3	4	5	6	7	8	9
6.0	.7782	.7789	.7796	.7803	.7810	.7818	.7825	.7832	.7839	.7846
6.1	.7853	.7860	.7868	.7875	.7882	.7889	.7896	.7903	.7910	.7917
6.2	.7924	.7931	.7938	.7945	.7952	.7959	.7966	.7973	.7980	.7987
6.3	.7993	.8000	.8007	.8014	.8021	.8028	.8035	.8041	.8048	.8055
6.4	.8062	.8069	.8075	.8082	.8089	.8096	.8102	.8109	.8116	.8122
6.5	.8129	.8136	.8142	.8149	.8156	.8162	.8169	.8176	.8182	.8189
6.6	.8195	.8202	.8209	.8215	.8222	.8228	.8235	.8241	.8248	.8254
6.7	.8261	.8267	.8274	.8280	.8287	.8293	.8299	.8306	.8312	.8319
6.8	.8325	.8331	.8338	.8344	.8351	.8357	.8363	.8370	.8376	.8382
6.9	.8388	.8395	.8401	.8407	.8414	.8420	.8426	.8432	.8439	.8445
7.0	.8451	.8457	.8463	.8470	.8476	.8482	.8488	.8494	.8500	.8506
7.1	.8513	.8519	.8525	.8531	.8537	.8543	.8549	.8555	.8561	.8567
7.2	.8573	.8579	.8585	.8591	.8597	.8602	.8609	.8615	.8621	.8627
7.3	.8633	.8639	.8645	.8651	.8657	.8663	.8669	.8675	.8681	.8686
7.4	.8692	.8698	.8704	.8710	.8716	.8722	.8727	.8733	.8739	.8745
7.5	.8751	.8756	.8762	.8768	.8774	.8779	.8785	.8791	.8797	.8802
7.6	.8808	.8814	.8820	.8825	.8831	.8837	.8842	.8848	.8854	.8859
7.6	.8865	.8871	.8876	.8882	.8887	.8893	.8899	.8904	.8910	.8915
7.8	.8921	.8927	.8932	.8938	.8943	.8949	.8954	.8960	.8965	.8971
7.9	.8976	.8982	.8987	.8993	.8998	.9004	.9009	.9015	.9020	.9026
8.0	.9031	.9036	.9042	.9047	.9053	.9058	.9063	.9069	.9074	.9079
8.1	.9085	.9090	.9096	.9101	.9106	.9112	.9117	.9122	.9128	.9133
8.2	.9138	.9143	.9149	.9154	.9159	.9165	.9170	.9175	.9180	.9186
8.3	.9191	.9196	.9201	.9206	.9212	.9217	.9222	.9227	.9232	.9238
8.4	.9243	.9248	.9253	.9258	.9263	.9269	.9274	.9279	.9284	.9289
8.5	.9294	.9299	.9304	.9309	.9315	.9320	.9325	.9330	.9335	.9340
8.6	.9345	.9350	.9355	.9360	.9365	.9370	.9375	.9380	.9385	.9390
8.7	.9395	.9400	.9405	.9410	.9415	.9420	.9425	.9430	.9435	.9440
8.8	.9445	.9450	.9455	.9460	.9465	.9469	.9474	.9479	.9484	.9489
8.9	.9494	.9499	.9504	.9509	.9513	.9518	.9523	.9528	.9533	.9538
9.0	.9542	.9547	.9552	.9557	.9562	.9566	.9571	.9576	.9581	.9586
9.1	.9590	.9595	.9600	.9605	.9609	.9614	.9619	.9624	.9628	.9633
9.2	.9638	.9643	.9647	.9652	.9657	.9661	.9666	.9671	.9675	.9680
9.3	.9685	.9689	.9694	.9699	.9703	.9708	.9713	.9717	.9722	.9727
9.4	.9731	.9736	.9741	.9745	.9750	.9754	.9759	.9763	.9768	.9773
9.5	.9777	.9782	.9786	.9791	.9795	.9800	.9805	.9809	.9814	.9818
9.6	.9823	.9827	.9832	.9836	.9841	.9845	.9850	.9854	.9859	.9863
9.7	.9868	.9872	.9877	.9881	.9886	.9890	.9894	.9899	.9903	.9908
9.8	.9912	.9917	.9921	.9926	.9930	.9934	.9939	.9943	.9948	.9952
9.9	.9956	.9961	.9965	.9969	.9974	.9978	.9983	.9987	.9991	.9996

Appendix IV
pH: The Way Chemists
Express the Strength of
Acidic and Basic Solutions

When chemists are interested in the strengths of aqueous (in water) solutions of acids and bases, what they are really concerned with is the concentration of H^+ ions. This may vary from as low a concentration as 10^{-14} M up to as much as 1 M. To make it easier to work with numbers that can vary over so many magnitudes, chemists have worked out a system that uses only numbers between 1 and 14. In this system, they express the concentration of H^+ ions as the exponent to which 10 is raised when the concentration is expressed in exponential notation. Recall from Section B in Chapter 1 that the logarithm to the base 10 is exactly this. If you have forgotten what logarithms mean and how to work with them, you had better check back to Chapter 1.

To simplify things still further, you give the log of the H^+ ion concentration a negative sign. When you take the logarithm of something and give it a negative sign, you call this a p function. The p means "$-\log$." Thus, if you are interested in the H^+ ion concentration, you can write

$$pH = -\log [H^+]$$

The [   ] is the conventional way of showing concentration in moles per liter.

In the same way, you can write

$$pOH = \underline{\hspace{2cm}} [OH^-].$$

$-\log$

Not only concentrations have p functions. Any number can have a "$-\log$." Thus you can express the dissociation constant for water, K_w, as a p function.

$$pK_w = -\log K_w$$

Similarly, you can express the dissociation constant of a weak acid, K_a, as a p function.

$$pK_a = \underline{\hspace{2cm}}$$

$-\log K_a$

Determining the pH
from the H⁺ Ion Concentration

Let's see how this system works with pH. Say that you have a solution in which the H⁺ ion concentration is 10^{-3} M. To get the pH, you write

10^{-3}

-3

3

$$pH = -\log [H^+] = -\log [\underline{\hspace{1cm}}]$$

$$= -(\underline{\hspace{1cm}})$$

$$= \underline{\hspace{0.5cm}}$$

(If you weren't able to get from log 10^{-3} to -3, go back to Section B, Chapter 1. You should also remember the sign convention. Two negatives multiplied or divided give a positive.)

If the H⁺ ion concentration were 0.00001, to get the pH you would write

$-\log [H^+]$

10^{-5}

-5

5

$$pH = \underline{\hspace{2cm}}$$

$$= -\log [\underline{\hspace{1cm}}]$$

$$= -(\underline{\hspace{1cm}})$$

$$= \underline{\hspace{1cm}}$$

This is pretty easy. You can check yourself by doing the following on a separate sheet of paper and checking your answers with those in the margin. Determine the pH for the following [H⁺].

	[H⁺]	pH
5	10^{-5}	_____
6	0.000001	_____
0	1	_____
13	10^{-13}	_____
1	0.1	_____

Now that you are able to get the pH easily when you know the H⁺ ion concentration of a solution, let's look at what these pH

1

values mean. If the [H⁺] is 0.1 M, the pH is _____.

2

If the [H⁺] is 10 times less, 0.01 M, the pH is _____.

3

If the pH is 100 times less, 0.001 M, the pH is _____.

4

And if the pH is 1000 times less, 0.0001 M, the pH is _____.

greater

The lower the [H⁺], the _____ the pH. Since the strength of an acid is directly related to the concentration of H⁺ ions, the

greater

weaker the acid solution, the _____ the pH. An acid with a pH

stronger

of 1 would be 10 times _____ than an acid with a pH of 2.

An acid solution with a pH of 6 would be _____ times weaker than one with a pH of 4.

100

So far we have considered only acid solutions whose H^+ ion concentrations can be expressed with the digits 0 and 1. We are now going to consider other concentrations. Your instructor may suggest that you stop at this point. If so, you may eliminate the rest of this section and start again when we are determining the H^+ ion concentration from the pH.

Now say that you want to know the pH for a solution whose $[H^+]$ is 0.050. You start by writing

$$pH = \underline{\hspace{1cm}} [H^+]$$

$-\log$

and you put in the value for $[H^+]$

$$pH = -\log [\underline{\hspace{1cm}}]$$

0.050

To get the logarithm, you have to express this in scientific notation.

$$pH = -\log [\underline{\hspace{2cm}}]$$

5.0×10^{-2}

As you did in Section B, Chapter 1, you break this up into the sum of two logarithms.

$$pH = -(\log 5.0 + \underline{\hspace{1.5cm}})$$

$\log 10^{-2}$

You can find the log of 5.0 in the table of logarithms in Appendix III,

and you know that $\log 10^{-2}$ equals _____.

-2

$$pH = -[\underline{\hspace{1cm}} + (-2)]$$

0.70

(Notice that you may write only two significant figures on the log because the 5.0 has just two significant figures.)

Making the algebraic addition gives you

$$pH = -(\underline{\hspace{1.5cm}})$$

-1.30

and the two $-$ signs cancel to give you

$$pH = \underline{\hspace{1cm}}$$

1.30

It should be obvious that the pH is between 1 and 2, since the

$[H^+]$ is less than 0.1 M, which has a pH of _____, but more than

1

0.01 M, which has a pH of _____. So the pH has to be between 1 and 2.

2

Here is another to make sure that you have this. You want the pH of an acid solution whose $[H^+]$ is 0.0000356. You always start by writing

$$pH = \underline{\hspace{2cm}}$$

$-\log [H^+]$

Writing the $[H^+]$ in scientific notation, you have

$$pH = -\log [\underline{\hspace{2cm}}]$$

3.56×10^{-5}

You then break this up into the sum of two logarithms

3.56, 10⁻⁵ -> $pH = -(\log\underline{\hspace{1cm}} + \log \underline{\hspace{1cm}})$

When you look log 3.56 up in the table of logarithms and put in the log 10⁻⁵, you have

0.551, −5 -> $pH = -[\underline{\hspace{1.5cm}} + (\underline{\hspace{1cm}})]$

(Notice the number of significant figures on log 3.56.)

Making the algebraic addition, you have

—4.449 -> $pH = -(\underline{\hspace{1.5cm}})$

which gives you the pH

4.449 -> $pH = \underline{\hspace{1.5cm}}$

To see that you are doing this correctly, determine the pH for the following H⁺ ion concentrations.

	[H⁺]	pH
0.815	0.153	_____
3.137	0.000730	_____
2.577	0.00265	_____

Determining the H⁺ Ion Concentration from the pH

Let's see how you can reverse this process. Instead of finding the pH from [H⁺], find the H⁺ ion concentration from the pH. Say that the pH of a solution is 3. You know that the relationship between pH and [H⁺] is

−log [H⁺] -> $pH = \underline{\hspace{2cm}}$

Therefore you can write

−log [H⁺] -> $3 = \underline{\hspace{2cm}}$

There is no way of dealing with negative logarithms, so what you do is multiply both sides of the equation by —1. This does not change the values, but it does change the signs. The + becomes a −, and the − becomes a +.

$$-3 = \log [H^+]$$

If you want to determine the number whose logarithm you know, you must find the antilogarithm. Once again, if you are rusty on any of this, refer to Chapter 1, Section B. There is an entire part that deals with antilogarithms. You write

$$\text{antilog } (-3) = [H^+]$$

The antilogarithm of -3 is simply _____. So you can write 10^{-3}

$$10^{-3} = [H^+]$$

It is as simple as that.

Here is another example. You want to know the H^+ ion concentration of a solution that has a pH of 11. You start by writing the relationship between pH and $[H^+]$.

pH = _____ $-\log [H^+]$

And you then put in the value for the pH from the problem.

_____ $= -\log [H^+]$ 11

Next you change the signs on both sides.

_____ $= \log [H^+]$ -11

Now all you have to do is to get the _____ of -11.   antilogarithm

It is _____. 10^{-11}

So you know

_____ $= [H^+]$ 10^{-11}

This is very easy as long as the pH is a whole number greater than 1. If it has something written to the right of the decimal, getting the antilogarithm is a little more difficult. Your instructor may suggest that you not continue this. If so, you can skip the rest of this section and go on to the section, "Calculating the pOH from the OH^- Ion Concentration." For the stout-hearted, let's go on with the more complicated cases.

Say that you want to know the H^+ ion concentration of a solution that has a pH of 3.40. You start in the usual way, by writing

pH = _____ $-\log [H^+]$

and you put in the value of the pH in the problem.

$3.40 =$ _____ $-\log [H^+]$

Changing the signs in both sides gives you

_____ $= -\log [H^+]$ -3.40

Since you cannot get the antilogarithm of a negative number, you must change the -3.40 to $(-4+0.60)$. Now you can write

$$\text{antilog } (-4+0.60) = [H^+]$$

Since the antilogarithm of two terms added together is the same as the antilogarithm of one times the antilogarithm of the other, you write

$$\text{antilog } -4 \times \text{antilog } 0.60 =$$ _____ $[H^+]$

10⁻⁴

You know that antilog −4 is _____, and you can look the antilog of

3.98

0.60 up in a table of logarithms. It is _____. So you can write

3.98 × 10⁻⁴

_____ = [H⁺]

It is better to express this as 4.0×10^{-4} to have the correct number of significant figures.

Here's another example. You want to know the H⁺ ion concentration that has a pH of 11.68. You always start by writing the relationship between pH and [H⁺].

−log [H⁺]

pH = _____

Putting in the value from the question, you have

11.68

_____ = −log [H⁺]

and then changing signs on both sides of the equation, you have

−11.68

_____ = log [H⁺]

Writing the −11.68 as a negative whole number and a positive number less than 1 gives you

−12 + 0.32

_____ = log [H⁺]

antilogarithm

You must determine the _____ of (−12+0.32) in order to get the value of [H⁺]. Since the antilogarithm of the sum of two numbers equals the antilogarithm of one _____ the antilogarithm of the other, you can write

times

×

antilog −12 ____ antilog .32 = [H⁺]

10⁻¹²

The antilog 12 equals _____, and when you look up antilog .32

2.09

in the table of logarithms you find that it is _____. So you can write

2.09 × 10⁻¹²

_____ = [H⁺]

It is best to express this as 2.1×10^{-12}, considering the significant figures.

So that you can check yourself and see that you are able to do these, determine the [H⁺] that is equivalent to the following pH values. The answers are given in the margin.

pH	[H⁺]
9.23	_____
6.37	_____
2.10	_____
1.54	_____

5.889 × 10⁻¹⁰

4.226 × 10⁻⁷

7.943 × 10⁻³

2.884 × 10⁻²

You should realize that these answers have more than the allowable number of significant figures. The nature of logarithms is such that it is best to give your answers to only two significant figures in all these cases. Thus the first is best shown as

5.9×10^{-10}, the second as 4.2×10^{-7}, the third as _____, and the fourth as 2.9×10^{-2}. There is no need for interpolation or even for getting the exact column in the three-place logarithm table. Simply determining the nearest number in the left-hand column is enough.

7.9×10^{-3}

Calculating the pOH from the OH⁻ Ion Concentration

When a base, such as NaOH, is dissolved in water, it breaks up into ions—a positive ion and an OH^- ion. The strength of the base depends on the concentration of the OH^- ions just as the strength of acids depends on the concentration of H^+ ions. And just as you used a p function with the $[H^+]$, you can use a p function to express the concentration of the OH^- ion, $[OH^-]$.

$$pOH = -\log [OH^-]$$

The calculations are identical. Thus, if you know that the $[OH^-]$ is 10^{-3} *M*, then you can calculate the pOH.

$$pOH = -\log [\_\_\_\_]$$
$$= -\log [\_\_\_\_]$$
$$= -(\_\_\_\_)$$
$$= \_\_\_$$

OH^-

10^{-3}

-3

3

If the OH^- concentration cannot be expressed simply as 10 to a whole-number power, you have to use a table of logarithms to get the pOH value. For example, if you want the pOH of a solution that has an OH^- ion concentration of 0.00065 *M*, you start by writing

$$pOH = _____$$
$$= _____ \text{ (in scientific notation)}$$
$$= -[\_\_\_\_ + (-4)]$$
$$= \_\_\_$$

$-\log [OH^-]$

$-\log [6.5 \times 10^{-4}]$

0.8129

3.19

If you can determine the pH from $[H^+]$, then you can just as easily determine pOH from $[OH^-]$.

Determining the OH⁻ Ion Concentration from the pOH

You use exactly the same method to get the OH⁻ ion concentration from the pOH that you used to get the H⁺ ion concentration from pH. For example, say that you want to know the OH⁻ ion concentration that gives you a pOH of 4. You start by writing the relationship between pOH and [OH⁻].

$-\log$ [OH⁻]

$$pOH = \underline{\hspace{2cm}}$$

Then you put in the value for the pOH.

$-\log$ [OH⁻]

$$4 = \underline{\hspace{2cm}}$$

change

To get the logarithm, you have to _____ the signs on both sides of the equation.

-4

$$\underline{\hspace{1.5cm}} = \log \text{[OH⁻]}$$

And now to get the value for [OH⁻], you must find the

antilogarithm

$$\underline{\hspace{3cm}} \text{ of } -4.$$

antilog

$$\underline{\hspace{1.5cm}} (-4) = \text{[OH⁻]}$$

This gives you

10^{-4}

$$\underline{\hspace{1.5cm}} = \text{[OH⁻]}$$

If your pOH is not a simple whole number, then you have to use the table of logarithms to get the antilogarithm of the pOH value, as you did with pH. We will not go through this procedure, as you have had plenty of practice with [H⁺] from pH.

The Water Equilibrium

Pure water itself breaks up very slightly into ions. One molecule of water produces one H⁺ ion and one OH⁻ ion. The equation for this is

$$H_2O \rightarrow H^+ + OH^-$$

This type of reaction is called a *dynamic equilibrium*, since as fast as the water molecules break up, H⁺ and OH⁻ ions are recombining to produce more water molecules. You can show this by using two arrows.

$$H_2O \rightleftarrows H^+ + OH^-$$

Since the ions are being used up as fast as they are being formed, their concentration remains constant. In pure water the concentration of H⁺ ions is 10^{-7} *M*. The concentration of OH⁻

10^{-7} *M*

ions is also _____, since as many H⁺ ions are produced as OH⁻

7

ions. Thus the pH of pure water is ____, and the pOH of pure water is also ___. The chemist says that pure water is *neutral*, that is,

neither acidic nor basic. A neutral solution therefore has a pH

of ___.

7

7

Because of the nature of dynamic equilibria, if the concentration of any of the substances in equilibrium is changed, the entire system responds to this change to minimize its effect. This is called the *Le Chatelier principle* or the *Law of Mass Action*. It is expressed mathematically by stating that the product of the concentrations remains constant. In the case of water, you write

$$K_w = 10^{-14} = [H^+] [OH^-]$$

K_w is called the dissociation constant for water, and its value is 10^{-14}. You can see how this works for pure water whose $[H^+]$ is

10^{-7} and $[OH^-]$ is also _____.

10^{-7}

$$K_w = 10^{-14} = [H^+] [OH^-]$$
$$= [\text{\_\_\_\_\_}] [\text{\_\_\_\_\_}]$$

10^{-7}, 10^{-7}

If you have an acid solution whose $[H^+]$ ion concentration is increased to 10^{-5} (remember, 10^{-5} is larger than 10^{-7}), you have

$$K_w = 10^{-14} = [H^+] [OH^-]$$
$$= [10^{-5}] [OH^-]$$

or

$$\frac{10^{-14}}{10^{-5}} = [OH^-]$$

$$\text{_____} = [OH^-]$$

10^{-9}

To have the product of the two concentrations still equal to 10^{-14}, the $[OH^-]$ must decrease to 10^{-9} M. Thus, increasing the $[H^+]$ decreases the $[OH^-]$. (Remember, 10^{-9} is smaller than 10^{-7}. As a matter of fact, if you have forgotten how to manipulate these powers of 10, check back to Section A in Chapter 1.)

Here is another example. You have a basic solution where the OH^- ion concentration is 10^{-2} M. You can calculate the H^+ ion concentration in the following way.

$$K_w = 10^{\text{\_\_\_\_}} = [H^+] [OH^-]$$
$$= [H^+] [\text{\_\_\_\_\_}]$$

-14

10^{-2}

or

$$\frac{\text{\_\_\_\_\_}}{10^{-2}} = [H^+]$$

$$\text{\_\_\_\_\_} = [H^+]$$

10^{-14}

10^{-12}

Let's consider these two examples when we express the [H⁺] as pH and the [OH⁻] as pOH for each case. Something interesting happens.

5   The acidic solution with an [H⁺] of 10^{-5} *M* has a pH of ___. When we calculated its [OH⁻], it worked out to be 10^{-9}. This

9   gives a pOH of ___.

2   In the basic solution with an [OH⁻] of 10^{-2} *M*, the pOH is ___.

12   Its [H⁺] worked out to be 10^{-12}, so its pH is ___.

14   In both cases, the sum of the pH and the pOH is ___. This is always true.

$$pH + pOH = 14$$

We can see how this comes about. Say that you use the water equilibrium expression

$$K_w = 10^{-14} = [H^+][OH^-]$$

and you take the p values for everything. Since the p value means "− log," you would have

$$-\log K_w = -\log 10^{-14} = -\log [H^+][OH^-]$$

The log [H⁺] [OH⁻] can be broken up into the sum of two logarithms

$$\log [H^+][OH^-] = \log [H^+] + \log [OH^-]$$

so you can write

$$-\log K_w = -\log 10^{-14} = -(\log [H^+] + \log [OH^-])$$

Moving the − sign inside the parentheses gives you

$$-\log K_w = -\log 10^{-14} = -\log [H^+] - \log [OH^-]$$

−14   But you know that log 10^{-14} equals ___, that −log [H⁺] is the

pOH   pH, and −log [OH⁻] is the ___. You also know that $-\log K_w$ equals pK_w. So we can write

$$pK_w = -(-14) = pH + pOH$$

or

14      $$pK_w = \underline{\hspace{1cm}} = pH + pOH$$

This is the reason that

$$pH + pOH = 14.$$

This is a very useful relationship. It allows you to go from pH to pOH for a given solution by a simple subtraction. For example, if a solution has a pOH of 3, you can find its pH by

$$pH + pOH = 14$$

$$pH + \underline{\quad} = 14 \qquad\qquad 3$$

Subtracting 3 from both sides of the equation, you have

$$pH + 3 - 3 = 14 - \underline{\quad} \qquad\qquad 3$$

and finally

$$pH = \underline{\quad}. \qquad\qquad 11$$

What this amounts to (if your algebra is okay, you know this already) is

$$pH = 14 - pOH$$

and in the same way

$$pOH = 14 - \underline{\quad}. \qquad\qquad pH$$

Try this one for practice. What is the pH for a solution whose pOH is 2.5? All you do is write

$$pH = 14 - \underline{\quad} \qquad\qquad pOH$$

$$= 14 - \underline{\quad} \qquad\qquad 2.5$$

$$= \underline{\quad} \qquad\qquad 11.5$$

Because it is so easy to get from pOH to pH, chemists rarely speak about pOH values. Instead, they give the solutions a pH value. However, when they are dealing with a basic solution, it is easiest to calculate the pOH. Then they can easily change the pOH to pH.

Here is how this is done. Say that you have a solution with an OH^- ion concentration of 10^{-5} M and you want the pH. The first thing you do is determine the pOH.

$$pOH = \underline{\qquad\qquad} \qquad\qquad -\log [OH^-]$$

$$pOH = \underline{\qquad\qquad} \qquad\qquad -\log [10^{-5}]$$

$$pOH = \underline{\quad} \qquad\qquad 5$$

Now to determine the pH, all you have to do is subtract 5 from____.   14

$$pH = \underline{\quad} - 5 \qquad\qquad 14$$

$$= \underline{\quad} \qquad\qquad 9$$

Here's another for practice. You want the pH for a solution that has an OH⁻ ion concentration of 0.000731 *M*. The first thing

pOH	you do is determine the _____.
[OH⁻]	$$pOH = -\log \underline{\hspace{2cm}}$$
[7.31 × 10⁻⁴]	$$= -\log \underline{\hspace{3cm}} \text{ (in scientific notation)}$$
3.1361	$$= \underline{\hspace{1.5cm}}$$
subtract	Now to get the pH, you _____ the pOH from 14.
3.1361	$$pH = 14 - \underline{\hspace{1.5cm}}$$
10.8639	$$= \underline{\hspace{1.5cm}}$$

According to the rules of significant figures, you are allowed only three places after the decimal, so the answer is better expressed as 10.864. Determining the number of significant figures to which you can express a logarithm is very tricky. The significant figures may vary with the number whose logarithm you are finding. If finding the significant figures is vital, you can determine two pH values. First you assume that the last digit in your [H⁺] or [OH⁻] is one *more* than the given value, and you calculate a pH. Then you assume that the last digit in the concentration is one *less* than the given value and you calculate the pH. Compare these two values and you will see which figure is uncertain.

In the case above, an [OH⁻] of 0.000732 has a pH of 10.8633. An [OH⁻] of 0.000730 has a pH value of 10.8645. It is the third digit after the decimal that varies, and this is the last that you can write. We will examine this again in the next problem.

You can also use the changing of pH to pOH when you have a pH value and you want to know the OH⁻ ion concentration. For example, let's say that you want to know [OH⁻] when the pH is 9.77. Since you are interested in [OH⁻], you change pH to pOH.

14	$$pOH = \underline{\hspace{1cm}} - pH$$
14	$$= \underline{\hspace{1cm}} - 9.77$$
4.23	$$= \underline{\hspace{1cm}}$$

Now you can calculate the OH⁻ ion concentration from the pOH.

[OH⁻]	$$pOH = -\log \underline{\hspace{1.5cm}}$$
4.23	$$\underline{\hspace{1.5cm}} = -\log [OH^-]$$
−4.23	$$\underline{\hspace{1.5cm}} = \log [OH^-]$$
−5	$$(\underline{\hspace{1cm}} + .77) = \log [OH^-]$$
antilog	$$\underline{\hspace{1.5cm}}(-5 + .77) = [OH^-]$$
5.889 × 10⁻⁵	$$\underline{\hspace{2cm}} = [OH^-]$$

We can check this answer to see how many significant figures it may have. What we do is determine the $[OH^-]$ for a pH of 9.76 (this is one less than the given value in its last place) and also for 9.78 (this is one more than the given value in its last place). These values are

pH 9.76 gives $[OH^-] = 5.755 \times 10^{-5}$

pH 9.78 gives $[OH^-] = 6.026 \times 10^{-5}$

In this case, the first place after the decimal is uncertain, so this is the last digit that you can show. Your answer to the correct number of significant figures is 5.9×10^{-5}.

Appendix V
Conversion Factors

Metric Units

Length		Mass		Volume	
Angstrom	$\dfrac{Å}{10^{-10}\,m}$	Microgram	$\dfrac{\mu g}{10^{-6}\,g}$	Milliliter	$\dfrac{ml}{10^{-3}\,liter}$
Micron	$\dfrac{\mu}{10^{-6}\,m}$	Milligram	$\dfrac{mg}{10^{-3}\,g}$	Cubic centimeter	$\dfrac{1\ ml}{cm^3}$
Millimeter	$\dfrac{mm}{10^{-3}\,m}$	Kilogram	$\dfrac{kg}{10^3\,g}$	Microliter	$\dfrac{\mu l}{10^{-6}\,liter}$
Centimeter	$\dfrac{cm}{10^{-2}\,m}$				
Kilometer	$\dfrac{km}{10^3\,m}$				

British Units

Length	Mass	Volume
$\dfrac{12\ in.}{ft}$	$\dfrac{16\ oz\ (avoir)}{lb\ (avoir)}$	$\dfrac{16\ oz\ (fluid)}{pt}$
$\dfrac{3\ ft}{yd}$	$\dfrac{2000\ lb}{ton}$	$\dfrac{2\ pt}{qt}$
$\dfrac{5280\ ft}{mi}$		$\dfrac{4\ qt}{gal}$

Metric to British Units

Length	Mass	Volume
$\dfrac{2.54\ cm}{in.}$	$\dfrac{454\ g}{lb}$	$\dfrac{946\ ml}{qt}$
$\dfrac{1.61\ km}{mi}$	$\dfrac{2.20\ lb}{kg}$	

Pressure

$\dfrac{760\ mm\ Hg}{atm}$	$\dfrac{760\ torr}{atm}$	$\dfrac{76\ cm\ Hg}{atm}$	$\dfrac{14.7\ psi}{atm}$

Temperature (not conversion factors; formulas)

$$°C = \frac{5}{9}(°F-32) \qquad °F = \frac{9}{5}(°C) + 32 \qquad °K = °C + 273$$

Avogadro's Number

$$\frac{6.02 \times 10^{23} \text{ molecules}}{\text{mole of molecules}} \qquad \frac{6.02 \times 10^{23} \text{ atoms}}{\text{mole atoms}}$$

Ideal Gas Constant

$$\frac{0.0820 \text{ (liter) (atm)}}{\text{(mole) (°K)}} \qquad \frac{22.4 \text{ liters (STP)}}{\text{mole}}$$

Heat

$$\frac{10^3 \text{ cal}}{\text{kcal}} \qquad \frac{4.184 \text{ joules}}{\text{cal}} \qquad \frac{1 \text{ cal}}{\text{(g water) (°C)}} \text{ (the specific heat of water)}$$

Electrical Units

$$\frac{1 \text{ faraday}}{\text{mole electrons}} \qquad \frac{1 \text{ faraday}}{96,500 \text{ coulombs}} \qquad \frac{1 \text{ coulomb}}{\text{(amp) (sec)}} \qquad \frac{1 \text{ volt}}{\text{(amp) (ohm)}} \qquad \frac{1 \text{ watt}}{\text{(amp) (volt)}}$$

Properties Expressed as Conversion Factors

Property	*Units*
Density	$\dfrac{\text{g}}{\text{ml}}$ (solids and liquids) $\dfrac{\text{g}}{\text{liter}}$ (gases)
Specific gravity	$\dfrac{\text{g sample}}{\text{g water}}$ occupying same volume as sample
	or $\dfrac{\text{density sample}}{\text{density water}}$
Percent	$\dfrac{\text{parts substance of interest}}{100 \text{ parts total mixture}}$
Weight percent	$\dfrac{\text{g substance of interest}}{100 \text{ g total mixture}}$
Volume percent	$\dfrac{\text{ml substance of interest}}{100 \text{ ml total mixture}}$
Percent purity	$\dfrac{\text{g pure component}}{100 \text{ g impure substance}}$

Chemical Properties Expressed as Conversion Factors

Property	*Units*
Molecular weight	$\dfrac{g}{mole}$
Atomic weight	$\dfrac{g}{mole}$
Equivalent weight	$\dfrac{g}{equivalent}$
Formula weight	$\dfrac{g}{mole}$
Molarity (M)	$\dfrac{mole\ of\ solute}{liter\ solution}$
Formality (F)	$\dfrac{formula\ weights\ of\ solute}{liter\ solution}$
Normality (N)	$\dfrac{equivalent\ of\ solute}{liter\ solution}$
Molality (m)	$\dfrac{mole\ of\ solute}{1000\ g\ solvent}$
Molal freezing point depression constant (K_f) or Molal boiling point elevation constant (K_b)	$\dfrac{(°C)\ (1000\ g\ solvent)}{mole\ solute}$

Thermodynamic Properties Expressed as Conversion Factors

Property	*Units*
Specific heat	$\dfrac{cal}{g°C}$
Heat of fusion or Heat of vaporization or Heat of sublimation	$\dfrac{cal}{g}$ or $\dfrac{kcal}{mole}$
Heat of reaction ($\Delta H°$) (at 25°C, 1.0 atm)	$\dfrac{kcal}{mole}$
Heat of formation ($\Delta H_f°$)	$\dfrac{kcal}{mole}$
Calorimeter constant	$\dfrac{cal}{°C}$

In 1971 an international organization, the International Bureau of Weights and Measures, reached an agreement on a system of units which will be used by all scientists throughout the world. This standard international system (the French call it Système Internationale), abbreviated as SI, is very much like the metric system. The SI system simply tries to reduce the number of different units by expressing everything in terms of seven "base" units.

For example, volume in the metric system is often expressed in liters. However, a volume can be merely the product of three lengths. Why not discard "liter" and replace it with the equivalent cube which is 10 cm on an edge (that is, $10 \text{ cm} \times 10 \text{ cm} \times 10 \text{ cm} = 10^3 \text{ cm}^3$).

Or, for a somewhat more complicated example, consider pressure, which is expressed in torr, mm Hg, or atm. Pressure is the force exerted per unit area, and force can be expressed as the product of the distance that a mass is moved in one sec per sec (i.e., accelerated). Area can be reduced to the product of two lengths, and force can be reduced to length, mass, and time. Thus it is possible to express pressure simply in terms of length, mass, and time.

Another example would be heat, which is expressed in calories in the metric system. Heat is simply energy, which can be expressed in terms of force and area; and once again it is easily reduced to units of length, mass, and time.

The seven "base" units needed are tabulated below. The last, candela, will not normally be of interest to chemists.

SI Base Units

Quantity	Name of unit	Abbreviation
Length	metre	m
Mass	kilogram	kg
Time	second	s
Electric current	ampere	A
Temperature (thermodynamic)	kelvin	K
Amount of substance	mole	mol
Luminous intensity	candela	cd

Very large and very small multiples of the base units can be expressed in the same way as in the metric system, by putting a prefix before the name to indicate to what power of 10 the base unit is multiplied. Recall, a kilometer was equal to 10^3 meters. In the SI, a kilometre is equal to 10^3 metres. Below is a partial list of these prefixes and the power of 10 they represent. (In discussing SI units, we use the French endings: *re* instead of *er*.)

SI Prefixes

Power of 10	Prefix	Symbol
10^3	kilo	k
10^1	deca	da
10^{-1}	deci	d
10^{-2}	centi	c
10^{-3}	milli	m
10^{-6}	micro	μ
10^{-9}	nano	n

Some units, which are in such common usage that it would be impractical to eliminate them, have been retained. Thus, although the base unit for time is seconds, the units for minutes, hours, and days have been kept. This is also true for atmospheres (pressure), liters (volume), and tonne (10^3 kg). Some familiar units have however been removed. Thus the angstrom (Å) (10^{-10}m or 10^{-8} cm), the torr or mm Hg, and the calorie have been deleted. Also the word and symbol degree (°) have been removed from temperatures when they are expressed in Kelvin notation. Thus water boils at 373 kelvins (373 K). However, the degree, word and symbol, has been kept if you express the temperature in Celsius (centigrade). Water boils at 100 degrees Celsius (100°C). The name micron (10^{-6} m) has been changed to micrometre.

Actually, you will find not too much difference between the metric system and SI. Those units which will be important to you as a chemist and which vary from the metric system are shown in the tables below. If in the future you need a source of all the SI units, there is a pamphlet available, National Bureau of Standards Special Publication 330, 1972 Edition, SD Catalog No. 13.10:330/2, from the U.S. Government Printing Office, Washington, D.C.

Conversion Factors, Metric to SI

	Length			Volume	
Angstrom (Å)	$\dfrac{Å}{10^{-10}\text{ m}} = \dfrac{Å}{0.1\text{ nm}}$		Liter (l)	$\dfrac{1}{10^{-3}\text{ m}^3} = \dfrac{1}{\text{dm}^3}$	
Millimicron (mμ)	$\dfrac{\text{m}\mu}{\text{nm}}$		(The liter and its multiples can still be used, but the use of m^3 for high precision is recommended)		

Pressure	*Temperature*
The SI unit is the pascal (Pa)	[The only change is that the degree (°) symbol is omitted in temperatures expressed in the Kelvin scale.]
mm Hg $\dfrac{\text{mm Hg}}{133.3 \text{ Pa}}$	
Torr $\dfrac{\text{torr}}{133.3 \text{ Pa}}$	$K = °C + 273$
Atmosphere (atm) $\dfrac{\text{atm}}{101.3 \times 10^3 \text{ Pa}}$	
(The atm can still be used)	

Ideal Gas Constant

$$\frac{0.0820 \text{ (liter) (atm)}}{\text{(mole) (°K)}} = \frac{8.31 \text{ (m}^3\text{) (Pa)}}{\text{(mol) (K)}} = \frac{8.31 \text{ J}}{\text{(mol) (K)}}$$

J (joule) is the unit for heat (energy), $\dfrac{\text{cal}}{4.184 \text{ J}}$

Electrical Units

Faraday is not considered. Simply use mole of electrons	Coulomb (C) $\dfrac{C}{(A)\ (s)}$
	(Only the symbol is changed)

Volt (V) $\dfrac{(V)\ (A)\ (s)}{J}$	Ohm (Ω) $\dfrac{(\Omega)\ (A)}{V}$	Watt (W) $\dfrac{(W)\ (s)}{J}$

Properties Expressed as Conversion Factors with SI Equivalent

	Metric	*SI*
Density	$\dfrac{g}{ml}$	$= \dfrac{kg}{m^3}$
Molecular weight	$\dfrac{g}{mole}$	$= \dfrac{10^{-3} \text{ kg}}{mol}$

Equivalent weight is not considered.

Molarity, the word, is not to be used in SI, but the concentration is still allowed, and for high accuracy work it should be expressed as

$$\frac{\text{mole}}{\text{liter}} = \frac{\text{mol}}{10^{-3} \text{ m}^3} = \frac{\text{mol}}{\text{dm}^3}.$$

Normality is not considered since the equivalent is not considered.